T0254206

AQUATIC PLANTS IN BRITAIN AND IRELAND

AQUATIC PLANTS IN BRITAIN AND IRELAND

A JOINT PROJECT OF THE ENVIRONMENT AGENCY,
INSTITUTE OF TERRESTRIAL ECOLOGY AND THE JOINT
NATURE CONSERVATION COMMITTEE

BY

C.D. PRESTON AND J.M. CROFT

ILLUSTRATED BY

G.M.S. EASY

BRILL

LEIDEN • BOSTON
2014

This paperback was originally published by Harley Books (B.H. & A. Harley Ltd),
Colchester, UK, in hardback in 1997 under ISBN 0 946589 55 0 and in paperback in 2001 under
ISBN 0 946589 69 0, for the Environment Agency, the Institute of Terrestrial Ecology and the
Joint Nature Conservation Committee.

THE ENVIRONMENT AGENCY is a non-departmental public body whose responsibilities
largely cover the management and regulation of the water environment, including the conservation
of its flora and fauna; the control of certain industrial processes and the management of waste
disposal. The Agency's responsibilities relate only to England and Wales.

THE INSTITUTE OF TERRESTRIAL ECOLOGY is a component of the Natural
Environment Research Council's Centre for Ecology and Hydrology. The Institute develops
long-term multidisciplinary research to advance the science of terrestrial ecology. The securing,
expansion and dissemination of ecological data provide the basis for impartial advice on
environmental protection, conservation and the sustainable use of natural resources.

THE JOINT NATURE CONSERVATION COMMITTEE is the forum through which the
three country nature conservation agencies – the Countryside Council for Wales (CCW), English
Nature (EN) and Scottish Natural Heritage (SNH) – deliver their statutory responsibilities for
Great Britain as a whole and internationally. These responsibilities, known as the special functions,
contribute to sustaining and enriching biological diversity, enhancing geological features and
sustaining natural systems.

Designed by Geoff Green. Text set in Ehrhardt by Saxon Graphics Ltd, Derby.

Library of Congress Control Number: 2014940955

ISBN 978-90-04-27729-8 (paperback)
ISBN 978-90-04-27730-4 (e-book)

CONTENTS

CONSERVATION DESIGNATION OF THE RARER SPECIES

FOREWORD

PLANTS are an essential part of all healthy stretches of water from the lochans high in the Cairngorms to the fenland drains near my home in East Anglia. They provide a diversity of structure and form, a nutrient recycling process, food sources and shelter for organisms of very many kinds. At the same time, some plants can cause problems by physically obstructing water courses, by de-oxygenating water or because they are poisonous to farm animals.

Sound knowledge of the status and distribution of freshwater plants and an understanding of their biology are essential if man is to live in harmony with nature – if our use of natural systems is to be at all sustainable.

This volume presents, for the first time, a thorough review of all freshwater plants in Britain and Ireland. It is both an atlas of distribution and a compendium of scientific information. As such, it will help to inform and improve decision-making for the management of our rivers, streams and lakes. It also establishes a unique benchmark against which future changes in the aquatic flora, and their wider environmental implications, can be assessed. Consequently, it is a fitting 'status report' for the end of this millenium.

I congratulate all those involved in the partnership that has produced this important publication. I am delighted, at the dawn of the new Environment Agency, to be able to commend this book to everyone with an interest or involvement in one of our greatest natural assets – the freshwater environment.

May 1996   Lord De Ramsey
                                            Chairman, Environment Agency

PREFACE

AQUATIC plants have never been popular with the majority of botanists. Their aesthetic appeal is understated rather than blatant, they grow in habitats which can be uncomfortable or even physically difficult to examine, and some of the most important genera are taxonomically difficult. However, biologists who overcome these initial discouragements find that aquatic plants possess a number of particularly interesting traits. Species from many different families have become adapted to the aquatic environment, and provide both remarkable examples of convergent evolution and striking illustrations of the fact that the same biological problem can be solved in several different ways. Reproduction by vegetative means is often more significant than the normal modes of sexual reproduction, and sterile hybrids may be ecologically important and very persistent. Many species are effective colonists of remote or newly created habitats, and some have spread at a dramatic rate following their introduction to areas outside their native range. These subjects have been treated in innumerable research papers and in two books which have become classics: Arber's *Water Plants* (1920) and Sculthorpe's *The Biology of Aquatic Vascular Plants* (1967).

Since the publication of Sculthorpe's book, it has become increasingly apparent that major changes in the distribution of aquatic plants in Europe have taken place in the last 50 years. Many species have declined in the face of habitat destruction, unsympathetic management of existing habitats, or the acidification or eutrophication of freshwaters. The long history of botanical recording in the British Isles allows us to document the effects of these changes. Other species have been introduced and spread, including several that originate in the New World. As the aquatic vegetation in many areas

of Britain, Ireland and mainland Europe has been progressively modified by the direct or indirect effect of human action, it has become increasingly important to recognize and protect those areas which still retain a rich aquatic flora.

In this book we have attempted to summarize the distribution, habitat and reproductive biology of the aquatic plants of the British Isles, and to provide a brief summary of their wider distribution. Many terrestrial groups are rather poorly represented in our area compared to mainland Europe, and they often reach maximum diversity in the high mountains of central Europe or the arid lands around the Mediterranean. By contrast, the aquatic flora of the British Isles is relatively rich. We have (for example) 21 of the 22 European species of *Potamogeton* and nine of the eleven species in *Ranunculus* subgenus *Batrachium*. Our wet climate, coupled with the effective dispersal of many aquatics, is probably responsible for this. An account of our aquatic plants should, therefore, be of more than parochial interest.

In his book *The Black Death*, Ziegler (1969) confesses that 'no-one...can be expert in this vast panorama and the spectacle of rival historians, each established in his fortress of specialised knowledge, waiting to destroy the unwary trespasser, is calculated to discourage even the most intrepid'. Although our field is more limited, we too have had to deal with many subjects about which we have no expert knowledge. We apologize for the errors and omissions which we must inevitably have made, and hope that the value of the book outweighs these faults. Above all, we hope that this work will encourage many more botanists to study these fascinating plants.

Monks Wood
April 1996

J. M. Croft
C. D. Preston

ACKNOWLEDGEMENTS

IN preparing this book we have been constantly encouraged by the enthusiasm with which many organisations and individuals have greeted the idea of an updated account of the aquatic plants of Britain and Ireland. We are grateful to many people for assistance in planning the project, and in the preparation of the maps and the text.

Professor R. W. Edwards first suggested to us how the project might be funded. The initial stages (1989–1991) were financed by the Institute of Terrestrial Ecology (ITE), a component body of the Natural Environment Research Council, the Nature Conservancy Council (NCC) and the Water Research Centre (WRC), acting under contract to the National Rivers Authority (NRA). Subsequently the Joint Nature Conservation Committee (JNCC) inherited this responsibility from NCC, and the NRA assumed direct responsibility for their part in the project. The NRA became part of the new Environment Agency (EA) in 1996. The co-operation of three funding bodies might have been fraught with difficulties; the fact that it has run smoothly has owed much to the commitment of the staff of all the organisations listed. We are particularly grateful to P. T. Harding (ITE), Mrs M. A. Palmer (JNCC) and J. Hogger (NRA/EA) for their constant support and their forbearance on those occasions when contract deadlines had to be extended. Others who have represented their organisations at project meetings, and who have always provided constructive support, are J. Hellawell and M. Gibson (NCC), M. E. Bramley, P. Raven, N. Smith and C. Spray (NRA/EA) and T ap Rheinallt and J. Gulson (WRC).

The project has also received support from other organizations, notably the Botanical Society of the British Isles, the Countryside Council for Wales, the Department of Agriculture (Northern Ireland), the Department of the Environment (Northern Ireland), English Nature, the National Parks and Wildlife Service (Department of Arts, Culture and the Gaeltacht, Republic of Ireland) and Scottish Natural Heritage.

Many people have generously provided records for inclusion in the database. We thank P. P. Abbott; J. Barneveld & T. H. Blackstock (Countryside Council for Wales); S. L. Bell, I. Butterfield, E. Clegg & O. Lassière (Scottish Natural Heritage); J. H. Bratton; A. O. Chater; A. P. Conolly; T. G. F. Curtis (National Parks and Wildlife Service); F. H. Dawson; J. J. Day; R. FitzGerald; C. E. Gibson, S. J. Smith & S. Wolfe-Murphy (Department of Agriculture, Northern Ireland); M. Gibson & S. J. J. Lambert (English Nature); R. Janes; L. A. & P. D. Livermore; L. J. Margetts; J. O. Mountford; D. A. Pearman; E. G. Philp; A. Spink; R. Surry (Dorset Environmental Records Centre); K. J. Watson; S. D. Webster; M. J. Wigginton (Joint Nature Conservation Committee). We are particularly grateful to N. F. Stewart, who not only provided his own numerous field records but also extracted records from survey and research reports held by the Research Branch of the National Parks and Wildlife Service, Republic of Ireland. We also thank T. G. F. Curtis of NPWS for permission to publish the data from these reports.

On visits to herbaria we have been assisted by D. Bolton (Exeter), M. Dowlen, J. R. Press & R. Vickery (Natural History Museum), J. R. Edmondson (Liverpool), R. G. Ellis, G. Hutchinson & A. D. Tipper (Cardiff), R. Goyder (Kew), D. R. McKean (Edinburgh), S. Marner (Oxford), J. G. Murrell & P. D. Sell (Cambridge) and M. J. P. Scannell & D. M. Synnott (Dublin).

At the Biological Records Centre V. J. Burton and W. A. Forrest have input records, A. Stewart and S. E. Yates dealt with records of aquatic plants received during the Scarce Species Project, and V. J. Appleby helped with word processing. H. R. Arnold and J. C. M. Dring have been responsible for database management, and we are particularly grateful to Henry Arnold for much advice and assistance. The maps have been plotted using the DMAP program kindly modified for us by A. J. Morton.

M. Palmer of the Monks Wood library has obtained copies of many publications for us. Information on particular genera or species has been provided by A. Brewis, K. J. Adams, J. R. Akeroyd, P. M. Benoit, A. O. Chater, T. G. F. Curtis, R. FitzGerald, V. Gordon, R. J. Gornall, G. Halliday, P. Hawkins, D. A. Pearman, D. A. Ratcliffe, T. C. G. Rich, W. Scott, P. D. Sell, D. A. Simpson, J. E. Smith, C. A. Stace, N. F. Stewart and G. T. D. Wilmore. We thank C. D. K. Cook, P. T. Harding, N. T. H. Holmes, J. O. Mountford and M. A. Palmer for reading and commenting on many of the species accounts, and H. R. Arnold for heroically proof-reading the entire manuscript.

We are most grateful to G. M. S. Easy for drawing the illustrations.

INTRODUCTION

THIS book summarizes the distribution, habitat and reproductive biology of the vascular plants which grow in freshwater in Britain and Ireland. It has been made possible by the co-operation of three funding bodies, the Environment Agency (formerly the National Rivers Authority), the Institute of Terrestrial Ecology and the Joint Nature Conservation Committee. We hope that it will be of use to all those with a professional or amateur interest in the aquatic plants of Britain and Ireland, and that it will introduce more botanists to the study of these fascinating, but often neglected, organisms.

There is a clear need for a new summary of the distribution of aquatic plants in Britain and Ireland. Maps of aquatic vascular plants in our area were published alongside those of terrest-rial taxa in the *Atlas of the British Flora* (Perring & Walters, 1962), and reprinted in 1976, 1982 and 1990 with only minor revision. We now know that these maps under-estimated the dis-tribution of many aquatic plants, not surpris-ingly in view of the fact that most of the fieldwork for the *Atlas* was carried out in five hectic seasons (1954–1958), often by botanists with little previous experience of aquatic plants. Since 1962 many areas have been surveyed in much greater detail, sometimes by specialist teams which have systematically surveyed aquatic habitats. Taxonomic studies have resulted in an improved understanding of some groups, notably *Ranunculus* subgenus *Batrachium*. In addition to these changes in our knowledge of plant distribution and taxonomy, there have also been major changes in the dis-tribution of the plants themselves. Some native species such as *Hydrocharis morsus-ranae*, *Potamogeton acutifolius* and *P. praelongus* are much less common than they were in the 1950s, whereas aliens such as *Crassula helmsii*, *Elodea*

nuttallii, *Lemna minuta* and *Myriophyllum aquaticum* were not then recorded in our area, or were very rare, but are now widespread. Revised distribution maps for some genera and species have appeared in a number of scattered publications. This book provides us with an opportunity to publish updated maps of all the aquatic species in a single volume. Further details of the mapped records are held on computer file in a database managed by the Biological Records Centre at ITE Monks Wood.

The maps are accompanied by accounts of the individual taxa. Scientific papers dealing with aquatic plants, like those dealing with other aspects of botany, have been published with increasing frequency in recent years. It is difficult to keep abreast of new literature, even if one has access to a large library and comput-erized abstracts of the major journals, and easy to forget older publications which contain much useful information. We have tried to provide an up-to-date account of the ecology, reproductive biology and distribution of the British and Irish aquatic plants, and references to sources of further information. The text and maps were completed in July 1995 but we were able to make minor changes to both in April 1996.

DEFINITION OF AQUATIC PLANTS

Aquatic vascular plants form an ecological rather than a taxonomic group, and cannot be defined with any degree of precision. The boundary between the land and the water fluc-tuates from day to day, season to season and year to year. The difficulties of definition are compounded by the fact that many species are adapted to life in the boundary zone, where predictable or unpredictable changes in water-level reduce the competition from larger but more specialized terrestrial or aquatic plants. Most authors of books on aquatic plants therefore begin by agonizing over their definition. All would agree with Aston (1973) that 'there is no firm boundary dividing aquatic from non-aquatic species'. One may attempt to draw the arbitrary line between aquatic and terrestrial species in different places and there are therefore many different definitions of an aquatic, some narrow and some broad. As Mason (1957) remarks, 'if we include only the wholly aquatic species...we eliminate the very important amphibious species. If we include also the amphibious species, we are drawn immediately up the shore, where the naturalness of communities and the overlapping of their species lead us farther and farther from the water.' Indeed, 'under some conditions almost any plant may be found in the water' (Fassett, 1940). Beal (1977) was driven to define aquatic and marsh plants as 'those plants which a typical terrestrial taxonomist does not collect for fear of getting his feet wet'!

We have attempted to include in this book those species which characteristically grow in water which persists throughout the year. The entire plant may be submerged or may float on the water surface, or the basal parts may be rooted in water but send out aerial leaves and inflorescences. We have included those species which usually grow in permanent water, and those such as *Juncus bulbosus* and *Persicaria amphibia* which often grow terrestrially, but have aquatic populations which are a significant feature of aquatic vegetation. We have excluded those which are only occasionally found, in small quantity, in water. We have also excluded species which grow in water for only part of the year, such as ephemerals which grow on damp mud in sites where the water recedes during the summer. As the boundary between aquatic and terrestrial plants is so blurred, we have sometimes bent the rules slightly to include a species if this allows us to cover an entire genus, sub-genus or group of related taxa. As Godfrey & Wooten (1979) predicted, 'criticism for having included certain plants, for not having included others, may confidently be expected'.

We have included accounts of those native or naturalized species listed by Stace (1991) which fulfil our criterion, together (if possible) with their component subspecies. We have also included accounts of the aquatic hybrids which are treated in full in Stace's Flora, i.e. those which sometimes occur in the absence of both parents. We have also provided accounts of *Hydrocotyle ranunculoides*, an alien which is now known to be well naturalized, *Cabomba caroliniana*, which is established (at least temporarily) in the Basingstoke Canal, and the widespread hybrid *Veronica × lackschewitzii*, which is not treated in full by Stace (1991) but often grows in the absence of both parents.

The three marine vascular plants in our area, *Zostera angustifolia*, *Z. marina* and *Z. noltii*, are not included here but comparable accounts of these species in Britain are provided by Stewart *et al.* (1994).

GEOGRAPHICAL COVERAGE

The geographical areas covered by this book are Great Britain, Ireland and the Channel Islands. The political units are the United Kingdom, the Republic of Ireland, the Isle of Man and the Bailiwicks of Jersey and Guernsey. We have used the phrases 'Britain and Ireland' and 'British Isles' to denote the entire area covered by this book. The northern and western regions of Britain and Ireland have much in common, and we have often referred to these as the 'north and west'; similarly we sometimes use the phrase 'south and east' to designate the southern and eastern areas of both Britain and Ireland.

SCIENTIFIC AND ENGLISH NAMES

The scientific and English names of the aquatic plant families, genera, species and subspecies follow Stace (1991), as does the order in which they appear. However, we have used the English name Yellow Flag for *Iris pseudacorus* rather than Yellow Iris, as Yellow Flag is the name in common use and we have encountered the name Yellow Iris only in books. The authors of the names of plants which are treated in full follow the standard forms in Brummitt & Powell (1992). Names of other British and Irish plants mentioned in the text also follow Stace (1991), and we have not cited the authors of these names. Authors are cited for species and infraspecific taxa which do not appear in Stace (1991).

SOURCES OF RECORDS

THIS chapter outlines the sources of the records plotted on the distribution maps. The maps are based on the records collected by the members of the Botanical Society of the British Isles for publication in the *Atlas of the British Flora* (Perring & Walters, 1962) and its *Critical Supplement* (Perring & Sell, 1968). These are held by the Biological Records Centre (BRC). To these have been added records received by BRC since the publication of these atlases, and a large number of records collected and incorporated into the BRC database specially for this publication.

A detailed account of the collection of records for the *Atlas of the British Flora* is given by Perring & Walters (1962). Most of the published records were collected between 1954 and 1958, when members of the BSBI attempted to record the flora in all the 10-km squares in Britain and Ireland. The remaining records were extracted from earlier field surveys or herbarium and literature sources. In the 1950s Irish records were collected on an extension of the British grid rather than on the current Irish grid. These have now been reallocated to the correct Irish grid square using the details of the locality; unlocalized records of commoner species have been assigned to the most appropriate Irish grid square.

Distribution maps of a selection of critical taxa were published in the *Critical Supplement to the Atlas of the British Flora* (Perring & Sell, 1968). These included the following taxa mapped as aquatics in this book: *Eleocharis palustris* subsp. *palustris*, *Glyceria* × *pedicellata*, *Lythrum portula* subsp. *longidentata*, *L. portula* subsp. *portula*, *Nuphar* × *spenneriana*, the *Potamogeton* hybrids, *Rorippa microphylla*, *R. nasturtium-aquaticum*, *R.* × *sterilis*, the subspecies of *Sparganium erectum*, *Utricularia australis* and *U. vulgaris*. Records of these taxa

were derived from herbarium specimens and (in some cases) from reliable literature sources and field recorders.

The records collected for the *Atlas of the British Flora* and its *Critical Supplement* formed the original basis for the BRC database. Records from numerous additional surveys have subsequently been added to them. These include the BSBI Monitoring Scheme, led by T. C. G. Rich, which resurveyed one in nine of the 10-km squares in Britain and Ireland in 1988 and 1989 (Palmer & Bratton, 1995). They also include records collected for the distribution maps in the *Atlas of Ferns of the British Isles* (Jermy *et al.*, 1978), covering all pteridophytes, and the BSBI handbooks *Sedges of the British Isles*, covering the genus *Carex* (Jermy *et al.*, 1982), and *Pondweeds of Great Britain and Ireland*, covering *Groenlandia*, *Potamogeton* and *Ruppia* (Preston, 1995). Records were also collected by the authors of papers on *Callitriche truncata* (Barry & Wade, 1986), *Elodea* spp. (Simpson, 1986) and *Ranunculus penicillatus* (Webster, 1988a). BRC has also received numerous individual records, often sent because they represented new vice-county records or additions to a particular 10-km square. The new vice-county records are published annually in the journal *Watsonia*.

CRITERIA FOR INCORPORATING NEW RECORDS

When we began to update the BRC database, we soon became aware that it was not possible to incorporate all the available records of aquatic plants, which were held by many different organizations and individuals, and in many different forms. It was necessary to decide which data to include, and we drew up the following criteria for assessing the available datasets.

1. Records should be made by botanists who are known to be competent in identifying aquatic plants.
2. Detailed surveys are preferable to broad-brush initial surveys of the same area.
3. Precisely localized records are preferable to records which provide less accurate information on the localities of the species recorded.
4. Recent records are preferable to old records, provided that sufficient old records have been obtained to identify major historical trends in the distribution of species.
5. Records from areas for which existing data are known to be inadequate are preferable to those which contain additional records from well-worked areas.
6. Records of species which are local or rare, or are good indicators of environmental conditions, are preferable to those of widespread species which occur in a broad range of habitats.

All datasets were assessed against these criteria, which are listed in descending order of importance. The datasets which met some or all of these criteria were then incorporated into the database if at all possible. Records which were already computerized were taken into the database without undue difficulty, and others were computerized by BRC during the course of the project. However, computerizing some desirable datasets would have been too labour-intensive, and we were unable to incorporate these into the database.

MAJOR SOURCES OF RECORDS INCORPORATED DURING THE PROJECT

The main datasets incorporated during the course of the project are listed below.

Scottish Loch Survey

The Scottish Loch Survey is a systematic survey of a large sample of Scottish lochs, which was started by the Nature Conservancy Council in 1984; by 1990 (when it was discontinued) it had covered a large area of northern Scotland.

These records were computerized at BRC and included in the database from which the maps in this volume were plotted. The Loch Survey was revived by Scottish Natural Heritage in 1993, but the only records which we have been able to incorporate from the 1993 and 1994 field seasons are those of *Potamogeton* species and hybrids.

Northern Ireland Lake Survey

This survey of a large sample of the lakes of Northern Ireland was carried out between 1988 and 1990 by a team based at the Freshwater Biological Investigation Unit of the Department of Agriculture (Northern Ireland), and funded by the Department of the Environment (Northern Ireland). Records from the survey were sent to us in electronic form.

Nature Conservancy Council surveys of lakes and other standing waters

Standing waters (including canals) were surveyed by C. Newbold, M. A. Palmer and other members of the former Nature Conservancy Council Chief Scientist Team (later renamed the Chief Scientist Directorate) between 1977 and 1988. Their records were compiled on the NCC Standing and Freshwater Habitats card (CST 0W2). We were allowed to copy these cards and we have computerized the records from these copies.

Other surveys of lakes and turloughs

Records have been extracted from published surveys of lakes, including the classic surveys of Scottish lochs by West (1905, 1910), Seddon's (1972) study of Welsh lakes and Stokoe's (1983) records from the lakes and tarns of Cumbria. The lochs surveyed by West (1910) in Fife and Kinross were resurveyed by Young & Stewart (1986) and we have also been able to include these records. Other sources which we have drawn upon include the surveys of lochs in the Outer Hebrides carried out by the Royal Botanic Garden, Edinburgh (Biagi *et al.*, 1985; Chamberlain *et al.*, 1984; Royal Botanic Garden, Edinburgh, [1983]) and an NCC survey of lochs and meres in Shropshire and Cheshire (Wigginton, 1989). Records from Irish lakes have been drawn from Heuff (1984), and records from the larger turloughs from Goodwillie (1992).

Surveys of individual canals

Numerous canals have been surveyed in recent years and the records summarized in unpublished reports. We have been able to incorporate records from the following canals into the database: Ballinamore & Ballyconnell Canal (Environmental Consultancy Services, 1989), Basingstoke Canal (Hall, 1988), Bude Canal (C. D. Preston, unpublished records), Chesterfield Canal (Alder, 1986), Exeter Canal (C. D. Preston, unpublished records), Forth & Clyde Canal (Watson & Murphy, 1988), Grand Canal (Dromey et al., 1992), Grand Western Canal (B. Benfield & L. J. Margetts, unpublished records), Grantham Canal (Candlish, 1975), Lancaster Canal (Livermore & Livermore, [1988]), Leven Canal (Kendall, 1987), Pocklington Canal (Tolhurst, 1987), Royal Canal (Caffrey, 1988; Dromey et al., 1991) and Union Canal (Anderson & Murphy, 1987). It is interesting that in Britain many of these surveys cover isolated canals; there appear to be fewer surveys from those areas where canals are most frequent.

Surveys of rivers by N. T. H. Holmes and others

The results of a major survey of British rivers by N. T. H. Holmes (Holmes, 1983) are held in a computer database by Scottish Natural Heritage. Records collected by a number of other surveyors have been added to them. Rivers and streams in the former South Region of Nature Conservancy Council were surveyed by P. D. Goriup in 1977 and 1978: these records have been computerized from copies of the NCC South Region Macrophyte Recording cards held by BRC or from listings formerly held by the Chief Scientist Directorate, NCC, Peterborough. Records from Irish rivers have come from the reports by Douglas & Lockhart (1983), Goodwillie et al. (1992), Heuff (1987) and McGough (1984).

Surveys of grazing marsh ditches

A pioneer survey of the grazing marsh ditches of Broadland was carried out by R. J. Driscoll

between 1972 and 1974 (Driscoll, 1975). The results of this survey were copied to BRC and have been incorporated into the database. The flora of grazing marsh ditches in a number of areas has been surveyed by teams from the Institute of Terrestrial Ecology led by J. O. Mountford (Mountford, 1994a): records from the Somerset Levels, the Romney and Walland Marshes and the Lower Idle Valley and Misson Levels have been added to the BRC database. Teams from the England Field Unit of the former Nature Conservancy Council also carried out important surveys of grazing marsh ditches between 1987 and 1989 (Palmer, 1991). We have received records from the Exminster Marshes, Derwent Ings, Broadland, and coastal grazing marshes in Essex, Suffolk and North Norfolk. The Gwent Levels have been surveyed under the auspices of the Countryside Council for Wales. Records from CCW, ITE and NCC (the latter now held by English Nature) were available in electronic form and were incorporated into the database. We also computerized records from a survey of dykes in the Deal–Sandwich area (Henderson, 1982).

Records from BSBI Vice-county Recorders

Records of some aquatic plant species have been obtained from the vice-county recorders appointed by the Botanical Society of the British Isles. These include records of *Groenlandia densa* and of all *Potamogeton* species and hybrids except the five commonest species (*P. crispus, P. natans, P. pectinatus, P. perfoliatus* and *P. polygonifolius*), which were obtained during the preparation of the BSBI Pondweeds Handbook (Preston, 1995). They also include records from Britain (but not Ireland) of those aquatic species which were covered by the joint BSBI/ITE/JNCC Scarce Plant Project (Stewart et al., 1994): *Callitriche hermaphroditica, C. truncata, Carex aquatilis, C. elata, Ceratophyllum submersum, Elatine hexandra, E. hydropiper, Eleocharis acicularis, Hydrocharis morsus-ranae, Isoetes echinospora, Luronium natans, Myriophyllum verticillatum, Najas flexilis, Nuphar pumila, Nymphoides peltata, Oenanthe fluviatilis, Pilularia globulifera, Potamogeton coloratus, P. compressus, P. filiformis, P. friesii, P. praelongus, P. trichoides, Ranunculus baudotii, R. tripartitus, Ruppia cirrhosa, Stratiotes aloides, Subularia aquatica* and

Wolffia arrhiza. We have also been able to make use of the records of the following nationally rare aquatic species in Britain which were contributed by vice-county recorders to JNCC for the proposed new edition of the *British Red Data Book*: *Alisma gramineum*, *Carex recta*, *Crassula aquatica*, *Damasonium alisma*, *Eriocaulon aquaticum*, *Hydrilla verticillata*, *Ludwigia palustris*, *Najas marina*, *Potamogeton epihydrus*, *P. nodosus*, *P. rutilus*, *Rumex aquaticus* and *Schoenoplectus triqueter*. However, the data collection for the Red Data Book project was still continuing when we went to press in July 1995.

We have not been able to obtain records of the remaining species from all vice-county recorders, partly because recorders were fully occupied with the Scarce Plants Project while we were gathering data for this book, and partly because we had not got the manpower needed to input any additional records. However, we did incorporate records from some vice-county recorders working in hitherto under-recorded areas, and they are listed in the next section.

Records from regional surveys

Records of aquatic plants in Dorset were provided by the Dorset Environmental Records Centre, and we are grateful to H. J. M. Bowen and D. A. Pearman for their critical scrutiny of these records. These records are particularly valuable as the Dorset records plotted in the *Atlas of the British Flora* were collected by R. d'O. Good in the 1930s, rather than during the main period of recording for the *Atlas* in the 1950s. Records from rivers, streams and major drainage ditches have been collected by the NRA Anglia region and are held in the Rivers Environmental Database (REDS). These have also been made available to us and incorporated into the BRC database. We have also received records from a number of other areas, many of them surveyed by the BSBI vice-county recorders, sometimes with the assistance of other local botanists. These areas include Co. Antrim (S. Beesley), Cardiganshire (A. O. Chater), Coll & Tiree (C. D. Preston & N. F. Stewart), Donegal (C. D. Preston & N. F. Stewart), Kent (E. G. Philp), the Lleyn peninsula of Caernarvonshire (A. P. Conolly), West Perth (N. F. Stewart), Wexford (R. FitzGerald) and Worcestershire (J. J. Day). The records from Kent include those published by Philp (1982), with later additions; those from West Perth were used in the compilation of the *Checklist of the Plants of Perthshire* (Smith *et al.*, 1992). We have also included records of aquatic plants made since 1986 in other, scattered localities in Britain and Ireland by C. D. Preston and N. F. Stewart.

Records from major herbaria

We have extracted details of herbarium specimens of selected taxa from a number of herbaria, notably those of the National Botanic Garden, Glasnevin, Dublin (DBN), National Museum of Wales, Cardiff (NMW), Natural History Museum, London (BM), Royal Albert Memorial Museum, Exeter (RAMM), Royal Botanic Garden, Edinburgh (E), Royal Botanic Gardens, Kew (K), and the Universities of Cambridge (CGE), Lancaster (LANC), Leicester (LTR) and Oxford (OXF). We have not been able to extract records of all the taxa covered in this book from any of these herbaria, but we have concentrated on extracting records of submerged and floating plants (which tend to be under-recorded) and, in particular, tried to obtain records of the more critical taxa. The visits to herbaria have allowed us to assess which taxa are frequently misidentified, and have helped us to assess field records of these plants.

SYSTEMATIC ACCOUNTS

EXPLANATORY NOTES ON THE TEXT

The introduction to each family and genus aims to set the aquatic representatives found in our area in the context of the family and genus as a whole. The number of species in the family and genus has usually been taken from two invaluable reference works, *The Plant-book* (Mabberley, 1987) and the *Aquatic Plant Book* (Cook, 1990a), on which we have also drawn freely for the rest of the introductory information. *The Families and Genera of Vascular Plants* (Kubitzki, 1990, 1993) is an even more useful work, but only the families from Isoetaceae to Polygonaceae in our account have been covered so far. In order to avoid repetition, we have included information which is applicable to all the aquatic members of a genus (or occasionally, of a family) in these introductory notes. **This means that the accounts of the family and the genus as well as that of the species have to be read to obtain the full information on any species.**

The first paragraph of the text for each species outlines its habitat in Britain and Ireland. In preparing these notes we have drawn on our own field experience, and on the publications of others. We have made particular use of a number of publications, including those of George (1992), Grime *et al.* (1988), Holmes (1983), Palmer *et al.* (1992), Rodwell (1995), Seddon (1972), Spence (1964), Stewart *et al.* (1994) and Wheeler (1980a,b,c). For information on the behaviour of species in western Ireland, we have often consulted Webb & Scannell (1983). We have sometimes drawn on these sources without citing them directly. The altitudinal limits of each taxon are cited. The term lowland is used for altitudes below 300 metres; in some cases where the species is

clearly confined to the lowlands (e.g. coastal taxa), we have not provided an explicit statement to this effect. The altitudinal range of vascular plants in Britain and Ireland has not received much attention recently, and Wilson (1956) is still the basic source of information. We have updated this where we have additional information, and G. Halliday has kindly checked the altitudinal limits against the records he has acquired during the Flora of Cumbria project and revised them where necessary. Nevertheless, there is scope for a more systematic updating of Wilson's records.

The second paragraph presents information on the reproductive biology of the taxon. Detailed information is available on the reproductive biology of a few species, but there are many about which we are surprisingly ignorant. We have, where necessary, included information in this paragraph derived from studies in mainland Europe or North America.

The third paragraph outlines trends in the distribution of the species in Britain and Ireland. Information on Quaternary history is taken from Godwin (1975). In describing more recent trends in the distribution of taxa we have tried to avoid unsubstantiated statements about the increase, or the decrease, of a species. We have instead drawn on the detailed local studies available in many county and other local floras to substantiate any claims we make. Few such floras are available for the Irish counties, and it has therefore proved more difficult to assess trends in the distribution of species in Ireland than in many parts of England. We have also noted in this paragraph those taxa which are believed to be under-recorded, and those which botanists are prone to misidentify.

The fourth paragraph outlines the world distribution of the taxon. This information is often based on the maps in Hultén & Fries (1986) or, for more southerly species, in Meusel & Jäger (1992) and Meusel *et al.* (1965, 1978). Although we have tried to assemble the world distributions as critically as possible, it is difficult to find reliable data on the world distributions of many species. Early botanists from Europe often applied the names of taxa they knew well at home to plants which they found when exploring other continents. Later it was often realized that the foreign plants were distinct, but the early, erroneous records may still persist in the literature.

Additional information may be given in a fifth paragraph.

The accounts in this book can provide only brief summaries of the information which is available for some species, especially widespread dominants which may be ecologically important throughout the northern hemisphere. We have therefore tried to cite the most important sources of additional information on each taxon.

EXPLANATORY NOTES ON THE MAPS

Available records are mapped in the 10-km squares of the Ordnance Survey National Grid in Britain, the Ordnance Survey/Suirbheireacht Ordonais National Grid in Ireland, and the Universal Transverse Mercator Grid in the Channel Islands. Orkney and Shetland, and the Channel Islands, are shown in insets. The following symbols are used on the maps:

■   Record of native population made in or after 1970.

●   Record of native population made between 1950 and 1969 (inclusive); no later record available.

○   Record of native population made before 1950 (or undated); no later (or dated) record available.

×   Record of introduced population made in or after 1950; no record of native population available.

+   Record of introduced population made before 1950 (or undated); no record of native population or later (or dated) record of introduced population available.

The total number of 10-km squares in each category in Britain (GB) and in Ireland (IR) is given in an inset to the left of the map. The figures for the Channel Islands are not included in these totals.

In interpreting the maps, it is important to appreciate that the original records collected for the *Atlas of the British Flora* are the only ones which result from a systematic survey which was aimed at covering all 10-km squares in Britain and Ireland. All the remaining sources are the result of partial surveys which were restricted to a sample of those squares, or particular habitats or areas. This has influenced our choice of the date categories to be mapped. The two symbols for native records from 1950 onwards [■, ●] together indicate the records made during or after the *Atlas* survey. However, we have distinguished records made in or after 1970. The absence of a later record may be significant for the rarer species or in well-recorded areas. The maps also indicate areas which still require survey work.

ISOETACEAE

The Isoetaceae is a small and taxonomically iso-lated family of pteridophytes. It contains a single genus which is divisible into two subgenera: the widespread subgenus *Isoetes* and subgenus *Stylites* (Amstutz) L. D. Gómez, which is endemic to the Andes (Kubitzki, 1990).

Isoetes L.

Isoetes is a genus of *c*.130 species, of which eleven are recognized in Europe. They have linear leaves (or sporophylls) arising from a corm. The base of the sporophyll is expanded and contains a sporan-gium. The large female megaspores and the smaller male microspores are borne in sporangia on separate sporophylls on the same plant. The spores are released as the sporophylls die and decay. Both male and female prothalli develop within the spore walls, the mobile male gametes emerging from the microspores to effect fertiliza-tion.

The simple structure of *Isoetes* offers few char-acters to the taxonomist, and this (coupled with extreme phenotypic plasticity) means the taxon-omy of the genus has always been difficult. Recent evidence indicates that interspecific hybridization can take place in the genus, and that some species have arisen as fertile allopolyploids (Hickey *et al.*, 1989). There are still a number of taxonomic problems to be resolved even in Europe. Three *Isoetes* species are known in Britain and Ireland, two aquatics (*I. echinospora*, *I. lacustris*) and one which is found in terrestrial or seasonally flooded habitats (*I. histrix*, a rare Mediterranean-Atlantic species restricted in our area to S.W. England and the Channel Islands). Plants which are apparently intermediate between *I. echinospora* and *I. lacustris* have recently been found in Britain (Camus *et al.*, 1988). Their taxonomic status has still to be eluci-dated and they are not considered further in this account. Cytological variation occurs within British *I. echinospora*, and both diploid (2n=22) and triploid (2n=33) plants occur at Cogra Moss (Rumsey *et al.*, 1993).

The aquatic *Isoetes* species exemplify the isoetid growth habit. Isoetids have a rosette of short stiff leaves (often with extensive internal air spaces) arising from a short stem. This growth habit is found in a number of other British and Irish species in unrelated families: *Subularia aquatica* (Brassicaceae), *Littorella uniflora* (Plantaginaceae), *Lobelia dortmanna* (Campanulaceae) and

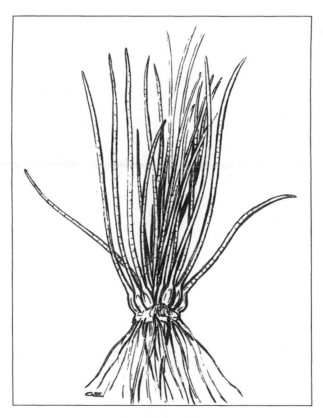

Isoetes lacustris L.

Eriocaulon aquaticum (Eriocaulaceae). These plants are so similar in habit that they are sometimes confused by field botanists. Isoetids are found in nutrient-poor habitats, often in areas which are disturbed by wave-action. They are slow-growing, evergreen plants. A high proportion of their biomass is accounted for by their roots, which are often mycorrhizal (Farmer & Spence, 1986). They are unable to utilize bicarbonate ions as a source of carbon dioxide, and are thus restricted to using CO_2 in the sediment in which they grow. It has recently been shown that *Isoetes* species and *Littorella uniflora* possess Crassulacean Acid Metabolism (CAM), a photosynthetic pathway more usually associated with desert succulents (Keeley, 1982; Keeley & Morton, 1982; Farmer & Spence, 1985). This pathway allows CO_2 to be accumulated at night in malic acid, which is then released in the day and used in photosynthesis. Desert plants are thus able to minimize water loss by opening the stomata at night. The CAM pathway allows *Isoetes* and *Littorella* to recycle CO_2 released into the leaf lacunae at night by respiration, a valuable adaptation for plants which characteristically grow in habitats where the inorganic carbon content is low (Boston, 1986).

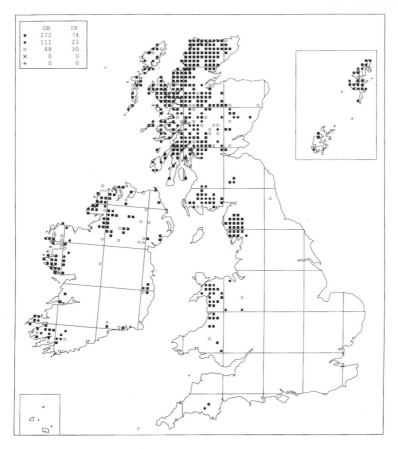

	GB	IR
■	272	74
●	111	23
○	48	30
×	0	0
+	0	0

Isoetes lacustris L.   Quillwort

I. lacustris is a characteristic species of oligotrophic lakes. It usually grows over rocky substrates with skeletal soils or over base-poor sands or clays, being absent from silty mud. Scattered plants often grow in shallow water at the edge of lakes, where they are frequently accompanied by *Littorella uniflora* and *Lobelia dortmanna*. However, the species is usually more frequent at greater depths and it can be the dominant (and sometimes the only) macrophyte at depths below 1.5–2.5m. It has been recorded in water as deep as 6m at Loch Lundie (West, 1905). The freshwater sponge *Spongilla lacustris* (L.) is often a conspicuous associate in these deep water stands (Spence, 1964). In addition to natural lakes, *I. lacustris* has also colonized artificial reservoirs and the Muirtown Basin of the Caledonian Canal. It is recorded from sea-level to 825m at Ffynnon Llyffant.

Little information is available about the reproduction of *I. lacustris* in Britain or Ireland, but the presence of small plants at the edge of lakes suggests that establishment from spores is not infrequent. In N. America neither megaspores nor microspores require a cold treatment for germination (Kott & Britton, 1982). A high proportion of megaspores germinate (an average of 87%), the maximum germination being reached in 45 days or less. Megaspores from dried plants fail to germinate. Plants also reproduce by gemmae which are formed at the base of the leaf (Kubitzki, 1990).

I. lacustris was clearly under-recorded in the *Atlas of the British Flora* (Perring & Walters, 1962), presumably because it is most abundant in deeper water and scattered plants are easily overlooked when growing with *Littorella uniflora* in the shallows. It is probably still under-recorded in areas which have not been covered by intensive surveys. There is little evidence for any major decline in *I. lacustris* in Britain or Ireland. Like other isoetids, it is extinct at some of its more easterly sites, which are now eutrophic, and other such populations are probably vulnerable (Farmer & Spence, 1986). It formerly grew, for example, at Bomere Pool but, like *Lobelia dortmanna*, it is now extinct there (Lockton & Whild, 1985). Although it has been suggested that *I. lacustris* and other isoetids have declined in Scandinavia as a result of acidification, Rørslett & Brettum (1989) conclude that the evidence for this is extremely scanty. There is much clearer evidence of a decline in the Netherlands, where acidification has followed the increased input of ammonium ions into water bodies in recent years. *Isoetes lacustris* and *Littorella uniflora* have often been replaced by *Juncus bulbosus* and *Sphagnum* spp., which lack the ability of the isoetids to use CO_2 from the sediment, but benefit from the increased concentrations of CO_2 in the water after acidification (Roelofs, 1983; Roelofs *et al.*, 1984).

I. lacustris is frequent in Europe north of *c.*53°N.; it extends south as scattered populations to N. Spain (Jalas & Suominen, 1972). It is also found at similar latitudes in eastern N. America. The American plants were formerly regarded as a separate species, *I. macrospora* Durieu, but are now recognized as conspecific with *I. lacustris* (Morin, 1993).

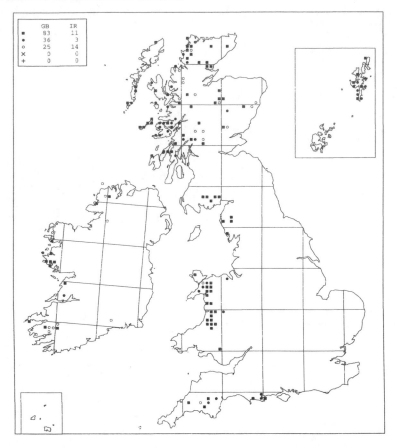

	GB	IR
■	83	11
●	36	3
○	25	14
×	0	0
+	0	0

Isoetes echinospora **Durieu**   **Spring Quillwort**

Over much of its British and Irish range *I. echinospora* grows, like *I. lacustris*, in oligotrophic lakes; it has also been found at the edges of slow-flowing rivers and in pools. It occurs over a wide range of nutrient-poor substrates including rocks and stones, base-poor sands and gravels, sandy mud or silt and peaty deposits. In Shetland it favours a finer substrate than *I. lacustris* and apparently prefers the sheltered southern and western sides of lochs (Scott & Palmer, 1987); this is probably true elsewhere in Britain. Although it usually grows in oligotrophic sites, *I. echinospora* has occasionally been found in eutrophic lakes, notably Llyn Llygerian and Llyn Llywenan in Anglesey (Seddon, 1965). Here it is found in areas where the growth of emergent species is restricted by rocky outcrops or by heavy wave action. *I. echinospora* has colonized artificial reservoirs in Cumbria and Wales, flooded gravel-pits at Capel Bangor and disused clay-pits in Dorset and Devon. It ascends from sea-level to 470m at Llyn Glas.

Little or no information is available about the reproduction of *I. echinospora* in Britain or Ireland. Experimental studies in N. America have shown that a cold period of 4°C for twelve weeks stimulates megaspore germination, but even with such a treatment total germination was very low (15%). Spores subjected to shorter pretreatments and control samples failed to germinate, as did megaspores from dried specimens.

Microspores germinated without a cold pretreatment but a 4°C pretreatment for twelve weeks resulted in a higher percentage germination (Kott & Britton, 1982).

Although *I. echinospora* often has more flaccid, gradually tapering leaves than *I. lacustris*, the two species can be identified with certainty only by microscopic examination of ripe megaspores. Although *I. echinospora* was first reported from Britain by Babington (1862, 1863), it is almost certainly under-recorded both because *Isoetes* plants are themselves often overlooked and because *I. echinospora* might be assumed to be *I. lacustris* or not noticed when the species are growing together. Fossil evidence indicates that *I. echinospora* has been present in Britain from the late Weichselian onwards, and long-term persistence through the Flandrian has been demonstrated at some sites (Godwin, 1975). The botanical records are insufficient to provide any evidence of change in its distribution since it was first reported in 1862.

I. echinospora has a similar European range to *I. lacustris*: it is frequent in N. Europe and present as very scattered populations south to N. Spain and Greece (Jalas & Suominen, 1972). *Isoetes echinospora* sensu lato is a very variable and taxonomically complex species which has a circumboreal distribution, with isolated occurrences in Barbados and Chile.

I. echinospora and *I. lacustris* are ecologically similar in Britain and Ireland, and can grow together (Stokoe, 1978). In Scandinavia *I. lacustris* 'occurs mainly in oligotrophic to mesotrophic lakes' whereas *I. echinospora* 'is a ubiquitous species of clayey riverside habitats and intermittently dried-out mudflats', also occurring in dystrophic lakes, clear-water lakes, turbid eutrophicated sites and in slightly brackish water (Rørslett & Brettum, 1989). This suggests that *I. echinospora* has a much wider habitat range in Scandinavia than in Britain or

Ireland, with a correspondingly clearer distinction between *I. echinospora* and *I. lacustris*. *I. echinospora* tends to occur in shallower water than *I. lacustris* in Scandinavia (although there is considerable overlap between the two species). Unfortunately, critical observations on the depth at which *I. echinospora* grows in the British Isles are not available.

EQUISETACEAE

This family contains a single extant genus, *Equisetum*. The species of *Equisetum* are the only living representatives of the Sphenopsida, a class which is known from fossil evidence to have been in existence for over 250 million years. In the Upper Carboniferous, members of the class included both tall trees which were co-dominant in the coal-measure swamp forests and herbaceous plants similar to modern *Equisetum* species.

Equisetum L.

A small genus containing 15 species (Hauke, 1974). The genus is widespread in both hemispheres, but is absent from Australia and New Zealand (Kubitzki, 1990). Cook (1990a) classifies two species as aquatics, and several others are plants of damp or wet terrestrial habitats. Descriptions of the ecology of the British and Irish taxa are given by Page (1982, 1988) and Page & Barker (1985) and their distributions are mapped by Jermy *et al.* (1978).

The genus is divided into two subgenera. The species in subgenus *Equisetum* have annual stems which develop in the spring and die back in autumn; the cones are obtuse. The species in subgenus *Hippochaete* (Milde) Baker usually have persistent stems and the cones are apiculate. Both subgenera have been monographed by Hauke (1963, 1978). Representatives of both subgenera occur in our area but the taxa treated here are in subgenus *Equisetum*.

Equisetum species are rhizomatous perennials. The jointed stems are the main photosynthetic structures; they often have whorls of branches at the nodes. The leaves are reduced to inconspicuous scales. The sporangia are borne in cones at the apex of the stems or, in some species, at the end of the branches. The vegetative and fertile stems are usually similar in appearance, but the cone-bearing stems of some species, including *E. arvense*, are pale brown and ephemeral. The plants are homosporous and the short-lived spores germinate to produce prothalli which are green and photosynthetic. In *Equisetum* subgenus *Equisetum* the prothalli are either male or bisexual; the bisexual individuals initially produce archegonia and later develop antheridia. Detailed studies of the reproductive biology of species in this subgenus have been published by Duckett (1970a,b,c, 1979)

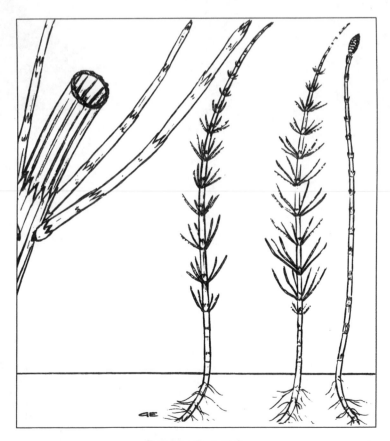

Equisetum fluviatile L.

and Duckett & Duckett (1980).

Hybrids have been recorded between species within a subgenus, but not between members of different subgenera. They normally become established on moist, freshly exposed soil surfaces. They are partially or totally sterile but can form persistent clones which spread by growth or frag-mentation of the rhizome. There is a particularly rich concentration of hybrids in western Britain and Ireland, where the chances of hybridization and the subsequent establishment of hybrid clones may be favoured by the oceanic climate (Page & Barker, 1985).

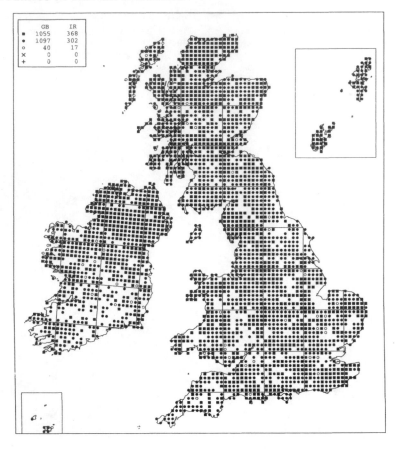

Equisetum fluviatile L.   Water Horsetail

E. fluviatile grows in a wide range of water bodies from
ditches and small ponds to large lakes and the sheltered
backwaters of rivers, growing in marshes and fens and as
an emergent in standing water. In fens it occurs as scattered stems in closed vegetation where the dominant
species are often tall sedges; common fen species such as
Galium palustre, *Menyanthes trifoliata*, *Potentilla palustris*
and *Calliergon cuspidatum* are amongst the most frequent
of its numerous associates. It can also be found in open
fen carr. In small lakes and the more sheltered bays of
larger lakes it is frequently found with *Carex rostrata* in
swamps where the water is at least 0.3m deep; in shallower water it is usually replaced by *C. rostrata*. It also
grows in open stands of other emergents including
Phragmites australis and *Schoenoplectus lacustris*. In more
exposed lakes it often grows in water up to 1.5m deep,
forming open stands in the absence of other emergents
but with a range of aquatics with submerged or floating
leaves including *Littorella uniflora*, *Myriophyllum alterniflorum*, *Potamogeton natans* and *P. gramineus*. *E. fluviatile*
is a species with a notably wide ecological amplitude, not
only tolerating a range of water depths but also found in
oligotrophic, mesotrophic and eutrophic water and over
a variety of aerobic to highly anaerobic substrates
including inorganic sands, clays, and muds with an
appreciable organic content. It extends from sea-level to
690m on Little Dun Fell and up to 900m in the

Breadalbanes (White, 1898). There is an unlocalized
record from 1050m (Grime *et al.*, 1988).

The cones of *E. fluviatile* are borne on green, photosynthetic stems. The spores are released as soon as the
cones are mature. The normal period of release is from
mid-May to the end of June; a few fertile shoots are
sometimes produced throughout the summer and spores
can continue to be released from these into September
(Duckett, 1970a). The spores are short-lived (Duckett,
1970b). The gametophytes require newly exposed, bare
mud on which to develop and are typically found around
the edge of reservoirs where the water-level fluctuates
markedly or on mud left when lakes have been drained.
Gametophytes develop into sporophytes only if the mud
remains exposed above the water during the subsequent
winter. For a detailed description of the ecology of the
gametophyte, see Duckett & Duckett (1980). Established
stands of *E. fluviatile* spread vegetatively by means of the
extensive rhizome system. Stems are often detached by
waterfowl or broken off by waves, and even short lengths
root from the nodes when floating on water or stranded
on wet mud (Praeger, 1934b).

E. fluviatile is a distinctive species. It has presumably
decreased in abundance in areas where wetlands have
been drained in historic times. It is widespread and
locally abundant in Britain and Ireland and in most of
our area there is no evidence to suggest that its distribution is changing. In the London area Burton (1983) suggests that it may have been eliminated from some sites

by *Typha latifolia*, which has become more frequent in recent years.

E. fluviatile is widespread in the boreal and temperate zones throughout the northern hemisphere. In Europe it is frequent from Iceland and N. Scandinavia south to France, N. Italy and Romania, but rare in the Mediterranean region (Jalas & Suominen, 1972).

The ecological requirements of *E. fluviatile* and *Schoenoplectus lacustris* are rather similar, but Spence (1964) comments that they can rarely be found together in Scottish lakes. The reasons for this are unclear.

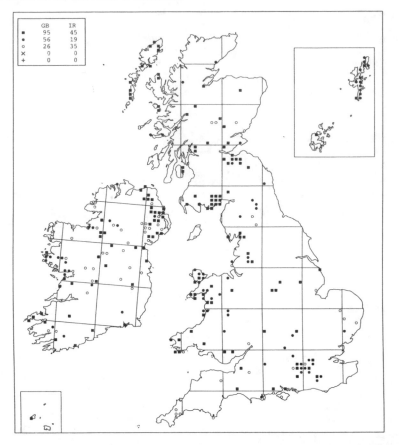

	GB	IR
■	95	45
●	56	19
○	26	35
×	0	0
+	0	0

***Equisetum* × *litorale* Kühlew. ex Rupr.**
(*E. arvense* L. × *E. fluviatile* L.)   Shore Horsetail

As a hybrid between a terrestrial and an aquatic species, *E.* × *litorale* can be found in a wide variety of moist lowland habitats, and occasionally in water up to 0.3m deep. Many of its sites are on marshy ground near bare or disturbed soil. It has been recorded from the edge of quarry pools, ponds, lakes, reservoirs, streams and rivers, flooded gravel-pits, the dried-out beds of canals, moist ditches, roadsides and railway cuttings, open peaty mud in bogs, damp pastures, open willow scrub and the edges of moist woods. In the Hebrides it is particularly frequent by the eroding edge of streams flowing over shingle or machair, where *Iris pseudacorus* is a characteristic associate (Page, 1988). This hybrid can form large and vigorous stands which may displace the parent species. Some colonies along streams and canals can extend for over 1.5km. Like the parents, it is found over a wide range of substrates from calcareous sand to acidic peat.

E. × *litorale* produces cones on green, photosynthetic stems, although cone-bearing stems are infrequent in some colonies. The spores are infertile. Clones spread by the growth of rhizomes, and perhaps also by waterborne shoot fragments. Experiments in N. America have shown that cut stems of the hybrid can become estab-

lished if they are buried in mud, and rhizome fragments can regenerate even if placed on the surface of the mud. The ability of the hybrid to regenerate in this way is greater than that of its parents (Peck, 1980).

E. × *litorale* has been known from Britain since 1886, when it was first reported (as a species) by Beeby (1886), but for many years it was regarded as a very rare plant. It was discovered in Ireland by Praeger (1917), who subsequently found it in numerous Irish sites. It is still overlooked by many botanists and is probably much more frequent than the map suggests. Like other *Equisetum* hybrids, it is probably more frequent in northern and western areas than in the south and east.

The distribution of *E. fluviatile* is described under that species; *E. arvense* is even more widespread. *E.* × *litorale* is the most frequent hybrid in the genus, having been recorded at many localities in Europe and N. America.

This is a variable hybrid, but the variation is thought to be environmentally induced (Duckett & Page, 1975). The hybrid nature of this taxon has been demonstrated by cytological studies (Manton, 1950) and by the artificial synthesis of the hybrid. In experimental studies the hybrid has only been synthesized with *E. fluviatile* as the female parent; attempts to produce the reciprocal hybrid have been unsuccessful (Duckett, 1979).

MARSILEACEAE

This is a small and taxonomically isolated family of aquatic ferns, containing three genera and some 72 species. *Marsilea* is the largest genus; the 65 species are found in tropical and warm-temperate areas north to southern Europe. *Pilularia* is the only genus in Britain and Ireland. For an account of the family, see Kubitzki (1990).

Pilularia L.

Two of the six *Pilularia* species occur in Europe, but only one is represented in the British Isles. The species are rhizomatous and have erect, thread-like leaves which lack a lamina. The sporangia are borne in spherical sporocarps (pills) which form in the axils of the leaves. Both megasporangia and microsporangia are found in the same sporocarp. The unit of dispersal is probably the sporocarp, which dehisces to release the spores.

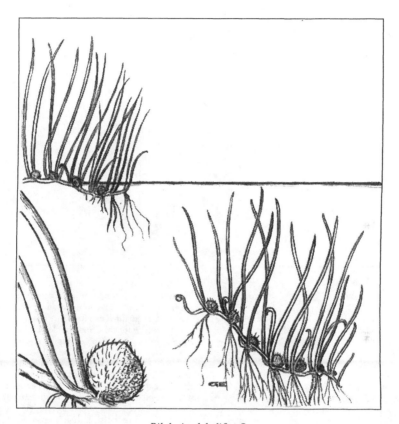

Pilularia globulifera L.

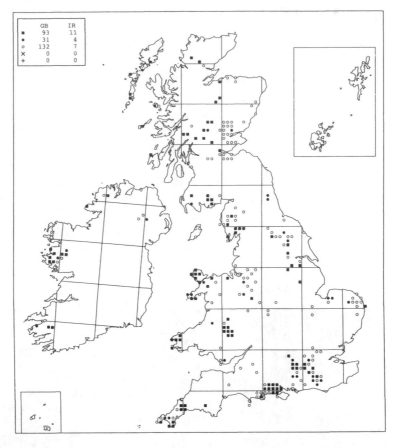

	GB	IR
■	93	11
●	31	4
○	132	7
×	0	0
+	0	0

Pilularia globulifera L.   **Pillwort**

This is a plant of lowland, non-calcareous silty, gravelly or peaty lake and pond margins, ditches and shallow pools. It is an opportunist species, which can rapidly colonize an open substrate exposed by falling water-levels or newly created in disused clay- and gravel-pits; it is even recorded from the swampy track of a disused railway (Brewis *et al.*, 1996). Characteristic associates include *Apium inundatum*, *Hydrocotyle vulgaris* and *Ranunculus flammula*. Some colonies on newly disturbed ground may be short-lived but the plant can persist in lakes and reservoirs, appearing in abundance when water-levels are low (cf. Stewart, 1988). Occasionally it maintains itself in denser plant communities as a submerged aquatic in base-poor pools, or in fens with *Carex diandra*, *Menyanthes trifoliata*, *Calliergon cuspidatum* and other pleurocarpous mosses. *P. globulifera* is not a competitive species, and at some of its sites suitable conditions are maintained by the trampling of cattle and horses. Most sites are at low altitudes, but the species is recorded up to 450m on Pant-y-llyn Hill.

P. globulifera is a perennial which dies back during the winter, but some 'sub-evergreen fragments' persist and regrow in the following spring (Page, 1982). Sporocarps

are formed in most populations except those in permanent deep water. Field observations suggest that the spores can remain viable for many years, but there is no hard evidence for this. Spores released from the sporocarp in late summer can develop through the gametophyte stage to produce new sporophytes within 17 days.

P. globulifera was lost from many sites before 1930, largely because of habitat destruction. Sites have continued to be destroyed since 1930, and the species has also disappeared from pools which have become overgrown following the cessation of grazing (Byfield & Pearman, 1994). The current distribution is probably stable: the species may disappear from some sites as a result of successional changes but these losses will perhaps be balanced by the colonization of new areas. It is an inconspicuous species which was recently the subject of a special survey led by A. C. Jermy (1984), but which may still be under-recorded in northern and western areas.

P. globulifera is endemic to western Europe, with lowland areas in Britain, France, N. Germany and S. Sweden containing the bulk of the populations and outliers extending to Portugal, Italy and Slovakia (Jalas & Suominen, 1972). It is decreasing in much of mainland Europe 'at an alarming rate' (Cook, 1983), and the sites in our area are therefore particularly significant.

AZOLLACEAE

The Azollaceae is a small family of pteridophytes which contains a single genus, *Azolla*.

Azolla Lam.

Azolla is a very distinctive genus of six rather similar species. The genus was formerly more diverse, and 30 species are known as fossils from the Cretaceous (Morin, 1993). The sporophytes are small free-floating plants with pinnately branched stems which bear scale-like leaves and pendulous roots. The leaves have an upper photosynthetic lobe, which is borne above the surface of the water, and a lower colourless lobe which rests on the water. The lower lobe is usually inhabited by the cyanobacterium *Anabaena azollae* Strasb. The presence of this nitrogen-fixing organism has led to the use of *Azolla* in E. Asia as a food for livestock and as a green manure for rice fields, a practice which originated in China at least 2000 years ago (Shi & Hall, 1988).

Sporocarps of two different sizes are produced on the lower surface of the mature frond. The smaller sporocarps contain microsporangia which produce the mobile male gametes. These fertilize the megaspores in the larger sporocarps. After fertilization the embryo develops within the wall of the megaspore until the first leaf grows, when it emerges as a free-floating sporophyte. For an illustrated account of the life-cycle of *Azolla* and a taxonomic review of the species, see Tan *et al.* (1986); further details of the morphology of the genus are provided by Kubitzki (1990).

Azolla caroliniana Willd. has been reported from Britain and Ireland, but all records are referable to *A. filiculoides*. *A. caroliniana* is 'the least understood species in the genus' (Tan *et al.*, 1986); plants from mainland Europe hitherto named *A. caroliniana* are referred to *A. mexicana* C. Presl by Tutin *et al.* (1993). For a detailed study of the differences between the two species naturalized in Europe, see Pieterse *et al.* (1977). A recent account in the *Flora of North America* (Morin, 1993) separates the species using different characters from those which have traditionally been employed, and European plants perhaps require re-investigation in the light of this new approach.

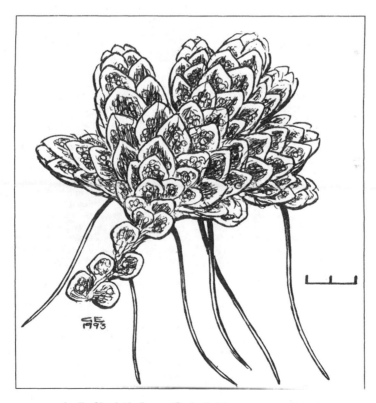

Azolla filiculoides Lam. (Scale divisions represent 1mm)

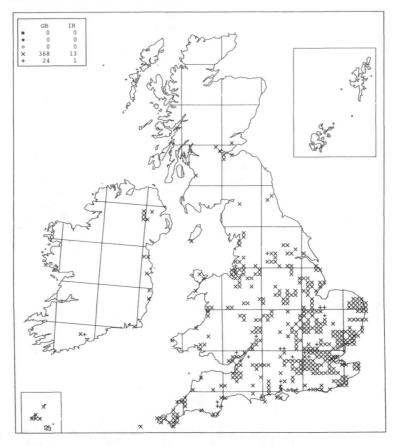

	GB	IR
■	0	0
●	0	0
○	0	0
×	368	13
+	24	1

Azolla filiculoides Lam.   Water Fern

A small but highly gregarious floating fern, which can occur in abundance on the surface of canals, ditches, ponds and small lakes or in the sheltered backwaters of larger water bodies. Dense stands exclude almost all competitors; less dense aggregations are often inter-mixed with *Lemna gibba*, *L. minor* and *L. minuta*. *Ceratophyllum demersum* and *Elodea canadensis* or *E. nuttallii* often grow under the *Azolla* layer. *A. filiculoides* is most frequent in eutrophic, calcareous waters and is often particularly frequent in ditches near the sea. It is most frequent in the lowlands, but ascends to 450m in a fellside pool above Rydal Water.

Plants are dispersed by water and perhaps also as frag-ments on the feet or feathers of birds or on drainage machinery (Mountford & Sheail, 1989). They increase very rapidly in the summer by clonal spread; colonies develop a striking red colour in autumn and persist into the early winter but then break up and disappear from view. Regeneration in the spring is thought to be from surviving fragments. However, examination of herbar-ium specimens and of established populations in the wild shows that sporocarps are regularly produced, sug-gesting that British and Irish populations may reproduce sexually.

A. filiculoides, like *Rhododendron ponticum*, was native to Britain and mainland Europe in previous interglacials (West, 1953; Tralau, 1959; Godwin, 1975) but in the present interglacial it occurs only as an introduction. In Britain it was first recorded at Pinner in Middlesex in 1886 (Odell, 1886), but did not become well established until some 25 years later. By 1914 it was well naturalized in the Thames valley and Broadland, and had been recorded at other scattered localities (Marsh, 1914). In Broadland its spread was assisted by floods in 1912. It is now sold as an ornamental plant in many garden centres and it has continued to spread in the wild, apparently from a combination of the continued release of unwanted garden stock and the subsequent dispersal of established populations. It can invade a region very rapidly, but although persistent at some sites it survives for only a few years at others or reappears after long intervals. There is much anecdotal evidence about the fluctuations of this species but an absence of critical study. Sporophytes have been eliminated from particu-lar sites by hard winters (cf. Muscott, 1983). It is not known whether populations can overwinter as spores even if mature sporophytes are killed. Warm summers apparently favour *A. filiculoides*.

A. filiculoides is a native of N. & S. America which is now widely naturalized elsewhere. In Europe it is wide-spread south of 55°N., but does not extend into colder latitudes (Jalas & Suominen, 1972).

NYMPHAEACEAE

This family includes the familiar water-lilies in the genera *Nuphar* and *Nymphaea*, as well as four small genera native to S. and E. Asia, S. America and Australia. Most members of the family have floating leaves and emergent flowers; many species also have delicate submerged leaves and in a few species these are the only leaves present. The non-European genera include the giant water-lily *Victoria amazonica* (Poepp.) J. C. Sowerby. For an account of the family, see Kubitzki (1993).

Mabberley (1987) has stressed the fact that the family, although classified in the dicotyledons, has many of the characters of the monocotyledons. The individual species possess characters which are not normally found in members of the same family (e.g. inferior and superior ovaries, orthotropous and anatropous ovules). Mabberley suggests that they are successful, specialized relics of a primitive stock which existed before the monocotyledons were distinct from the dicotyledons. They were able to penetrate temperate latitudes from a presumably tropical origin as they are rhizomatous plants which can overwinter in the buffered underwater environment.

For an account of the British water-lilies, see Heslop-Harrison (1955c). The cultivated species and hybrids are described by Swindells (1983).

Nymphaea L.

A genus of *c*.40 species, of which 4 are native to Europe. There is no recent monograph of the genus, and even in Europe there are unsolved taxonomic problems. These arise partly from the variability of the species, and the apparent occurrence of natural hybrids, and partly from the artificial hybridization of species by horticulturists followed by the planting of such hybrids in ornamental waters or semi-natural habitats. In addition to the native *N. alba* treated below, plants of horticultural origin have been planted in Britain in disused gravel-pits and other waters managed by anglers. They can usually be recognized by their pink-tinged or yellow flowers. Stace (1991) states that most are hybrids of *N. alba*, but suggests that they need further study. A N. American species, *N. odorata* Aiton, is established at two localities in S.E. England, Bolder Mere and Trilakes (Clement & Foster, 1994).

A detailed account of the ecology of *N. alba* in Britain and Ireland is provided by Heslop-Harrison (1955b).

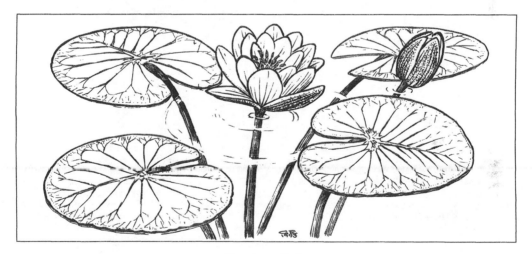

Nymphaea alba L.

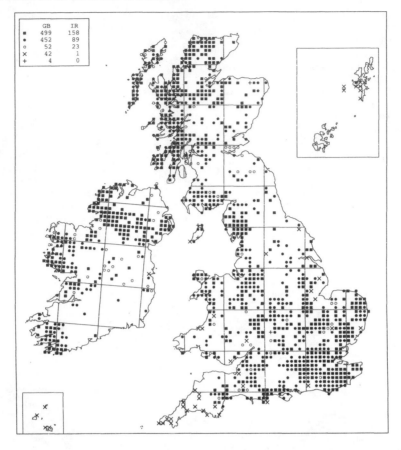

	GB	IR
■	499	158
●	452	89
○	52	23
×	42	1
+	4	0

Nymphaea alba L.   **White Water-lily**

N. alba often grows in abundance in the still water of lakes, ponds and pits, and in slowly flowing large ditches and the backwaters of lowland rivers. It is usually found at depths of 0.5–1.5m in open water or in very sparse stands of emergents, but it can grow in very shallow water, and is even found with aerial leaves in swamps in western Ireland (Praeger, 1934a) and Broadland (Wheeler, 1980a). In lakes stands of *N. alba* are often bordered on the landward side by emergents, and in deeper water by *Potamogeton natans* or submerged species. Stands of *N. alba* are usually species-poor, and few species occur as associates with any degree of constancy (Heslop-Harrison, 1955b; Spence, 1964). The species tolerates a wide range of water chemistry from oligotrophic and mesotrophic lakes to eutrophic lakes and rivers. It is, however, notably intolerant of disturbed and turbid water, and in windswept areas it is usually confined to small lakes and sheltered bays. It is also absent from rivers and canals with appreciable boat traffic. It occurs over a range of fine-grained substrates including peat, sand, silt, marl, mud and clay. Although primarily a lowland species, it grows at 405m at Dock Tarn and was formerly found at 456m at Angle Tarn, Place Fell.

N. alba is a rhizomatous perennial which, when mature, possesses only floating leaves. It sheds its leaves in autumn, coming into leaf in spring and flowering from early June to late August. The flowers close at night and open during the day. They are visited by insects but pollen is shed directly on to the stigmas and cross-pollination is probably not essential for fertilization. After fertilization the peduncles coil and draw the fruits under water, where they mature and eventually burst. The arillate seeds can float for 1–3 days (often in a mass resembling frog's-spawn) before sinking. Floating seeds are often eaten by birds and fish, which digest them completely and are therefore unlikely to act as agents of dispersal. The seeds have no adaptations to facilitate transport by animals and are also unable to withstand 28 days' desiccation. Natural dispersal is probably dependent on transport by water. The seeds do not germinate in the season in which they are shed. Thereafter germination is erratic, and in favourable years large crops of seedlings can be found. Germination is possible in anaerobic conditions. For details of the reproductive biology of the species see Heslop-Harrison (1955b), Smits & Wetzels (1986), Smits *et al.* (1989) and Velde (1986).

N. alba is certainly native in both Britain and Ireland; pollen evidence shows that *Nymphaea* was present throughout the late Weichselian and Flandrian. Its exact native distribution is more difficult to determine. It is clearly native in much of W. Scotland, N.W. England and W. Ireland, although even in Shetland the situation has been confused by introductions (Scott & Palmer, 1987). Its status in S.E. England has been greatly obscured by the drastic modification of natural river systems and the popularity of the species as an ornamental plant. It is probably less frequent as a native plant than the map suggests. In Broadland it initially increased in response to rising nutrient levels, and it was abundant in many sites between 1900 and 1950, but it subsequently declined as the water became excessively eutrophic (George, 1992). The species has also been lost from other sites because of eutrophication, but this decline has been counterbalanced by the planting of *N. alba*, or related hybrids, in new localities.

N. alba is widespread in Europe, reaching 66°N. in Norway; it is, however, scarce in the Mediterranean region. It is also recorded from N. Africa and W. Asia. In N.E. Europe and much of Asia it is replaced by the closely related *N. candida* C. Presl, with which it hybridizes. For maps of both taxa, see Jalas & Suominen (1989) and Hultén & Fries (1986).

This is a variable species, and populations in nearby but unconnected water bodies may differ in characters such as ovary shape or the number of petals, stamens or carpels. The degree of variation within populations differs from site to site (Heslop-Harrison, 1955c). Small populations in oligotrophic sites in N. and W. Scotland and W. Ireland are genetically distinct and have been distinguished as *N. alba* subsp. *occidentalis*. The consensus of opinion is that these plants do not merit taxonomic recognition (Heslop-Harrison, 1953a, 1955b,c; Tutin *et al.*, 1993), and few observers have recorded subsp. *occidentalis* in recent years. However, the subspecies has recently been taken up by Stace (1991). The map excludes records known to be of cultivated waterlilies doubtfully referable to *N. alba*, but some of these will almost certainly have been reported as *N. alba* and will consequently appear on the map.

Nuphar Sm.

The taxonomy of this genus is more problematical than that of *Nymphaea*, and Cook (1990a) could estimate the number of species only as between 7 and 20. Most authors, including Stace (1991) and Tutin *et al.* (1993), recognize two native species in Britain and Europe, *N. lutea* and *N. pumila*, with the hybrid *N.* × *spenneriana*. In addition, the alien

N. advena has recently been recorded from Britain. Beal (1956) considers that both *N. pumila* and *N. advena* are subspecies of a very broadly defined *N. lutea*.

For an account of the ecology of all three native taxa in Britain and Ireland, see Heslop-Harrison (1955a).

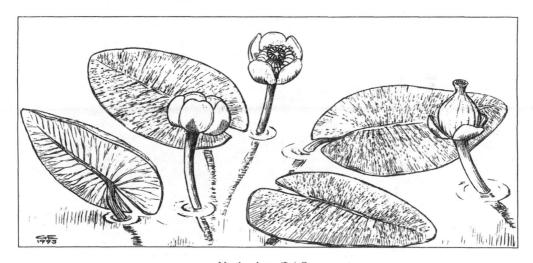

Nuphar lutea (L.) Sm.

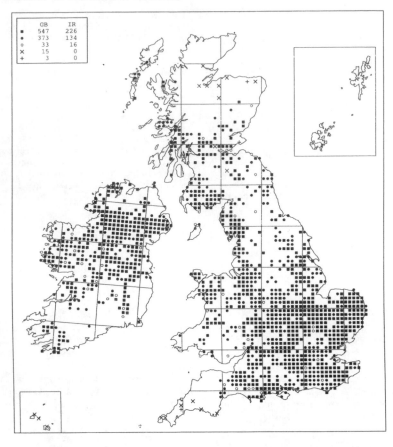

Nuphar lutea (L.) Sm.   Yellow Water-lily

N. lutea is a plant of mildly acidic or basic, mesotrophic or eutrophic water in lakes and in slowly flowing lowland rivers, canals and large drains and ditches. It is normally found in open water at depths of 0.5–2.5m, but it has been recorded as deep as 3.6m. Stands of *N. lutea* are at least as species-poor as those of *Nymphaea alba*; *Elodea canadensis*, the aquatic form of *Juncus bulbosus*, *Littorella uniflora* and *Potamogeton obtusifolius* are the most frequent associates recorded in Scottish lochs (Spence, 1964). Although it is tolerant of a range of water chemistry, *N. lutea* does not extend into sites which are as base-poor and oligotrophic as those tolerated by *Nymphaea alba*. Consequently it is rare in N.W. Scotland and the Hebrides. In richer lakes the species can often be found at the same site, with *N. lutea* usually growing in slightly deeper water than *Nymphaea alba*. *N. lutea* is notably more tolerant of mechanical disturbance than *Nymphaea alba*, and it can often be found along the edge of rivers and canals which carry considerable boat traffic. In very disturbed sites it persists as plants with submerged leaves which rarely produce floating leaves or flowers. *N. lutea* is usually found over mud or silt and rarely over peat; in rivers it is particularly characteristic of clay substrates (Holmes, 1983). Although primarily a lowland plant, it ascends to 510m at Llyn Crugnant.

N. lutea overwinters as a rhizome, producing translucent submerged leaves in early spring and coriaceous floating leaves in April and May. It flowers from June to August. The flowers, which remain open at night, are not automatically self-pollinated, but are visited by insects which may effect either self- or cross-pollination. The fruits ripen above the water. The seeds are embedded in the spongy tissue of the carpel, which normally sinks within a day of its release. The seeds resemble those of *Nymphaea alba* in lacking morphological adaptations for animal dispersal, and in being killed by desiccation and completely digested when eaten by birds or fish. The main chance of dispersal must be by water movement. Otherwise long-distance dispersal must be a rare event: Heslop-Harrison (1955a) found viable seed in the excreta of a heron which had presumably eaten a fish that had been recently feeding on *N. lutea*! A proportion of seeds can germinate in the year they are shed. Germination can take place in anaerobic conditions, but is erratic. In some years enormous numbers of seedlings can be found in favourable localities, but few survive to maturity. For further details of the reproductive biology of the species, see Heslop-Harrison (1955a), Smits & Wetzels (1986), Smits *et al.* (1989) and Velde (1986).

The fossil history of *N. lutea* is well documented, and Godwin (1975) concludes that it has probably persisted in the British Isles right through the Pleistocene, though with a restricted distribution in glacial periods. Its distribution has doubtless been somewhat modified by planting, but it is now less frequently introduced into semi-natural habitats than *Nymphaea alba*, probably because it is less suited to the shallow waters of ponds and gravel-pits. It is still a frequent plant in S. England.

N. lutea sensu stricto has a similar distribution in Europe to *Nymphaea alba*, being widespread up to 66° N. but scarce in the Mediterranean region (Jalas & Suominen, 1989). It is more frequent in N.E. Europe than *Nymphaea alba*, occurring in the areas where that species is replaced by *Nymphaea candida* C. Presl, and extends much further east in Asia. Related taxa occur in N. America (Hultén & Fries, 1986).

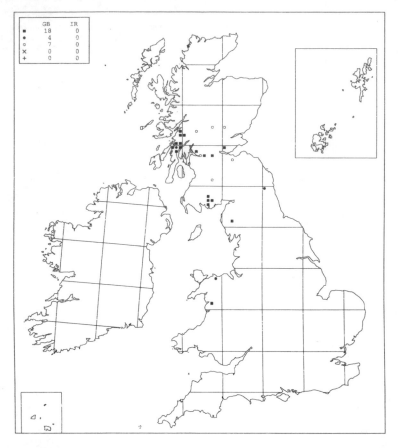

Nuphar × *spenneriana* Gaudin (*N. lutea* (L.) Sm. ×
N. pumila (Timm) DC.)   Hybrid Water-lily

This hybrid is found in oligotrophic and mesotrophic
lakes, pools, rivers and streams. It is recorded from
water up to *c*.2m deep over both peat and mineral soil.
At some sites it can occur in abundance in the absence of
both parents, a notable example being Chartners Lough
where *N.* × *spenneriana* covers almost the whole area of
the lake and 'the bottom ... in the area of the *Nuphar* is
entirely covered with entangled rhizomes' (Heslop-
Harrison, 1953b). Although most localities are lowland,
it is found at 476m in Blea Tarn.

   N. × *spenneriana*, like its parents, is capable of a limited degree of vegetative spread by the growth of its
rhizomes. Heslop–Harrison (1953b) suggests that some
colonies, such as that at Chartners Lough, may even
consist of a single enormous individual of great age. The
hybrid is less fertile than its parents, with 85% of the
pollen grains and 80% of the ovules being sterile.
Nevertheless, some ripe seed is set every year. In cultivation, seeds of the hybrid appear to germinate more
rapidly than those of the parent species, and develop
into vigorous young plants. Seedlings have not been
observed in the wild. British and European populations
show a surprising lack of variation, and may all be first

generation hybrids (Heslop-Harrison, 1955c).

   Colonies of *N.* × *spenneriana* are persistent: the hybrid
is still present at Chartners Lough, where it was first
found by Sir John Trevelyan before 1824. It can occur in
the absence of both parents and some colonies are a considerable distance from the rarer parent, *N. pumila*. The
seeds of *Nuphar* are not adapted to long-distance dispersal, and it seems likely that isolated colonies of *N.* ×
spenneriana are relic populations dating from a period
(perhaps in the early Flandrian) when the parents grew
in closer proximity. Unfortunately, direct evidence for
this hypothesis from pollen analysis is lacking.

   N. × *spenneriana* is widespread in N. Europe, and
extends into Asia. As in Britain, it can occur in pure
populations in the absence of both parents and in some
areas of Lapland it is more frequent than either.

   Heslop-Harrison (1953b) investigated the taxonomy
of *Nuphar* × *spenneriana* in detail, concluding that it was
a partially sterile hybrid between *N. lutea* and *N. pumila*.
The mapped records from N. England and Scotland are
likely to be correct, but the hybrid may well have been
overlooked at other localities or reported as one of its
parents. Heslop-Harrison (1975) queried records from
Wales, but the record from the southerly locality, Tal-y-
llyn Lake, has subsequently been confirmed by Benoit
(1984).

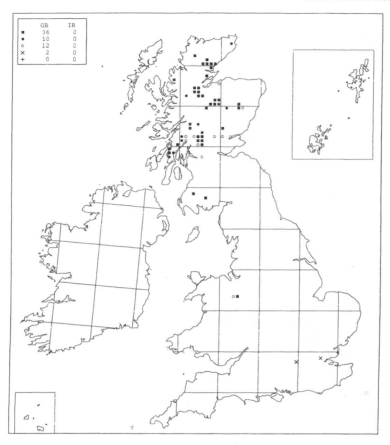

Nuphar pumila (Timm) DC.   Least Water-lily

This species grows in water 0.6–2.4m deep in sheltered lakes, the oxbows and sheltered bays of rivers, ditches, and in pools in marshes and bogs. In Scotland it is found in oligotrophic or, more frequently, mesotrophic sites, some of which receive base-rich drainage water. At the isolated sites in Shropshire it is found with *N. lutea* and *Persicaria amphibia* in sheltered bays in base-rich, eutrophic meres over glacial drift. These meres might, however, have been less eutrophic in former times. The species has been planted and is now established at Cow Pond, Windsor Great Park. It is found over a range of substrates including mud, silt and peat. It ranges from near sea level to 300m; a record from 520m near Lochan nan Damh (White, 1898) is probably correct but requires confirmation.

The reproductive biology of *N. pumila* is similar to that of *N. lutea*. Most populations are fertile, but a population at Avielochan shows reduced fertility, probably because the plants have suffered introgression with *N. lutea*. The seeds are unable to survive drying for 28 days. Little is known of their germination, and seedlings have not been seen in the wild.

Although there is no direct fossil evidence of the history of *N. pumila* in Britain, it seems likely that it was widespread in the early Flandrian and has since retreated to its current range. This is consistent with the occurrence of populations of *N.* × *spenneriana* in the absence of *N. pumila*. Although there are hints in the literature that the Shropshire plants are not native, there seems to be no

evidence to support this and the suggestion probably arose from botanists who were unable to accept the possibility of such a disjunct distribution. The Shropshire meres have a rich flora, which formerly included the demonstrably relict *Scheuchzeria palustris*, and Sinker's (1962) conclusion that *N. pumila* is 'a probable survivor from early post-glacial times' is the most likely explanation of its presence there. The plant is now restricted to a single site in Shropshire, having been recorded in two others from which it has probably been lost because of eutrophication. There is no evidence of a marked decline elsewhere, but this would be difficult to obtain because of the paucity of reliable historical records.

N. pumila is a circumboreal species. It is frequent in N. Europe; further south it is known from the Alps and from other very scattered localities south to 43°N. in Spain. It extends eastwards at similar latitudes through much of Asia and also occurs in eastern N. America. For a map of its European distribution, see Jalas & Suominen (1989).

The widespread occurrence of the hybrid *N.* × *spenneriana* was not appreciated by British botanists until the publication of Heslop-Harrison's (1953b) paper. Many old records of *N. pumila* are therefore dubious. It is difficult to provide an adequate revision on the basis of herbarium material, as specimens are not available for many localities and are less useful than a survey of a population in the field. The mapped distribution is probably broadly accurate, but it may incorporate some errors. A detailed field survey of the British populations of *N. pumila* and *N.* × *spenneriana* would be worthwhile.

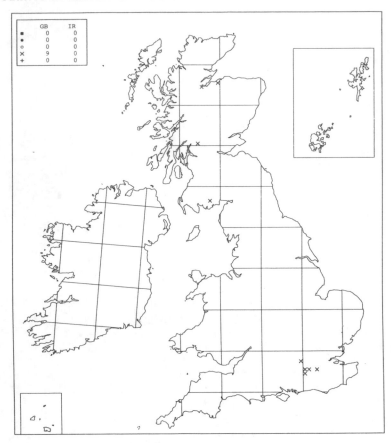

	GB	IR
■	0	0
●	0	0
○	0	0
×	9	0
+	0	0

Nuphar advena (Aiton) W. T. Aiton   Spatter-dock

This species is established in a number of lowland lakes and ponds, where it is found in water up to 1.5m deep. At most of the sites where it has been recorded it has clearly originated from planted stock, but the presence of a large clump of *N. advena* with *Nuphar lutea* and *Nymphaea alba* at Loch Ard is more difficult to explain.

N. advena is a rhizomatous perennial. Little is known of its reproductive biology in Britain, but it has presumably occupied most of its known sites by vegetative spread after planting. Riemer (1985) found that the seeds germinated slowly and sporadically over a long period of time; they were still germinating when he put a stop to his experiments one year after they had begun.

Specimens of *N. advena* were first collected by Miss M. McCallum Webster from an old curling pond at Auchnagairn House, Kirkhill, in 1963, but not identified until 1990. Miss McCallum Webster also noted that the same plant grew in a small loch in the grounds of Darnaway Castle. From 1979 onwards the plant was noted at sites in Surrey (in some of which it had clearly been established for a long time), and in 1990 the species was recognized in Loch Ard, where an unidentified exotic *Nuphar* had been noted eight years previously. In 1992 it was reported from Carlingwark Loch. It is unlikely to be overlooked, as the leaves do not float but are held above the water surface, but it may be under-recorded as botanists might have failed to identify it or considered that as an introduced species it was not worth noting.

N. advena is a native of N. America, where it extends from Wisconsin and Maine south to Mexico and Cuba. In its native range it grows in lakes, ponds, pools in marshes and swamps, ditches, canals and sluggish streams.

This species is related to *N. lutea* and some recent American authors, including Beal (1956) and Wiersema & Hellquist (1994), treat it as *N. lutea* subsp. *advena* (Aiton) Kartesz & Gandhi.

CABOMBACEAE

This is a small family with two genera, *Cabomba* and *Brasenia*, both of them aquatic (Kubitzki, 1993).

Cabomba Aubl.

Species of *Cabomba* have dissected submerged leaves and sometimes also produce a few laminar floating leaves. The conspicuous flowers are insect-pollinated, and have petals which may be white with a yellow base, or purple. There is therefore a similarity in habit between this genus and *Ranunculus* subgenus *Batrachium*. Ørgaard (1991) recognized five species, all native to the warm temperate and tropical regions of N. & S. America. *C. caroliniana* is often grown in aquaria and has been recorded in canals in Britain.

Cabomba caroliniana A. Gray

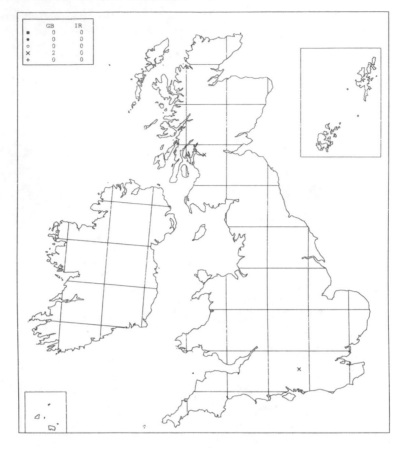

Cabomba caroliniana A. Gray   Fanwort

This species is known from two British canals. It formerly grew in a stretch of the Forth & Clyde Canal which was heated by water discharged from a nearby factory. Maitland (1971) reported 'several fine stands' growing with luxuriant *Myriophyllum alterniflorum*. The only other British records are from the Basingstoke Canal, where *C. caroliniana* currently grows in unheated, calcareous and eutrophic water with *Elodea nuttallii*, *Sagittaria sagittifolia* and *Sparganium emersum*.

The British populations of *C. caroliniana* consist of submerged, vegetative stems. The stems are rather fragile and the plant spreads from detached fragments. In N. America both rooted plants and floating shoots will flower and set seed. The flowers open by day; at night the petals close and the flowers fall below the surface of the water, rising again the next day. The flowers are usually cross-pollinated by flies or honey bees; flowers which are not visited by pollinating insects, and any which open beneath the surface of the water, do not set seed. After fertilization the peduncles curl down into the water. Plants will reproduce vegetatively from a section

of stem consisting of only one node and a single pair of leaves (Tarver & Sanders, 1977).

This species was first discovered in the Forth & Clyde Canal by the fish biologist P. S. Maitland in 1969 (Maitland, 1971). Although it was well established at the time of its discovery, it no longer survives at this site. It was found in the Basingstoke Canal by C. R. Hall in 1990; it was present in increased quantity in 1991 and still there in 1995 (Brewis *et al.*, 1996). *C. caroliniana* is a popular aquarium plant and there is little doubt that it was released into both canals by aquarists. Indeed, Maitland's attention was drawn to the stretch of the Forth & Clyde Canal in which *C. caroliniana* grew because it also harboured a thriving population of goldfish!

C. caroliniana is a native of eastern N. America, where it extends from Texas and Florida north to Illinois and Virginia and (as a naturalized alien) to Massachusetts. It has the most northerly native distribution of any *Cabomba* species. It also occurs as a native in S. America (Argentina, Brazil, Paraguay and Uruguay) and it is naturalized in India, Japan, New Guinea and elsewhere. In its native range it grows in ponds, lakes and quiet rivers, chiefly in the coastal plain (Godfrey & Wooten, 1981).

CERATOPHYLLACEAE

This is a small but distinctive family with only one genus, *Ceratophyllum* (Kubitzki, 1993). All the species are morphologically reduced and highly specialized aquatics, and the affinities of the family are difficult to assess. Cronquist (1981) interprets the Ceratophyllaceae as a specialized member of the Nymphaeales, the order which includes Nymphaeaceae and Cabombaceae. Les (1988a), however, considers that it shows no close affinity to any other angiosperm group. He places the family in a new order, Ceratophyllales, and suggests that the species are 'living fossils' which belong to a lineage which is even more ancient than the Nymphaeales. Fossil evidence suggests that the Ceratophyllaceae was formerly more diverse, comprising at least two genera.

Ceratophyllum L.

Ceratophyllum lacks roots; the stems float in the water or are anchored to the substrate by buried branches. The species are monoecious, and small flowers develop in the leaf axils. The male flowers release anthers which float just below the water surface, shedding pollen into the water. Pollination is only effective in standing waters; in streams and rivers the stigmas point downstream and are unlikely to make contact with pollen grains (Kubitzki, 1993). The fruits are small, hard nuts which may be dispersed by waterfowl. Little is known of the germination requirements of either species. Vegetative reproduction is probably frequent as plants are brittle and are able to regenerate from fragments. In winter plants sink, regrowing in spring from unspecialized buds.

The genus *Ceratophyllum* has a cosmopolitan distribution. Two recent taxonomic revisions have delimited the species in different ways. Wilmot-Dear (1985), in a traditional taxonomic study, recognized only two species, *C. demersum* and *C. submersum*. Each of them shows extraordinary variation in the surface and armature of the fruits, variation that is well illustrated in Wilmot-Dear's paper. Wilmot-Dear treats many of these variants as infraspecific taxa. Les (1986, 1988b,c, 1989), applying methods of multivariate statistical analysis, recognized six species (including *C. demersum* and *C. submersum*) and three subspecies. He regarded the species as morphologically distinct, and readily separable by statistical analysis.

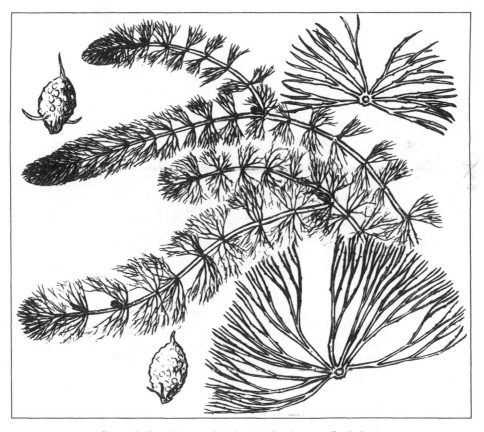

Ceratophyllum demersum L. (above); *C. submersum* L. (below).

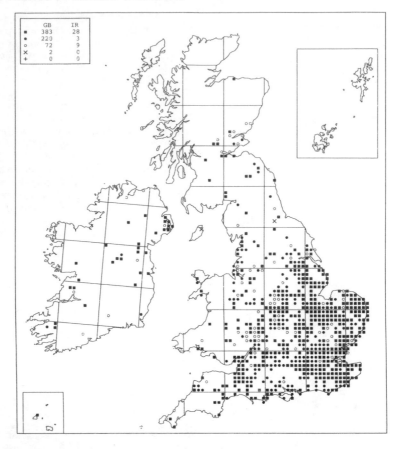

	GB	IR
■	383	28
●	220	3
○	72	9
×	2	0
+	0	0

Ceratophyllum demersum L. **Rigid Hornwort**

C. demersum is frequent and sometimes abundant in a wide range of still or slowly flowing lowland water bodies including lakes, ponds, canals, ditches and sluggish streams and rivers. As it lacks roots it is unable to become established in rapidly flowing water. Murphy & Eaton (1983) noted that in canals it was particularly frequent where boat traffic was low, suggesting that its fragile stems were vulnerable to mechanical damage in more heavily used channels. Its strong preference for eutrophic waters is well illustrated by its history in Broadland (George, 1992). It colonized the broads as nutrient levels rose in the post-war years, and in some broads it became very abundant. As eutrophication increased further, populations crashed and the broads became dominated by phytoplankton. At Alderfen Broad, for example, the excessive abundance of *C. demersum* was causing concern in 1963 but five years later the plant had virtually disappeared. When the flow of eutrophic water was diverted from Alderfen Broad, *C. demersum* again became abundant. *C. demersum* is shade-tolerant, and small quantities can persist in ditches under a heavy carpet of *Azolla filiculoides* or *Lemna* species.

C. *demersum* flowers freely. Many observers have noted that *C. demersum* rarely produces mature fruit (e.g. Burton, 1983; Norman, 1978; Scannell, 1976a; De Tabley, 1899). In a detailed study of the species in the London area Guppy (1894) failed to find fruiting plants in three successive summers. In the hot summer of 1893

fruits matured in shallow water but not in the cooler water of rivers, although it has been suggested that the failure to fruit in rivers may reflect the lack of successful pollination in a strong current (Kubitzki, 1993). If fruits were kept moist only 15% germinated in the first year, but this increased to 80% if the fruits were dried for 2–3 months. The main method of spread is by vegetative fragmentation, and the spread of the species in canals may be facilitated by clearing operations which break the fragile stems (Scannell, 1971). *C. demersum* is sold as a plant for garden ponds and may be released into the wild when overgrown ponds are cleared out.

The two species recognized in Britain and Ireland were confused until Sandwith (1927) demonstrated that the leaves of *C. demersum* branch once or twice whereas those of *C. submersum* branch three or four times. *C. demersum* has fluctuated in abundance at some sites in response to changes in the nutrient status of particular water bodies. There is little evidence to suggest that there is any change in its overall distribution or abundance in Britain or Ireland. However, an increase has been noted in Hampshire this century, where Brewis *et al.* (1996) suggest that it may have been introduced into the wild from garden ponds, and the results of the BSBI Monitoring Scheme suggest that the species has increased in frequency in Scotland since 1960 (Rich & Woodruff, 1996).

C. *demersum* is widespread in Europe. It is more frequent in Scandinavia than *C. submersum*, and almost reaches 70°N. in N. Norway (Jalas & Suominen, 1989).

It has a virtually cosmopolitan world distribution, being found in all continents except Antarctica. For a map of the world distribution of *C. demersum* sensu lato, see Wilmot-Dear (1985). Les (1988c, 1989) splits plants from Europe into two closely related species, *C. demersum* and *C. platyacanthum* Cham. The latter is characterized by a flattened, spiny wing to the fruit and spines on the faces of both fruit surfaces as well as at the base and apex. All British and Irish plants appear to be referable to *C. demersum* sensu stricto.

Electrophoretic evidence suggests that many American populations of *C. demersum* consist of a single clone. Morphological variation occurs between, but not within, populations (Les, 1991).

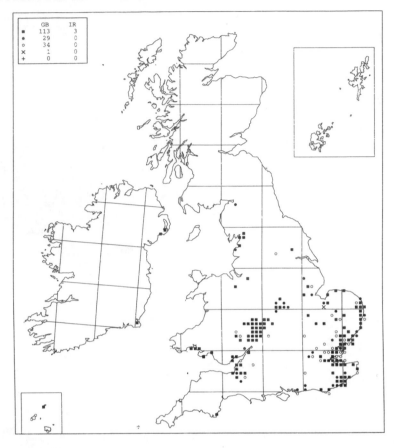

	GB	IR
■	113	3
●	29	0
○	34	0
×	1	0
+	0	0

Ceratophyllum submersum L.   Soft Hornwort

This is a plant of shallow, sheltered, lowland waters. Like *C. demersum*, it grows as a floating mass or lightly anchored to the substrate by buried stems. It is found in eutrophic, often base-rich or slightly brackish water. In coastal sites its most characteristic habitats are shallow ponds and ditches (including grazing marsh ditches), where it grows with other plants which tolerate slightly brackish conditions (e.g. *Myriophyllum spicatum, Potamogeton pectinatus, Ranunculus baudotii, Zannichellia palustris*) as well as species of wider habitat range. At inland localities it grows in similar habitats to the commoner *C. demersum*, including small field ponds, lakes and flooded marl- and gravel-pits. The two species can sometimes be found growing together. *C. submersum* is shade-tolerant, and can be found in ditches or small ponds surrounded by tall emergents or trees. In both coastal and inland sites it can be so abundant that it virtually excludes all competitors.

Plants of *C. submersum* flower freely. It has been suggested that this species fruits more regularly than *C. demersum* in S. England (Norman, 1978). Plants survive the winter as sunken stems, but are unable to tolerate prolonged freezing.

It is difficult to assess changes in the distribution of the species, as the simple vegetative distinction between *C. submersum* and *C. demersum* was not understood until it was described by Sandwith (1927). Older records of *C. submersum* are unreliable, and have been accepted only if supported by herbarium specimens. *C. submersum* is being recorded with increasing frequency in inland sites, and it was first found in Ireland in 1989 (Smith & Wolfe-Murphy, 1991). It is not clear if this species has increased or whether it was previously overlooked. At some sites it has certainly increased in abundance in recent years, and it now fills entire ponds which formerly had a richer flora. In Switzerland, *C. submersum* is becoming commoner and *C. demersum* rarer (C. D. K. Cook, pers. comm.).

C. submersum sensu lato is widespread in Europe, north to S. Scandinavia (Jalas & Suominen, 1989). It is also found in Asia, Africa and N. and S. America, but it is absent from Australasia. For a map of its world distribution, see Wilmot-Dear (1985). Les (1988b, 1989) regards the American plants as a separate but closely related species, *C. echinatum* A. Gray.

RANUNCULACEAE

A large family of 59 genera and some 2,500 species, most of which occur in temperate and boreal areas. Members of the family are very variable in vegetative form, floral structure and fruits (Kubitzki, 1993). Only two of the genera contain aquatic species, *Caltha* and *Ranunculus*.

There is one true aquatic species of *Caltha*, *C. natans* Pall. The other species grow in a variety of wet habitats. The only European representative is *C. palustris*, which is found in our area in sites which are flooded in the winter and where the water-table remains at or near the soil surface in summer but is absent from permanently flooded ground.

Ranunculus L.

This is the largest genus in the family, with 580 species. There are three rather distinct subgenera in Britain and Ireland. Subgenus *Ranunculus* and subgenus *Ficaria* include species with yellow petals; they are predominantly terrestrial but two species in subgenus *Ranunculus* (*R. flammula* and *R. reptans*) and their hybrid are treated here. The species in these subgenera are relatively straightforward taxonomically; few of them hybridize and the main taxonomic problems concern the treatment of the considerable intraspecific variation shown by some widespread species.

The species in subgenus *Batrachium* have white petals. They all grow in association with water, but vary in the extent to which they are adapted to the aquatic environment. All species are treated in the accounts below. Some (*R. hederaceus*, *R. omiophyllus*) have laminar leaves which cannot survive submersion, and these species are restricted to terrestrial habitats at the water's edge. Other species (*R. circinatus*, *R. fluitans*, *R. penicillatus* subsp.

pseudofluitans, *R. trichophyllus*) have capillary leaves which can survive only if submerged, and these are obligate aquatics. The largest group of taxa is heterophyllous, with the capacity to produce both capillary and laminar leaves (*R. aquatilis*, *R. baudotii*, *R. peltatus*, *R. penicillatus* subsp. *penicillatus*, *R. tripartitus*; the three hybrids treated below also fall into this category). These taxa usually grow in shallow water, with the capillary leaves submerged and the laminar leaves floating on the surface, but can sometimes develop in water as young plants with capillary leaves and then grow terrestrially with laminar leaves when they are mature. Both the species with capillary leaves and the heterophyllous species can grow as small, terrestrial forms if exposed above the water-level. The leaves of these forms resemble flattened, short capillary leaves. Terrestrial forms can survive on damp substrates during the summer but they are frost-sensitive and survive the winter only if they are flooded and therefore revert to the capillary-leaved form before the onset of cold weather.

Taxonomically, subgenus *Batrachium* is perhaps the most difficult group of aquatic plants in the British Isles. All available evidence indicates that the species are closely related. Like many aquatic plants they have undergone morphological reduction, and this limits the possible taxonomic characters which can be used to separate the species. Terrestrial forms with flattened, capillary leaves may be the only plants present at a site, but they cannot be identified with certainty. Barriers to hybridization within the subgenus are slight, and Cook (1970) suggests that all species may potentially be capable of exchanging genetic material. Both fertile and sterile hybrids can be found as persistent populations in the wild and have been synthesized in cultivation. *Ranunculus penicillatus* is particularly critical taxonomically. It is

Ranunculus circinatus Sibth.; *R. hederaceus* L.; *R. flammula* L.

believed to have evolved by hybridization between *R. fluitans* and other species, and is almost certainly polyphyletic in origin. Cook (1970) has suggested that some other species (including *R. aquatilis*, *R. peltatus* and *R. trichophyllus*) may also have a polytopic origin and perhaps represent no more than particular combinations of characters which are well adapted for survival in the aquatic or amphibious environment.

The breeding system of the species in subgenus *Batrachium* is similar for all species, and is outlined here. The flowers are protogynous and the stigmas are ready to receive pollen 2–6 days before the anthers dehisce. Self-fertilization takes place once the anthers dehisce and most species are self-compatible. At least some of the ovules in the species with smaller flowers tend to be self-pollinated before the flower buds open, especially if flowering is delayed by cold weather (Cook, 1966a; Hong, 1991). Little is known of the longevity of the seeds. Witztum (1986) reports that seeds of one species from a pond in Israel germinated after they had been stored for 10 years in dry conditions; Witztum identified his plant as *R. aquatilis* but C. D. K. Cook (pers. comm.) considers that it must have been *R. peltatus* or *R. trichophyllus*.

The modern taxonomic treatment of subgenus *Batrachium* is based on a revision by Cook (1966a); older records of the more critical species are of little value unless supported by herbarium specimens. A recently published revision of the subgenus in the Iberian peninsula (Pizarro, 1995) includes illustrations of all the species in our area except *R. circinatus*, as well as updated maps of their distribution in W. Europe. Guides to the identification of the British and Irish species have been prepared by Holmes (1979) and Webster (1988b). The treatment of the subgenus in Stace (1991) follows Cook (1966a), with modifications for the *R. penicillatus* group suggested by Webster (1988a). In mainland Europe there is little consensus about the taxonomic rank or even the definition of the taxa in the subgenus; the main differences of opinion are summarized by Jalas & Suominen (1989).

Much of the information in the accounts of *R. flammula*, *R. reptans* and *R. × levenensis* is taken from papers by Padmore (1957), Gibbs & Gornall (1976) and Gornall (1987). The accounts of the taxa in subgenus *Batrachium* are based on the work of Cook (1966a,b, 1970, 1975a) and Webster (1988a,b, 1990, 1991).

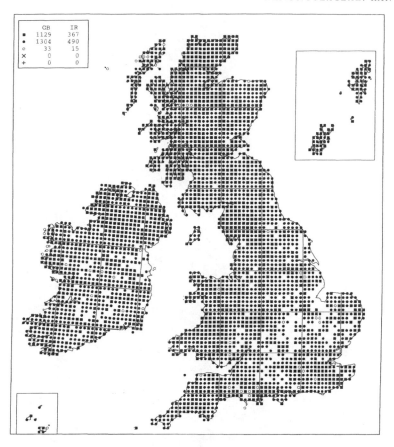

Ranunculus flammula L. subsp. flammula
Lesser Spearwort

R. flammula subsp. *flammula* is found in a wide range of moist, wet or seasonally flooded habitats, but only occasionally occurs in permanently flooded sites. It is restricted to areas where tall vegetation is unable to become established because of shallow or infertile soils, disturbance or grazing (*R. flammula* is unpalatable to stock). Typical habitats include stony or gravelly soil at the edge of pools and lakes, where it sometimes grows in shallow water, streamsides, springs and flushes, marshes, damp hollows in heathland and moorland, dune-slacks, ditches, damp grassland and deeply rutted woodland rides. It is sometimes found as a colonist of old gravel- and clay-pits. It is usually found over neutral or moderately acidic substrates associated with oligotrophic or mesotrophic water, and occurs on both peaty and mineral soils. In the Sheffield area it generally occurs where the soil pH is between 4.5 and 6.5, but there (as elsewhere) it is occasionally found in highly calcareous sites (Grime *et al.*, 1988). It is recorded from sea-level to 930m on Carnedd Llewelyn, and perhaps ascends to higher altitudes in the Scottish highlands (Wilson, 1956).

This is a perennial which overwinters as very small rosettes. It is a predominantly self-incompatible species, which scarcely sets any seed if self-pollinated (Gibbs & Gornall, 1976). The seeds lack buoyancy, but may be dispersed by animals as they have survived ingestion by

horses (Ridley, 1930). There is a persistent seed bank in the soil, and germination occurs in spring (Grime *et al.*, 1988). The seed of American plants only germinates sporadically when fresh but rapidly after subjection to alternate freezing and thawing (Cook & Johnson, 1968). Prostrate plants root at the nodes and if their shoots are accidentally detached they also may give rise to new plants.

R. flammula is frequent in suitable habitats in much of northern and western Britain and Ireland. In the south and east it is widespread but local, and in some counties such as Cambridgeshire it has decreased markedly because of drainage (Perring *et al.*, 1964). Subsp. *flammula* is much commoner than subsp. *minimus* or subsp. *scoticus*, and all records except those definitely attributable to the rarer subspecies have been included on the map.

R. flammula subsp. *flammula* is widespread in Europe, although absent from the extreme north and rare in the Mediterranean region (Jalas & Suominen, 1989). It extends south to N. Africa and east to western Asia; it is also naturalized in New Zealand. Plants in the *R. flammula* aggregate (which includes *R. reptans*) are widespread in N. America, and some of those in the western and eastern parts of its range may be referable to *R. flammula* sensu stricto, but the relationship of the American plants to those in Europe requires investigation.

This is a polymorphic taxon, with variants which range from procumbent phenotypes which resemble *R. reptans* to erect, large-leaved genotypes (var. *ovatus* Pers.) which can be mistaken for the fenland species *R. lingua*.

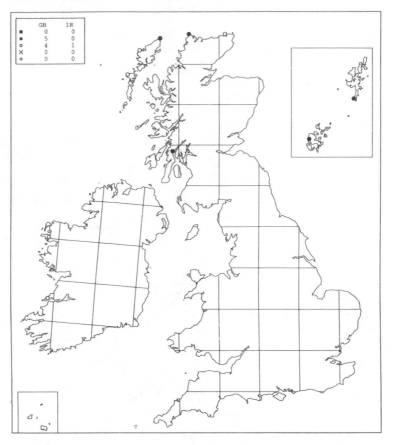

Ranunculus flammula subsp. minimus (A. Benn.) Padmore   Lesser Spearwort

This dwarf, fleshy, semi-prostrate subspecies of *R. flammula* grows in dense mats on exposed sea-cliffs. It is recorded from dry moorland, short, damp turf and wet flushes. All localities lie below 300m.

This is a fertile perennial which sets viable seed. Its reproduction in the wild has not been studied.

R. flammula subsp. *minimus* is a little-known taxon. It was described (as forma *minima*) by Bennett (1904), and has subsequently been collected from a number of scattered localities in Scotland and Ireland. Some observers have confused this subspecies with dwarf variants of subsp. *flammula* (Silverside, 1984); only records based on reliably determined herbarium specimens have been accepted for the map. One such specimen was collected from North Rona by F. Fraser Darling in 1939: this locality lies north of the Outer Hebrides, outside the border of the map. Silverside (1984) notes that subsp. *minimus* has been searched for but not refound recently at two of its known localities, Cape Wrath and Holborn Head, and Scott & Palmer (1987) have also failed to relocate it in Shetland.

The only confirmed records of this subspecies are from Britain and Ireland.

Well-marked specimens of subsp. *minimus* are very distinctive, and retain their characters in cultivation (Padmore, 1957). This suggests that the plant merits its subspecific status, but the taxonomy of the *R. flammula* aggregate in the northern hemisphere requires revision. Cook (1983) notes that he has seen herbarium specimens which appear to be this plant from America and perhaps elsewhere, but gives no further details. The current status of the British and Irish populations also requires investigation, but this will not be easy to achieve in view of the remoteness of the sites.

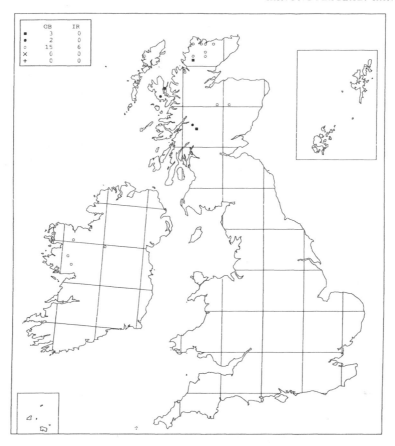

**Ranunculus flammula subsp. scoticus (E. S. Marshall)
A. R. Clapham**   **Lesser Spearwort**

This plant usually grows in water up to 0.6m deep at the edges of lakes. It is sometimes found exposed on stony or, more rarely, peaty shores. It is recorded at altitudes up to 485m at Loch Builg.

R. flammula subsp. *scoticus* is a fertile perennial which sets well-formed and apparently viable seed. Little is known of its reproduction in the wild.

This plant was discovered and described by Marshall (1889, 1892, 1898) but was subsequently misinterpreted by many observers, both before and after Padmore's (1957) account. The map is based on reliably determined herbarium specimens. Few botanists are familiar with the plant, and it may be more frequent in Scotland and Ireland than the map suggests.

The only confirmed records of this subspecies are from Britain and Ireland. Unconfirmed records from the Faeroes are reported by Jalas & Suominen (1989) but not taken up by Tutin *et al.* (1993).

R. flammula subsp. *scoticus* retains its characters in cultivation (Padmore, 1957). At a loch near Tongue it grows at a level below subsp. *flammula*, with intermediate plants between them (Silverside, 1984). Plants which are intermediate between the subspecies occur at most or perhaps all of the sites for subsp. *scoticus*, and this has contributed to the taxonomic confusion between the plants. Typical subsp. *scoticus* is illustrated by Marshall (1892) and the distinguishing characters discussed by Padmore (1957) and Silverside (1984). Cook (1983) suggests that this subspecies, like subsp. *minimus*, may occur in America and perhaps elsewhere, but gives no further details.

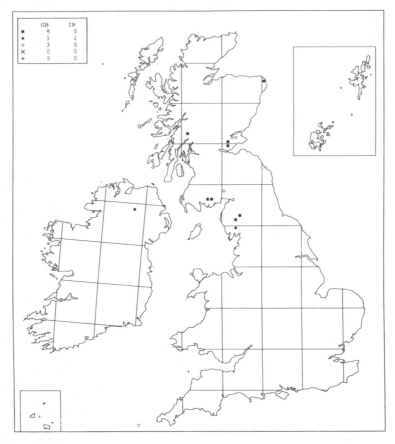

	GB	IR
■	8	0
●	1	1
○	3	0
×	0	0
+	0	0

Ranunculus × levenensis Druce ex Gornall (*R. flammula* L. × *R. reptans* L.)   Loch Leven Spearwort

This hybrid grows on lake shores, where it is usually found in a zone near the low water-level where the shore is damp and occasionally flooded in summer. It occurs on a variety of substrates including stones, gravel, silt-covered gravel and sand. *Carex* spp., *Juncus bulbosus* and *Littorella uniflora* are recorded as associates. All sites are below 300m.

R. × *levenensis* is a perennial which creeps and roots at the nodes. It is fertile and some hybrid plants are partially self-compatible. The hybrid produces a full complement of well-formed fruit.

R. × *levenensis* is known to have been present at its best-known British localities for many years. Plants from Loch Leven illustrated on the title page of Lightfoot's *Flora Scotica* (1777) are clearly referable to this hybrid (Gornall, 1987), and the plant is still present at this site. It also persists at Ullswater, where it was first collected in 1880. These populations were first reported as *R. reptans* but are now known to be of hybrid origin (Padmore, 1957; Gibbs & Gornall, 1976).

There are records of R. × *levenensis* from southern Scandinavia and the Alps, the main areas where the distributions of the parents overlap. The hybrid is also recorded from northern Germany which, like the sites in Britain and Ireland, is outside the main range of *R. reptans*.

Ranunculus flammula and *R. reptans* are completely interfertile, and there is an argument for treating them as a subspecies of a single species (Gibbs & Gornall, 1976). Plants of R. × *levenensis* in the wild are similar to synthesized hybrids. The hybrid is very variable, both genotypically and phenotypically. Genotypes in Britain range from those which closely resemble *R. reptans* to those which are very difficult to distinguish from dwarf phenotypes of *R. flammula* unless they are taken into cultivation. British populations probably consist for the most part of backcrosses with *R. flammula* rather than F1 hybrids. At Loch Leven and Ullswater, hybrids along the water's edge tend to be closest to *R. reptans* and above this they give way to hybrids closer to *R. flammula*, prostrate phenotypes of *R. flammula* and, above the high-water mark, to erect *R. flammula* (Padmore, 1957; Gibbs & Gornall, 1976).

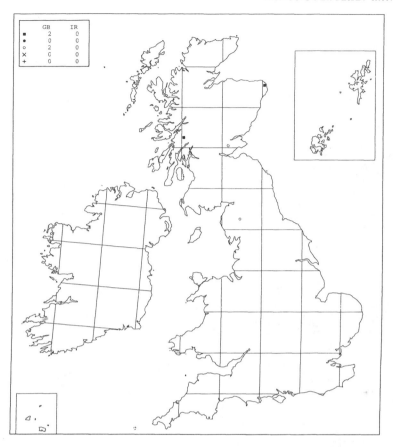

	GB	IR
■	2	0
●	0	0
○	2	0
×	0	0
+	0	0

Ranunculus reptans L.   **Creeping Spearwort**

Plants which are referable to *R. reptans* are known from four localities in Britain: the gravelly shores of Ullswater and Loch of Strathbeg, where they have been recorded with *R. × levenensis*, the shore of Loch Leven, where they were almost certainly associated with the same hybrid, and Loch Awe, where they have recently been found on silty sand near the top of the shore.

This species is a stoloniferous perennial, which creeps and roots at the nodes. It is partially self-compatible.

R. reptans appears to be a transient member of the British flora. It was collected at Ullswater in 1887 and from 1911 to 1917, and at Loch Leven in 1869, 1896 and 1935. It is currently known from Loch of Strathbeg, where it was collected in 1876, 1900 and 1989 and was still present in 1994, and Loch Awe, where it was collected in 1992. Gornall (1987) suggests that it is introduced repeatedly by waterfowl migrating in autumn from Iceland, Scandinavia or northern Russia. Mallard, pink-footed geese and wigeon appear to be the most

likely species to have carried the seeds to Loch Leven. Plants of *R. reptans* hybridize with *R. flammula* and give rise to persistent populations of the hybrid *R. × levenensis*. The reason that *R. reptans* itself fails to survive is not clear, but Gibbs & Gornall (1976) have shown that it is less phenotypically plastic than the hybrid and may therefore be at a disadvantage on lake shores where the water-level fluctuates. *R. × levenensis* has been recorded from sites in north-east Scotland and northern Ireland where *R. reptans* has never been reliably recorded, but where it was presumably once present.

R. reptans has a circumpolar distribution, being widespread in the boreal zone and extending southwards in the mountains of Europe, Asia and N. America. In Europe its main range is in Scandinavia and the north-east, but it is found at scattered sites south to the Alps, C. Italy and Bulgaria (Jalas & Suominen, 1989).

The distinction between *R. reptans* and the variable hybrid *R. × levenensis* is difficult and somewhat arbitrary. This account is based on specimens determined by R. J. Gornall or P. D. Sell.

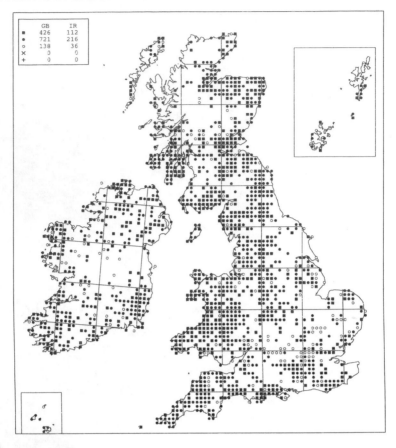

	GB	IR
■	426	112
●	721	216
○	138	36
×	0	0
+	0	0

Ranunculus hederaceus L.   Ivy-leaved Crowfoot

R. hederaceus is usually found at the edge of small water bodies such as ponds, ditches, streams, springs and flushes, although it can also occur by the sheltered back-waters of larger rivers. It is a low-growing species which is often found in areas where wet soil has been dis-turbed; characteristic sites include the cattle-poached edges of streams and ponds, recently cleared roadside ditches, places where trickles of water flow over foot-paths and ruts on unsurfaced tracks. It also grows in very shallow water as a form with floating leaves. It can colonize a range of substrates from gravel to mud, is found in both acidic and calcareous sites and it can also be found over slightly saline soil at the upper edge of salt-marshes. It tolerates a range of nutrient levels, and often grows in rather nitrophilous situations. It is found from sea-level to 770m (Little Dun Fell).

Plants of *R. hederaceus* are annuals or short-lived perennials. The seeds germinate very erratically if they are kept wet, but if dried then rewetted almost all will germinate simultaneously (Cook, 1966b). In the field, seeds germinate in autumn or spring and plants begin to flower from March to May. Pollination usually takes place within a flower bud before the flower opens, and after fertilization the pedicel bends down, forcing the fruits into the substratum. Growth of the parent plants continues until they are smothered by competing species such as *Lythrum portula*, *Montia fontana* or *Ranunculus*

sceleratus, or are killed by desiccation. Individuals which survive these hazards overwinter as small, tight cushions.

R. hederaceus is one of the most distinctive members of *Ranunculus* subgenus *Batrachium*, and the distribution map is therefore much more complete and reliable than those of the more difficult taxa. The species, like many other plants of similar habitat, has declined in south-east England. The decline is partly due to habitat destruc-tion: streams have been canalized, ponds filled in and tracks surfaced. Those habitats that remain are often unsuitable as they are no longer used for watering stock, and consequently the vegetation around them has become overgrown. In northern and western England and in much of Wales, Scotland and Ireland the species remains frequent.

In Europe this species has a western distribution. It extends from Portugal and Spain through France north-wards and eastwards to C. Germany, Denmark and S. Sweden. It is extinct in Norway, where it formerly occurred near Trondheim, and in Latvia. *R. hederaceus* also occurs on the east coast of N. America, where it may have been introduced. The European localities are mapped by Jalas & Suominen (1989) and Pizarro (1995) and the American sites by Cook (1966a,b). On the European mainland, as in Britain, the species has declined in the eastern part of its range (De Sloover *et al.*, 1977; Cook, 1983).

A detailed account of the ecology of *R. hederaceus* in the Low Countries is provided by Segal (1967).

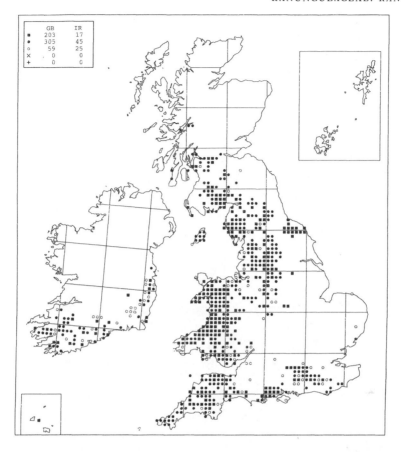

	GB	IR
■	203	17
●	305	45
○	59	25
×	0	0
+	0	0

Ranunculus omiophyllus Ten.
Round-leaved Crowfoot

The habitat requirements of this species are similar to those of *R. hederaceus*. It is found in open communities on disturbed soil at the edges of ponds and streams, in ditches, flushes and damp depressions in pastures and by the sheltered backwaters of lakes and rivers. It also grows in shallow water as a form with floating leaves. Unlike *R. hederaceus*, *R. omiophyllus* is confined to acidic, mesotrophic or oligotrophic soils. It often occurs over acid sands or peat and it is much more frequent than *R. hederaceus* in heathland pools. *Montia fontana* is a characteristic associate. *R. omiophyllus* is recorded from sea-level to 1000m on Carnedd Llewelyn.

R. omiophyllus has a similar life-history to *R. hederaceus*. It can behave as a winter or summer annual or, if conditions are suitable, persist as a short-lived perennial. Fresh seeds germinate erratically but most will germinate if they are dried then rewetted. Plants may begin to flower as early as January and the flowering period can extend until November. Flowers are usually self-pollinated before the flower buds open and the fruits are deposited in the substrate by the recurved pedicels. The spreading summer growth form, like that of *R. hederaceus*, is susceptible to desiccation or shade but the small compact overwintering plants are more tolerant of shade, desiccation and freezing.

This is a distinctive species and the map probably presents an accurate summary of its distribution since 1950. As a plant of acid soils it has always had a restricted distribution in south-east England, and it has declined in some areas for similar reasons to *R. hederaceus*. It is still frequent in its western strongholds.

R. omiophyllus is virtually endemic to Europe, where its distribution is even more restricted than that of *R. hederaceus*. In western Europe it is found in Portugal, Spain, France, Britain and Ireland; it is extinct in the Netherlands. There are also outlying populations in southern Italy, Sicily and N. Africa.

R. hederaceus and *R. omiophyllus* are the two species in subgenus *Batrachium* which lack the ability to produce capillary submerged leaves; they are therefore found in more terrestrial habitats than the others. Cook (1966b) suggests that they have evolved from ancestors which had the capacity to produce such leaves, and that they represent two homozygous strains that have arisen from a common polymorphic stock. The ecological requirements of the two species overlap, and they can occasionally be found growing together. Cook (1966b) has suggested that *R. omiophyllus* is excluded from nutrient-rich sites by competition with *R. hederaceus*. In Sicily, where *R. hederaceus* does not occur, *R. omiophyllus* is found on disturbed soil which is calcareous and eutrophic. Hybrids between the two species have not been found in the wild but a highly sterile hybrid has been synthesized artificially by Cook (1966b).

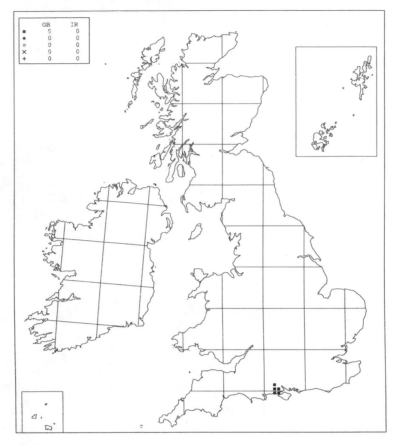

	GB	IR
■	5	0
●	0	0
□	0	0
×	0	0
+	0	0

Ranunculus × *novae-forestae* S. D. Webster (*R. omio-phyllus* Ten. × *R. tripartitus* DC.)

New Forest Crowfoot

This plant grows in shallow water or on damp clay at the edge of ponds, by small streams and drainage channels and in flushes and seasonally flooded hollows, sometimes in sites poached by cattle. It can be found with *R. omiophyllus*, but it appears to favour less acidic soils and is even found on calcareous clay in the southern New Forest.

There have been few detailed observations on the life-history of this hybrid. It normally behaves as an annual, but it may persist for over a year if submerged (Salisbury, 1974). It is variable in fertility, but some well-formed fruits usually develop from each flower. Like the parents, it presumably reproduces by seed rather than by vegetative spread. It was particularly plentiful in 1977, after widespread flooding in the winter of 1976–77 (Brewis *et al.*, 1996).

Cook (1966a) pointed out that material from the New Forest which had traditionally been ascribed to *R. tri-partitus* or *R. lutarius* by British botanists was actually the hybrid between *R. omiophyllus* and *R. tripartitus*. The hybrid was given a binomial by Webster (1990). It is still widespread in the New Forest.

R. × *novae-forestae* is known with certainty only from the New Forest and France.

Artificially synthesized hybrids are morphologically similar to the plant which grows in the New Forest. Both synthesized and wild hybrids are pentaploid, as would be expected from a hybrid between tetraploid (*R. omiophyllus*) and hexaploid (*R. tripartitus*) parents. In experimental studies both *R. omiophyllus* and *R. triparti-tus* can be backcrossed to hybrids which have been collected in the wild. The fertility of some wild plants greatly exceeds that of synthesized F1 hybrids, but highly fertile strains can be selected from the F2 generation or the backcrosses. The New Forest populations of *R.* × *novae-forestae* are very variable, and some are very similar to *R. tripartitus*. Cytological examination of these plants is needed to establish whether any of these are actually *R. tripartitus*, which has not been recorded from the area with certainty.

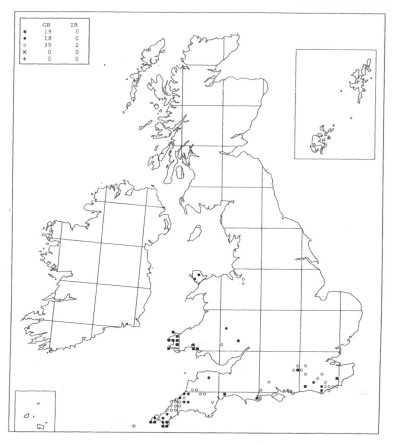

	GB	IR
■	19	0
●	18	0
○	39	2
×	0	0
+	0	0

Ranunculus tripartitus DC.   Three-lobed Crowfoot

R. tripartitus grows in shallow lowland pools which are flooded in winter but often dry out in the summer. Typical sites include ditches, heathland ponds, flushes in heaths and on sea cliffs, ruts in unsurfaced tracks and muddy depressions in field gateways. The species is a calcifuge which is found over a variety of soils ranging from peat to fine-grained loess. It is intolerant of competition and persists only where the development of a closed sward is prevented by fluctuating water-levels, poaching by livestock or disturbance by vehicles. Typical associates include *Apium inundatum, Eleogiton fluitans, Lythrum portula, Montia fontana* and *Nitella flexilis*.

In cultivation *R. tripartitus* is perennial but in the field it behaves as a winter annual, germinating in autumn on damp mud or under water. Individuals can overwinter as aquatics with capillary leaves, or as terrestrial plants with laminar leaves. They flower from March to May. If the habitat dries out the plants are then killed by desiccation; if it remains moist they are overwhelmed by more vigorous species.

R. tripartitus has undoubtedly undergone a marked decline in Britain. The reasons for this include the destruction of heathland, the draining or infilling of ponds, the improvement of rutted tracks and the reduction or cessation of grazing which has led to the development of rank vegetation or even willow carr around suitable habitats. The exact extent of the decline is difficult to assess as the plant has probably been overlooked in both the past and in recent years because of its early flowering season, and perhaps because of its similarity to *R. × novae-forestae*. The conservation of this species will require the protection and appropriate management of existing colonies, and perhaps the restoration of suitable habitat at former sites where the plant may still be present as dormant seed. *R. tripartitus* may be extinct in Ireland, where it has not been seen since 1897 in spite of much searching; recent records are now known to be based on misidentifications (T. G. F. Curtis, pers. comm.).

This species has an Atlantic distribution, being found in Morocco and in western Europe from Spain and Portugal to the Netherlands and northern Germany. It has also been reported recently from the Aegean island of Mykonos (Jalas & Suominen, 1989; Dahlgren, 1991). The species is declining in the northern part of its range; it is extinct in Belgium and it was last seen in the Netherlands in 1951 (Weeda *et al.*, 1990). There is little published information on its current distribution in southern Europe.

R. tripartitus is clearly adapted to an oceanic climate: it can be cultivated without difficulty in botanic gardens in western Europe but only with much effort in central Europe, where the transition between winter and summer appears to be too abrupt (Cook, 1976).

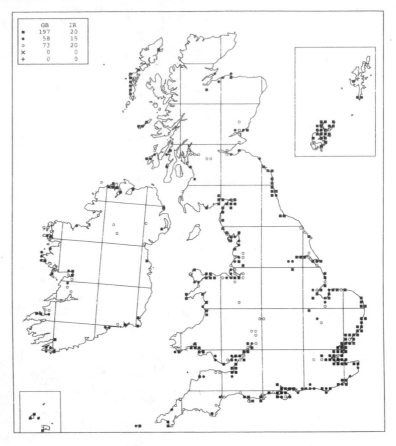

	GB	IR
■	197	20
●	58	15
○	73	20
×	0	0
+	0	0

Ranunculus baudotii Godr.
Brackish Water-crowfoot

This is a species of still, or very slowly flowing, brackish water. It is confined to water less than 1m deep, and usually found at depths shallower than 50cm. It grows in a range of coastal habitats including lagoons, the seaward side of machair lochs, ditches, pools, dune slacks and borrow-pits. It is a characteristic species of unfenced ditches in grazing marshes, where grazing and trampling by livestock prevent the growth of a dense sward of emergent vegetation. It is often found as a colonist of newly dug pools or pits, and even flooded ruts in unsurfaced tracks. It is often associated with species such as *Myriophyllum spicatum*, *Potamogeton pectinatus*, *Ruppia maritima* and *Zannichellia palustris* in water which is sufficiently saline to exclude most macrophytes but not brackish enough to support a dense sward of *Ruppia*. Exceptionally, it can be found in much more brackish situations, including sites which are flooded at high tide. There are a few inland records from canals, gravel-pits and clay-pits, where it grows in fresh water and in sites which receive saline drainage water from mines.

R. baudotii can behave as an annual or a perennial, depending on the particular habitat in which it grows. Perennial plants persist through the winter in the capillary leaf state. In summer the species can persist as a small, terrestrial form on damp mud. Plants are well adapted to self-pollination (Hong, 1991). The usual method of reproduction is probably by seed, but detached stems can also root and become established.

It is difficult to assess trends in the distribution of this species, which has been under-recorded in the past and is probably still overlooked in some areas. The species has almost certainly been lost from many sites in southeast England in recent years because of the conversion of much coastal grazing marsh to arable land, but these losses do not show at the scale of the map, and may have been partly compensated for by the ability of the species to colonize new sites.

R. baudotii is frequent along the Atlantic coast of Europe from Portugal to Denmark, and in the Baltic. It is apparently rarer along the Mediterranean coast and occurs at scattered inland localities in Europe (Jalas & Suominen, 1989). Outside Europe it is found in Morocco and the Canaries.

R. baudotii is a variable species; much of the variation is phenotypic but some has a genetic basis (Cook, 1966a). Typical plants are not difficult to identify, but not all plants attributable to this species show the full range of diagnostic characters and because of this the map is likely to show some erroneous records. *R. baudotii* and *R. peltatus* are difficult to distinguish in southern Europe, where the morphological gap between them is bridged by *R. peltatus* subsp. *saniculifolius* (Viv.) C. D. K. Cook. Cook (1984) has recently reduced *R. baudotii* to *R. peltatus* subsp. *baudotii* (Godr.) Meikle ex C. D. K. Cook, a treatment followed in *Flora Europaea* (Tutin *et al.*, 1993). However, in some areas, including the Outer Hebrides, *R. baudotii* is more difficult to distinguish from *R. trichophyllus*. It may be best to retain the plant at specific rank, while recognizing that several species of batrachian *Ranunculus* are less distinct than one would like them to be.

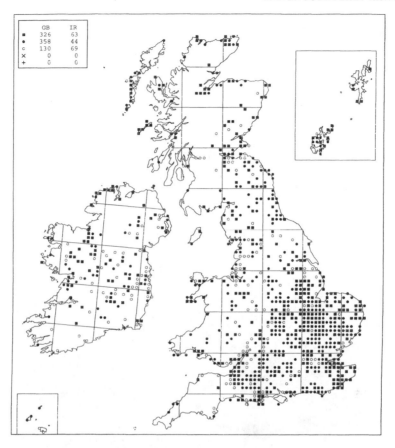

	GB	IR
■	326	63
●	358	44
○	130	69
×	0	0
+	0	0

Ranunculus trichophyllus Chaix
Thread-leaved Water-crowfoot

R. trichophyllus is a plant of shallow, still or very slowly flowing water. It is most frequently found in small water bodies such as ponds, dune slacks or drainage ditches, but it is also recorded from a wide range of other habitats including lakes, slowly flowing streams, canals and turloughs. It is an early colonist of new habitats or of sites which have been disturbed, where it is often accompanied by *Ranunculus aquatilis* and *Chara vulgaris*. It cannot persist (except perhaps as buried seed) when the vegetation becomes dominated by more competitive macrophytes or emergents. It tolerates a wide range of water chemistry but is most frequent in mesotrophic or eutrophic water and is absent from the most oligotrophic sites; it is sometimes found in brackish habitats near the sea. It is a mainly lowland species, which is recorded up to 310m on Alston Moor; the maximum altitude of 550m (Widdale Fell) given by Wilson (1956) is probably correct but the identity of the plant requires confirmation.

Like the other smaller members of subgenus *Batrachium*, *R. trichophyllus* can behave as an annual or a perennial. It often begins to flower in March or April, being the first member of the subgenus to flower in most parts of its British and Irish range. Plants are often self-pollinated before the flowers open. When the habitat dries out *R. trichophyllus* can persist on damp mud in a terrestrial phenotype, but these plants do not survive severe desiccation. Individuals can overwinter as submerged plants with capillary leaves but the terrestrial forms are killed by frost, and survive only if re-submerged before the onset of cold weather. Reproduction is by seed, which can germinate on damp mud or under water.

This is the most frequent batrachian *Ranunculus* in many lowland areas, and there is no evidence to suggest that it has decreased to any significant extent.

This species occurs throughout the northern hemisphere in arctic, boreal and temperate areas. It is also recorded from S.E. Australia and Tasmania. The plant in the British Isles is subsp. *trichophyllus*, which is replaced in arctic and montane areas by subsp. *eradicatus* (Laest.) C. D. K. Cook.

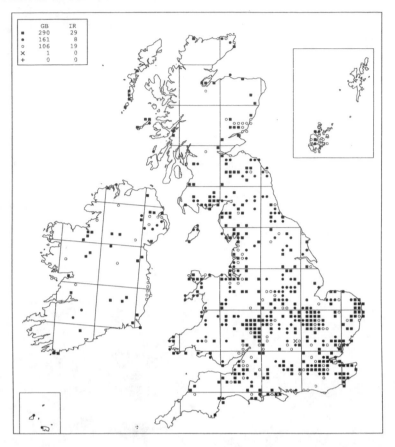

	GB	IR
■	290	29
●	161	8
○	106	19
×	1	0
+	0	0

Ranunculus aquatilis L.
Common Water-crowfoot

R. aquatilis has similar ecological requirements to *R. trichophyllus* with which it often grows. It is a plant of shallow water in marshes, ponds and ditches, the sheltered edges of lakes and the margins of slow-flowing streams. It appears to favour eutrophic and often rather base-rich water, and perhaps for this reason it is often recorded from farm ponds. Like many of the smaller species in subgenus *Batrachium*, it is favoured by disturbance and is often found in muddy water at the edge of streams and ponds where cattle congregate. *Callitriche obtusangula*, *Glyceria fluitans* and *Lemna minor* are characteristic associates. It is a primarily lowland species which is recorded up to 445m at Small Water.

R. aquatilis is an annual or a short-lived perennial. Most plants flower in May and June, but the flowering period can extend into the late summer. Like *R. trichophyllus*, plants are often self-pollinated and isolated plants set a full complement of seed. If the water dries up plants can persist through the summer on damp mud. However, the species produces only laminar leaves in long days; in short winter days only capillary leaves are produced and plants therefore overwinter under water or perish (Cook, 1969).

Our knowledge of the distribution of this species is incomplete as most records collected for the *Atlas of the British Flora* (Perring & Walters, 1962) under the name *R. aquatilis* are only attributable to an aggregate which also includes *R. peltatus* and *R. penicillatus*, and are not plotted on the map. Because the taxonomy of this group has only recently been clarified, and because *R. aquatilis* is under-recorded, it is difficult to assess whether there have been any changes in its distribution.

R. aquatilis is widespread in Europe from southern Scandinavia southwards, but is rare in the Mediterranean region. It is recorded from N. Africa and in scattered localities across Asia to Japan. It is also found in western N. and western S. America.

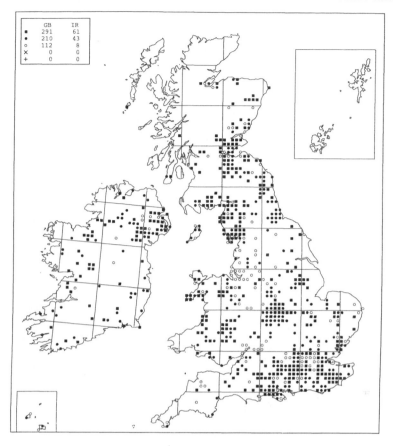

	GB	IR
■	291	61
●	210	43
○	112	8
×	0	0
+	0	0

Ranunculus peltatus Schrank
Pond Water-crowfoot

In southern England *R. peltatus* is often found in the upper reaches of highly calcareous rivers flowing from a chalk aquifer, in stretches which often dry out during the summer months. Further downstream it is replaced by *R. penicillatus* subsp. *pseudofluitans*. In Ireland it is the characteristic batrachian *Ranunculus* of streams and rivers over Carboniferous limestone. Away from calcareous bedrock it is found in a variety of other habitats, including slow-flowing, acidic streams and rivers over peat or clay, coastal lagoons, the shallow edges of lakes and nearby inflow streams, ditches, ponds and dune slacks. It can colonize newly-created habitats such as gravel-pits and in established vegetation it often benefits from disturbance. In many areas it shows some preference for base-poor water, but its trophic range seems to be broad and it has been recorded from both nutrient-rich and nutrient-poor localities. It is primarily a lowland species, which reaches 500m at Dogber Tarn. A record from 500m at Llyn Anafon (Wilson, 1956) is dubious as recent collections from this site have been determined as a sterile hybrid of doubtful parentage.

In permanently moist habitats *R. peltatus* is a perennial, which overwinters as an aquatic and persists through the summer in water or in a terrestrial form on damp mud. When the habitat dries out in summer it can behave as an annual. Cook (1966a) described the species

as self-compatible and adapted to self-pollination. However, the flowers are large and Hong (1991) concluded that the plants he studied were well adapted for outbreeding, and were pollinated by beetles. Their anthers did not dehisce in the buds, and there was some evidence that plants were self-incompatible.

R. peltatus has been recorded systematically only since the publication of Cook's monograph in 1966, and is still under-recorded. The patchy distribution shown on the map probably reflects this, but the species appears to be genuinely absent from some areas within its overall range, including the vice-county of Cambridgeshire. Some of the records plotted may be erroneous, as the species can be confused with *R. aquatilis*, *R. baudotii* or *R. penicillatus*. In the absence of adequate historical records it is impossible to assess whether its distribution has changed. A population growing at 460m in the R. Tees was destroyed when the site was flooded by Cow Green Reservoir (Proctor, 1978).

R. peltatus is widespread in Europe from the Mediterranean countries north to northern Scandinavia (Jalas & Suominen, 1989). Elsewhere it is known only from N. Africa. The closely related *R. sphaerospermus* Boiss. & Blanche extends from the Balkans through western Asia to the Himalaya.

In its ecological preferences *R. peltatus* is intermediate between the smaller species in subgenus *Batrachium*, which are plants of shallow, still or very slowly flowing water, and the larger riparian species.

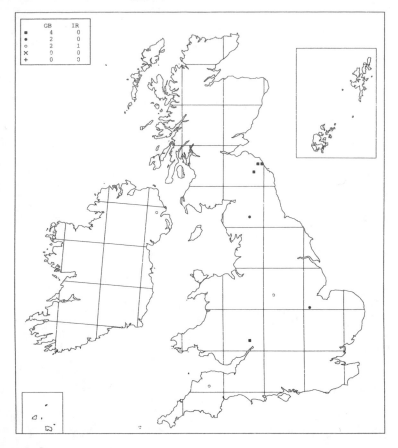

Ranunculus × *kelchoensis* S. D. Webster (*R. fluitans* Lam. × *R. peltatus* Schrank)

Kelso Water-crowfoot

This hybrid has been recorded from a number of major lowland rivers, most of which have a meso-eutrophic flora and are moderately base-rich. In the R. Teviot and the Whiteadder Water it is locally very abundant, forming large beds in shallows where the rivers flow rapidly over a stony substrate. Neither parent has been recorded in the vicinity of these stands, and the beds of *R.* × *kelchoensis* are normally sufficiently vigorous to exclude other species, but *R. penicillatus* subsp. *pseudofluitans* is sometimes present. The hybrid was also recorded from the R. Welland, where the water is more calcareous and probably more eutrophic than at these sites, but little is known of the exact habitat or the associated species in this river.

This is a robust perennial. Most populations are highly sterile, but the stems root at the nodes and the plant can presumably reproduce and spread by vegetative fragments. Sterile populations are very persistent:

the hybrid was collected in the Whiteadder Water at Allanton in 1841 and and the R. Teviot at Roxburgh Castle in 1876, and refound in both sites in 1991. Plants from the R. Wye which are also referable to *R.* × *kelchoensis* set well-formed fruit.

Brief accounts of this hybrid were given by Cook (1966a, 1975a) but it was not until Webster (1990) gave it the binomial '*kelchoensis*' that it was brought to the attention of most British and Irish botanists. The hybrid has not yet been searched for in all the rivers from which old herbarium specimens are known. It will probably be rediscovered in some localities from which only pre-1950 records are available, and found in new stations elsewhere. However, it has not been possible to refind the plant at sites in the R. Welland and its tributaries where it was present in the 1950s (Webster, 1990).

In addition to its British and Irish localities. *R.* × *kelchoensis* is known from France and Portugal.

Ranunculus penicillatus subsp. *penicillatus* is believed to be an amphidiploid derived from hybrids similar to *R.* × *kelchoensis*.

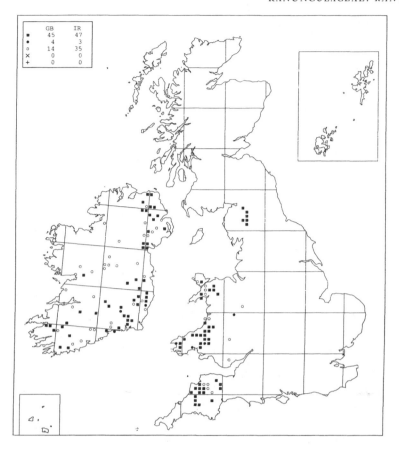

	GB	IR
■	45	47
●	4	3
○	14	35
×	0	0
+	0	0

Ranunculus penicillatus (Dumort.) Bab. subsp.
penicillatus   **Stream Water-crowfoot**

This is primarily a plant of rivers and streams, where it is
sometimes found in dense stands which dominate the
entire channel. In some rivers it grows in torrent conditions; in others it occurs in sections which are usually
more slow-flowing but where the current velocity is high
during floods (Holmes, 1980). In Britain it is confined to
base-poor and usually more or less mesotrophic waters
flowing over Palaeozoic or igneous rock. In Ireland it has a
considerably greater ecological range, being found in both
base-rich and base-poor rivers and streams over a range of
parent materials including basalt and Carboniferous limestone. It is also recorded from some Irish lakes and from
the Grand Canal in Dublin. It is not recorded above 300m.

This is a fertile perennial which overwinters as shoots
with capillary leaves. Its reproduction has not been
studied. Plants can presumably spread by vegetative
fragments, as lengths of stem are often detached and
washed downstream. The degree to which it reproduces
by seed is unknown.

R. penicillatus subsp. *penicillatus* has only been recognized as a distinct taxon in recent years, and is likely to
be under-recorded, particularly in Ireland.

The distribution of *R. penicillatus* and its constituent
subspecies is unclear because of taxonomic confusion.
The species appears to be widespread in the middle latitudes of Europe, but absent from both northern countries and much of the Mediterranean region (Jalas &
Suominen, 1989). The confirmed records of subsp. *penicillatus* are from W. and C. Europe, extending from the
British Isles, Denmark and St Petersburg southwards to
Portugal, Spain and the mountains of Sardinia (Cook,
1966a).

In Britain *R. penicillatus* subsp. *penicillatus* is replaced
by subsp. *pseudofluitans* in base-rich rivers. In Ireland
subsp. *pseudofluitans* is known from only one site, and
Webster (1991) suggests that this has allowed subsp.
penicillatus to occupy a wider range of habitats.

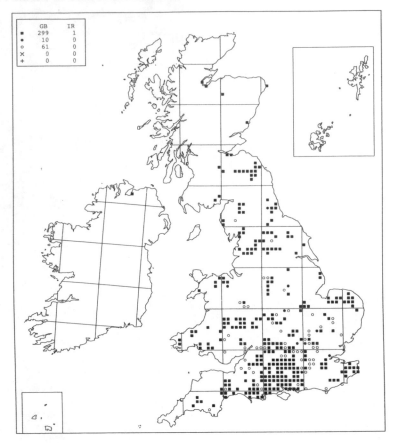

	GB	IR
■	299	1
●	10	0
○	61	0
×	0	0
+	0	0

Ranunculus penicillatus subsp. *pseudofluitans*
(Syme) S. D. Webster   **Stream Water-crowfoot**

This is the dominant macrophyte in many British low-land, base-rich streams and rivers, where it often forms dense beds which span the entire channel. It is found in moderately or rapidly flowing, mesotrophic or meso-eutrophic water over a wide range of geological strata, but it is most frequent in calcareous water arising from chalk or limestone aquifers. It is also found in some rivers which are only moderately base-rich. In eutrophic or sluggish rivers it is usually replaced by other aquatics, including *Myriophyllum spicatum* and *Potamogeton* species. It also grows in clear calcareous water in canals, pools and ditches. In Ireland it is confined to a single, rather base-poor river which flows over schist and gneiss, the usual habitat of subsp. *penicillatus*. Subsp. *pseudofluitans* is virtually confined to the lowlands, reaching 310m in the Carlcroft Burn.

Like subsp. *penicillatus*, subsp. *pseudofluitans* is a fertile perennial which overwinters as actively growing leafy shoots. Dawson (1980) noted that the first plants to flower in the R. Piddle were those near the source of the river, and there was then a progression of flowering downstream until the plants near the tidal limit flowered 2–3 months later. Material transplanted to other parts of the river retained its original flowering date. The extent to which subsp. *pseudofluitans* reproduces by seed or by the spread of vegetative fragments is unknown; reproduction by seed may be rare.

This species is so abundant in some rivers that it impedes the flow of water, and has to be removed by mechanical or chemical means. It is, however, vulnerable to pollution and it is, for example, much rarer in the R. Colne than it once was (Kent, 1975). However, it has also spread very rapidly in some rivers such as the R. Wear, which was formerly polluted by the coal industry and by sewage. It was first noted in the Wear in 1970 and it was perhaps the most abundant submerged flowering plant by 1976, having spread from a small population in a tributary (Holmes & Whitton, 1977a; Graham, 1988).

The distribution of *R. penicillatus* subsp. *pseudofluitans* is obscured by taxonomic confusion. It is recorded from W. & C. Europe by Cook (1966a), and has subsequently been reported from Greece.

Subsp. *pseudofluitans* differs from subsp. *penicillatus* in never having laminar leaves. The sole Irish population of subsp. *pseudofluitans* may actually be a variant of subsp. *penicillatus* which has lost the capacity to produce such leaves, and is therefore morphologically identical to subsp. *pseudofluitans* (Webster, 1991). Two varieties of subsp. *pseudofluitans* are recognized in Britain. Var. *pseudofluitans* is the commoner plant and is usually found in rivers, whereas var. *vertumnus* C. D. K. Cook can grow in rivers but is also found in streams, canals, ditches and pools. The distribution of both varieties is mapped by Webster (1988a), who also summarizes the extensive ecological literature on var. *pseudofluitans*. Murrell & Sell (1990) suggest that the varieties should each be recognized as species distinct from *R. penicillatus*.

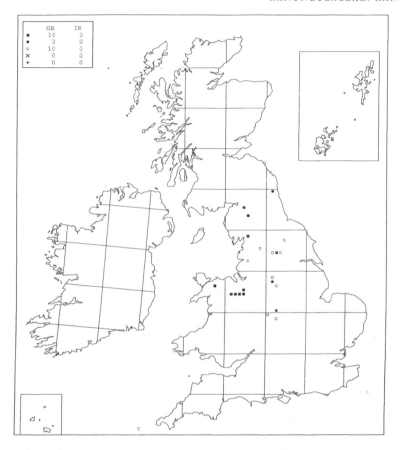

	GB	IR
■	10	0
●	3	0
○	10	0
×	0	0
+	0	0

***Ranunculus* × *bachii* Wirtg. (*R. aquatilis* L. × *R. fluitans* Lam. & *R. fluitans* Lam. × *R. trichophyllus* Chaix)** Wirtgen's Water-crowfoot

R. × *bachii* is known from a number of lowland rivers and streams, but most records are derived from herbarium specimens and there is little information about the nature of the sites in which it grows. In the R. Wharfe *R.* × *bachii* has recently been recorded in shallow, stony stretches of the river, growing with *R. penicillatus* subsp. *pseudofluitans* var. *vertumnus*, *Myriophyllum spicatum*, *Potamogeton crispus* and *P. perfoliatus*. The hybrid is known from rivers such as the Eden from which neither *R. aquatilis* nor *R. trichophyllus* has been recorded. The occurrence of the hybrid in lakes requires confirmation: plants collected as *R. cambricus* A. Benn. in Llyn Coron last century may be this hybrid, but they could be *R. fluitans* growing in an unfavourable habitat (Webster, 1990); they are not plotted on the map.

R. × *bachii* is a robust, highly sterile hybrid which can form large populations. Plants continue to grow through the winter months. Reproduction must be by vegetative spread, which presumably occurs when stems become detached and are carried downstream to become estab-

lished at new localities. Established populations can be very persistent: the hybrid was first collected in the R. Wharfe in 1904, and is still present there.

Cook (1966a) confirmed the presence of *R.* × *bachii* in Britain, but few observers are familiar with it and it is probably more frequent than the map suggests.

Plants attributable to *R.* × *bachii* have been recorded from a number of countries in W. and C. Europe, and, as in Britain, there is evidence that they have been established in some river systems for many years.

The name *R.* × *bachii* is used to cover the hybrids with *R. fluitans* as one parent and *R. aquatilis* or *R. trichophyllus* as the other. The type of *R.* × *bachii* is probably *R. fluitans* × *trichophyllus*, and the name is used in this restricted sense by Kent (1992). However, the two hybrids are genotypically and phenotypically variable and they cannot be distinguished with certainty on the basis of their morphology, although they differ in chromosome number. The population of *R.* × *bachii* at Monsal Dale has been identified as *R. aquatilis* × *fluitans* on the basis of a chromosome count. Hybrids similar to *R.* × *bachii* sensu lato are believed to have given rise to the fertile amphidiploid *R. penicillatus* subsp. *pseudofluitans*.

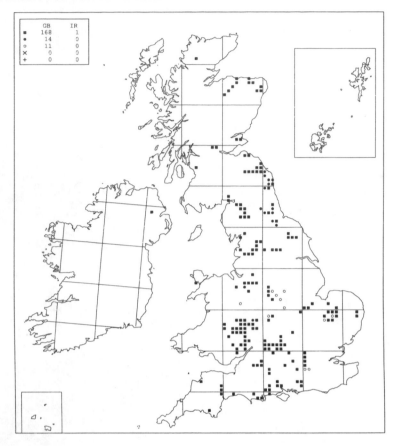

Ranunculus fluitans Lam.   River Water-crowfoot

R. fluitans is primarily a species of large rivers, where it usually grows in rapidly flowing water which is base-rich and meso-eutrophic. Its detailed distribution is often controlled by its requirement for a stable substrate. In Derbyshire, for example, it is present where rivers flow over non-calcareous rocks which weather to smooth pebbles around which silt collects, but is replaced by *R. penicillatus* subsp. *pseudofluitans* in those stretches of the same rivers which flow over limestone, and where the substrate is much less stable. In many northern rivers, including the Tees and the Swale, *R. penicillatus* subsp. *pseudofluitans* is usually the dominant species in the upper reaches but is replaced by *R. fluitans* downstream (Holmes & Whitton, 1977b,c); in the R. Dove the zone of overlap is less than 0.5km (Clapham, 1969). It is not recorded above 300m.

   R. fluitans is a fertile perennial. It overwinters as slow-growing, compact and prostrate patches of stems which root at the nodes. In April or May these clumps give rise to rapidly elongating and slightly hollow stems which lack the ability to root, and which flower from June onwards. Little is known of the relative importance of vegetative reproduction as opposed to reproduction by seed, but the species can spread fairly rapidly, as evinced by its colonization of the R. Wear following the deliberate introduction of plants by a water bailiff in 1959 (Graham, 1988).

In Britain this species has been confused with other taxa, particularly *R. penicillatus* and sterile hybrids which may have *R. fluitans* as one parent. Only records from rivers where the presence of the species has been confirmed by experts have been accepted for the map. Nevertheless, some records are probably erroneous and in other areas the species is probably under-recorded. Cook (1966a) describes it as fairly tolerant of pollution as long as the water remains clear, but soon dying if the water becomes cloudy or shaded. He suggests that it has become extinct or rare in many of the large rivers in the English Midlands. The species was eliminated by industrial pollution last century from that part of the Six Mile Water where it was first discovered in Ireland (Hackney, 1992). In recent years *R. fluitans* appears to have decreased in abundance in some rivers, such as the R. Lark in Suffolk, where it was frequent in the 1950s. A critical reappraisal of its past and present distribution would be worthwhile.

   R. fluitans is endemic to Europe. Most records are from W. & C. Europe, where the species extends from the Pyrenees north to southern Sweden, but there are scattered localities southwards and eastwards to Sardinia, Italy and Romania (Jalas & Suominen, 1989). In Switzerland *R. fluitans* has recently become a serious pest in some rivers and has to be removed by hydro-electricity operators; this increase has been attributed to a rise in phosphate levels in the water (C. D. K. Cook, pers. comm.).

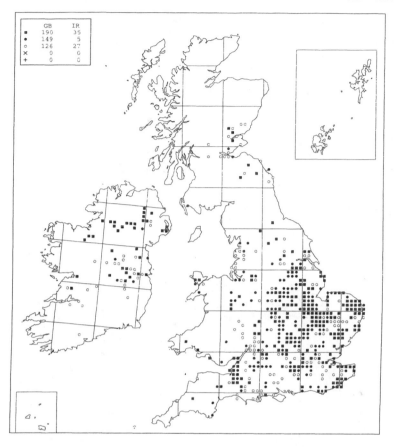

	GB	IR
■	190	35
●	149	5
○	126	27
×	0	0
+	0	0

Ranunculus circinatus Sibth.
Fan-leaved Water-crowfoot

This delicate species is found in standing or very slowly flowing, clear, base-rich, meso-eutrophic or eutrophic water. It is normally found in water 1–3m deep; it can grow in more shallow sites but only if the water is permanent, as it is intolerant of desiccation. Typical habitats include grazing marsh and fenland ditches, ponds, canals and major fenland drains, natural and ornamental lakes, reservoirs and sluggish streams and rivers. It is also a frequent colonist of gravel-pits, and it occasionally invades flooded quarries and newly-dug garden pools. It often grows in species-rich assemblages of aquatic plants; some of the more frequent of its many associates are *Elodea canadensis*, *Myriophyllum spicatum*, *Potamogeton crispus*, *P. pectinatus*, *P. pusillus* and, in larger waters, *Nuphar lutea*, *Potamogeton lucens* and *P. perfoliatus*. It is confined to the lowlands.

R. circinatus is a perennial; Salisbury (1960) suggested that it might behave as an annual but there is little or no evidence to support this view. Plants overwinter as prostrate stems which root at the nodes and bear rather flaccid leaves. In summer erect shoots grow upwards towards the water surface. Populations often fail to flower or flower only sparingly, but the plant can flower profusely on the surface of a ditch, lake or reservoir (cf. West, 1910; Church, 1925). Flowering sometimes begins before the stems reach the water surface, and these submerged flowers are often cleistogamous. It has little resistance to desiccation and in cultivation the terrestrial form can be grown only in moist conditions with long photoperiods. Cultivated terrestrial forms grow slowly and do not flower. It is not known whether the plant colonizes new sites by seed or by vegetative fragments. In Finland, where it rarely flowers, it is dispersed by fragments which become attached to boats (Cook, 1966a) and in Britain it might similarly be dispersed by anglers.

R. circinatus is the most distinctive aquatic *Ranunculus*, and it is therefore relatively well recorded. The results of the BSBI Monitoring Scheme indicated that it has declined in both England and Scotland since 1960 (Rich & Woodruff, 1996). It has been lost from many sites because of habitat destruction, including the infilling of farm ponds and the loss of drainage ditches in areas where pastoral land has been converted to arable. Driscoll (1985a), for example, noted a very marked reduction in its frequency when an area of grazing marsh in Norfolk was ploughed up and the drainage system modified. *R. circinatus* is intolerant of very eutrophic conditions, and it has therefore disappeared from sites which have become too eutrophic, including its northernmost localities (Balgavies Loch, Loch of Forfar and Rescobie Loch). The species has, however, colonized newly available habitats and in some counties such as Bedfordshire (Dony, 1976) and Essex (Jermyn, 1974) it is confined to gravel-pits or is more frequent in this habitat than in any other.

R. circinatus occurs in temperate latitudes from western Europe to eastern Asia; in Europe it is virtually absent from the Mediterranean region, and reaches its northern limit around the Gulf of Bothnia (Jalas & Suominen, 1989). Closely related taxa occur at similar latitudes in N. America.

POLYGONACEAE

This large family of 43 genera and over 1,100 species is found throughout the world, but is particularly well represented in the temperate northern hemisphere. It includes a wide range of life-forms (Kubitzki, 1993), and even in Britain the native and naturalized species include small, annual herbs (e.g. *Koenigia islandica*), giant perennial herbs (e.g. *Fallopia sachalinensis*), lianes (e.g. *F. baldschuanica*) and intricately branched shrubs (e.g. *Muehlenbeckia complexa*). An illustrated account of the British and Irish species is available (Lousley & Kent, 1981). The only genera to contain aquatic species are *Persicaria* and *Rumex*, both of which include true aquatics, emergents, plants of seasonally flooded ground and annuals which grow on mud exposed by falling water-levels, as well as many terrestrial species.

Persicaria Mill.

The species traditionally included in the genus *Polygonum* by British botanists are now classified in the three segregate genera *Fallopia*, *Persicaria* and *Polygonum* sensu stricto, following a revision of generic boundaries by Ronse Decraene & Akeroyd (1988). Some 150 species formerly included in the genus *Polygonum* probably belong to *Persicaria*, including the aquatic *Persicaria amphibia*.

Persicaria amphibia (L.) Gray

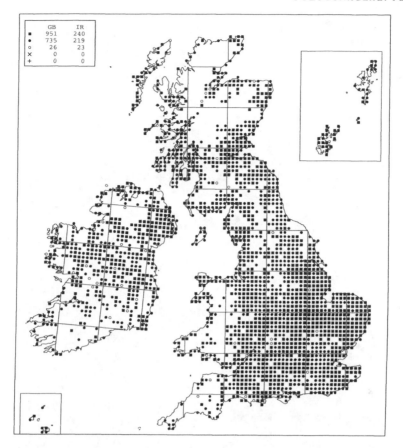

	GB	IR
■	951	240
●	735	219
○	26	23
×	0	0
+	0	0

Persicaria amphibia (L.) Gray   Amphibious Bistort

As an aquatic *P. amphibia* grows in standing or slowly moving water in lakes, reservoirs, canals, ditches, large fenland drains, and sluggish streams and rivers. In lakes it can form large, virtually pure stands in water from 0.5 to at least 2m deep; these stands often appear to represent a single clone. It is confined to mesotrophic or eutrophic water and is tolerant of turbidity caused by disturbed sediment or algal blooms, perhaps because it lacks submerged leaves. Colonies are found over a range of fine-textured substrates. *P. amphibia* also grows as a terrestrial plant in swamps and in herbaceous vegetation on ditchbanks, by canal towpaths, in wet meadows (where it is often dominant in winter-flooded hollows) or as a weed in arable fields near rivers or fenland drains. It can be abundant in sites such as reservoir margins, dune-slacks and the Breckland meres, where the water-level fluctuates greatly. It is virtually confined to the lowlands; the only record above 300m is from 572m at Blind Tarn, Coniston (Stokoe, 1983).

P. amphibia is a rhizomatous perennial. The shoots of a single plant measured by Arber (1920) were 13m long. The species can spread in water by regeneration from detached shoots and on land by growth from rhizome fragments following disturbance; in many such sites vegetative spread is probably more important than sexual reproduction. Aquatic populations often flower freely but even the most vigorous terrestrial stands usually produce only a few short inflorescences. The flowers have nectaries and are insect pollinated. The species is self-incompatible and seed is not set by those populations which consist of a single clone (C. D. K. Cook, pers. comm.). The fruits are eaten by surface-feeding duck (Mabbott, 1920). The seeds are dormant when they are first shed, not germinating even if the pericarp is removed. Under experimental conditions they do not respond to treatments which stimulate germination in many other *Persicaria* and *Polygonum* species, but 75–85% germination can be obtained by storing intact seeds for 4–6 months, then removing the pericarp and keeping the naked seeds in moist conditions at 2–4°C for a further 4–5 months (Justice, 1941). In the wild, seeds germinate in spring (Grime *et al.*, 1988).

P. amphibia is a frequent plant in lowland Britain and Ireland, and the results of the BSBI Monitoring Scheme suggested that it has increased in frequency in England since 1960 (Rich & Woodruff, 1996). It is common in suitable habitats even in the densely populated London area (Burton, 1983, Kent, 1975). Its capacity to grow both terrestrially and as an aquatic perhaps enables it to withstand all but the most drastic habitat change.

The variable *P. amphibia* is found at temperate latitudes throughout the northern hemisphere, and has also been introduced to areas south of its native range. In Europe it is frequent except in the far north and in the Mediterranean region (Jalas & Suominen, 1979).

The capacity of *P. amphibia* to exist as two markedly different phenotypes has been recognized since the 17th century. The aquatic plant has glabrous, floating leaves with a long petiole whereas the terrestrial plant has pubescent, short-stalked leaves. One phenotype may change to the other in response to changes in water-level.

Rumex L.

This is a genus of some 200, predominantly terrestrial species. The thirteen species native to the British Isles fall into three subgenera, which are sufficiently distinct to be treated as genera by some European botanists. Both subgenus *Acetosella* (Meisn.) Rech. f. and subgenus *Acetosa* (Mill.) Rech. f. are represented by a single terrestrial species in our area, but subgenus *Rumex* includes two aquatics.

Rumex hydrolapathum Huds.

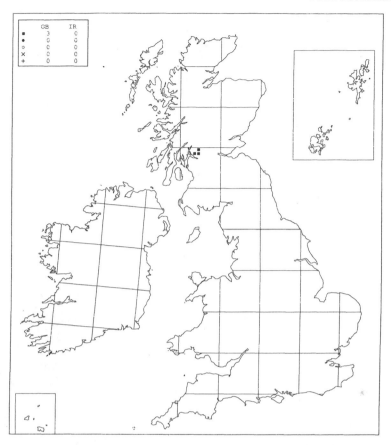

Rumex aquaticus L.   Scottish Dock

This robust species is found in lakeside vegetation, by the side of rivers or ditches, or on riverine sand banks. In these localities it grows amongst tall, emergent vegetation; its most frequent associates are *Carex vesicaria, Deschampsia cespitosa, Eleocharis palustris, Equisetum fluviatile, Filipendula ulmaria, Phalaris arundinacea* and *Sparganium erectum*. There is also a large colony in a clearing in a swampy alder wood. The plants grow on eutrophic, sandy, silty or muddy soils; at a lakeside site it has been recorded near a sewage outflow and some colonies receive nutrient-enriched water from farmland. All sites are at altitudes below 20m.

R. *aquaticus* is a perennial which, like other *Rumex* species, is believed to be wind-pollinated. In the largest British colony at least 89% of the seed is viable; in most other colonies the percentage of viable seed ranges from 55 to 76 (Hull & Nicholl, 1982). Mature plants with ripe fruit tend to topple over like wind-blown trees. Little is known of the conditions under which seeds germinate and seedlings become established in the wild.

R. *aquaticus* is known from fossil evidence to have been present in the late glacial period in Cornwall and N. Wales (Godwin, 1975). The only records of the living plant are from the shore of Loch Lomond, and by the Endrick Water which flows into the south-east corner of that loch. Most colonies are found on the flood plain of the Endrick, in areas which were drained for agriculture

in the 18th century, but subsequently reverted to marsh and damp woodland. The species was not discovered in Britain until 1935 (Lousley, 1939, 1944b). Its distribution around the south-east of Loch Lomond was mapped in 1967 by Idle (1968), and has not changed markedly in the last 25 years. However, two additional populations have been discovered, one on the west shore of Loch Lomond and the other by the Endrick 10 km upstream of the lake (Mitchell, 1982, 1993). The known populations of the species do not appear to be threatened, and the relatively recent discovery of the plant in Britain suggests that it may be found elsewhere in northern England or lowland Scotland.

R. *aquaticus* is a circumboreal species. It has a rather continental distribution in Europe, extending from the arctic to Scotland, C. France, the Alps and Bulgaria (Jalas & Suominen, 1979). It is extinct in the Netherlands (Weeda *et al.*, 1990). The species is polymorphic, and several subspecies are recognized. The closely related R. *aquaticiformis* Rech. f. is found in Patagonia.

In Scotland R. *aquaticus* hybridizes with R. *obtusifolius*, a terrestrial species. The hybrids tend to occur on drier ground than that normally occupied by R. *aquaticus*. The reduction in seed viability noted in some populations of R. *aquaticus* may be due to hybridity. Detailed accounts of R. *aquaticus*, from which much of the above information has been taken, are provided by Idle (1968) and Mitchell (1983).

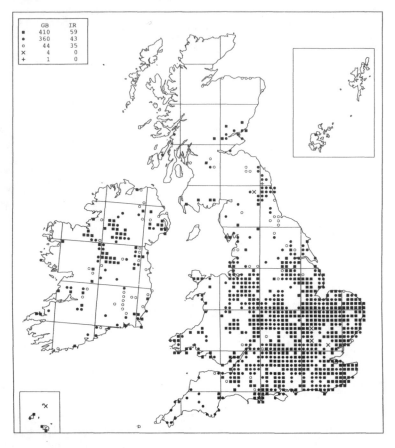

	GB	IR
■	410	59
●	360	43
○	44	35
×	4	0
+	1	0

Rumex hydrolapathum Huds. Water Dock

This species grows by still or slowly flowing, base-rich lowland waters, including lakes, slowly flowing rivers and associated ox-bows, canals and ditches. It is a species of eutrophic situations, and flourishes by lakes and ditches where submerged and floating species have been reduced in number or even eliminated by eutrophication. It is often found as scattered plants growing where stands of emergent vegetation join the open water; these emergents can include *Acorus calamus*, *Glyceria maxima*, *Iris pseudacorus*, *Phragmites australis* and *Typha latifolia*. *R. hydrolapathum* is less frequently found as a colonist of bare ground in marshes and fens, where it can temporarily occur in abundance: a 'great expanse' is described by Griffiths (1932) in an area of marshland which had previously been open water but had been partially drained 100 years earlier. However, it fails to persist in closed fen vegetation as established plants are outcompeted by surrounding vegetation and seeds are unable to germinate.

R. hydrolapathum is a wind-pollinated perennial. At Woodwalton Fen a plant of average size produces 16,000 fruits; more than 95% contain viable seed and this viability is not reduced by storage at room temperature for one year. Seeds at this site germinate in spring in areas of bare ground (Duffey, 1968). The fruits float for long periods, and they may also be dispersed by wind (Guppy, 1906; Lousley & Kent, 1981). Under experimental conditions the germination requirements of *R. hydrolapathum* are similar to those of more terrestrial

Rumex species: germination and growth is better in freely drained soils and soils where the water-table lies 10cm below the surface than in fully waterlogged conditions (Harper & Chancellor, 1959). Plants do not flower in their first year.

In southern England *R. hydrolapathum* has been lost from some sites because of the drainage of wetlands for agricultural use or for building, but it remains frequent in suitable habitats. In northern England and Scotland it has apparently extended its range in the last 100 years, probably by spread from sites where it was originally planted as an ornamental species. In Northumberland it was not recorded before 1890 but is now locally plentiful and possibly still spreading (Swan, 1993); in Angus it is apparently increasing (Ingram & Noltie, 1981).

R. hydrolapathum is virtually confined to Europe, occurring elsewhere only at sites in N.W. Anatolia and the Caucasus. In Europe it extends from southern Scandinavia to Sardinia, Italy and the Balkans, but it becomes rare towards the southern limit of its range (Jalas & Suominen, 1979).

This species is the foodplant of the extinct British subspecies of the Large Copper butterfly, and of the closely related Dutch subspecies which has been introduced to Woodwalton Fen. The butterfly lays its eggs on plants which grow in fen vegetation away from the water's edge, but are not hidden by surrounding vegetation. These are not the optimum conditions for *R. hydrolapathum* and suitable conditions for the butterfly can be maintained only by management practices such as cattle grazing or peat cutting (Duffey, 1968, 1977).

ELATINACEAE

A small family of two genera, *Bergia* and *Elatine*, and 36 species. The genus *Bergia* includes both aquatic and terrestrial species, most of which are robust and conspicuous. They are found in warm-temperate and tropical areas, although *B. capensis* L. is naturalized in Spain.

Elatine L.

This is a small genus of twelve species, all of which are small and inconspicuous. All the species are aquatic in the sense that they grow in sites which are at least seasonally submerged. Two of them are found in Britain and Ireland.

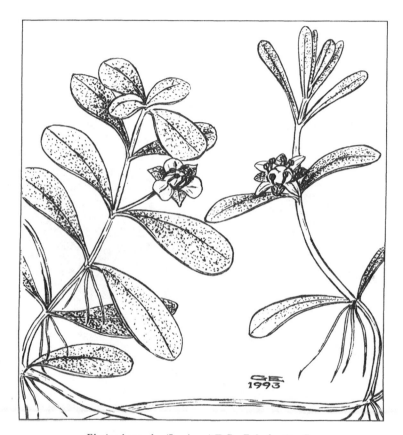

Elatine hexandra (Lapierre) DC.; *E. hydropiper* L.

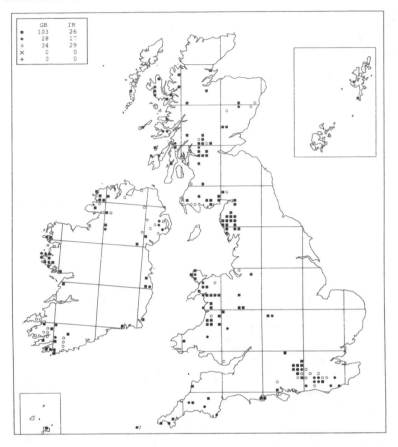

	GB	IR
■	103	26
●	28	17
○	34	29
×	0	0
+	0	0

Elatine hexandra (Lapierre) DC.
Six-stamened Waterwort

E. hexandra can be found at the edge of lakes, reservoirs, ponds and flooded sand- and gravel-pits. It is usually found on damp mud exposed at the water's edge, or in shallow water less than 0.5m deep, but it can sometimes be found in water at depths of 2m or more (West, 1910). It tolerates a wide range of water chemistry from oligotrophic lakes in the uplands to eutrophic lowland sites; it is, however, absent from highly calcareous waters. This species can grow with the rarer *E. hydropiper*; other associates include *Eleocharis acicularis*, *Littorella uniflora* and (where the water is peaty) *Baldellia ranunculoides* and *Juncus bulbosus*. Most sites are in the lowlands but *E. hexandra* ascends to 425m at Llyn Gynon; a record from nearly 500m in the Scottish Highlands (Hooker, 1884) has not been localized.

This species usually behaves as a summer annual on mud, or as a submerged, short-lived perennial. Individuals spread by the growth of the prostrate stem, but most reproduction is by seed. Plants in deep water do not flower, but flowers develop in shallow water and, more freely, on exposed plants. Submerged flowers are cleistogamous and produce fewer seeds than the open but self-fertilized terrestrial flowers. Seeds are dispersed by water and probably also by birds, as when wet they readily adhere to most surfaces. Seed which ripens in summer can germinate immediately, giving rise to a second generation of plants in a single season. Populations of *E. hexandra* vary in size from year to year. When water-levels are low and much substrate is exposed enormous numbers of plants can arise from dormant seed: one reservoir pool near Devil's Bridge contained an estimated 12 million plants in 1984 (A. O. Chater, pers. comm.). For a detailed account of the reproductive biology of this species, see Salisbury (1967).

The inconspicuousness of *E. hexandra*, coupled with its tendency to vary in numbers from year to year, means that it is easily overlooked. In recent years it has been found at many new sites in Wales, N. England, Scotland and W. Ireland. This almost certainly reflects the increasing amount of systematic fieldwork rather than any increase in the frequency of the plant itself. *E. hexandra* has, however, disappeared from some of its old sites in south-east England; the decline is particularly marked in Sussex (Hall, 1980) but has been offset in other counties by the appearance of new colonies in gravel-pits.

E. hexandra is widespread in W. & C. Europe, from southern Scandinavia south to Portugal, N. Spain and N. Italy. Elsewhere it is only known from N.W. Africa (Cook, 1983).

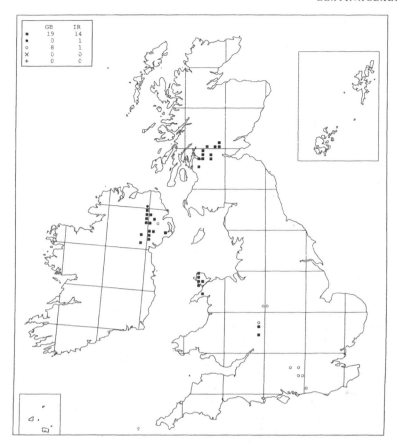

	GB	IR
■	19	14
●	0	1
○	8	1
×	0	0
+	0	0

Elatine hydropiper L. Eight-stamened Waterwort

This species is found in shallow water (up to 1m deep) at the edges of lowland lakes, reservoirs, ponds and old sand- and gravel-pits, or on damp mud or silty sand around the water's edge. It is therefore found in the same habitat as *E. hexandra*, with which it often grows, but unlike that species it is virtually restricted to mesotrophic or eutrophic sites. Characteristic associates include *Callitriche hermaphroditica*, *C. stagnalis*, *Eleocharis acicularis*, *Littorella uniflora*, *Lythrum portula* and *Persicaria amphibia*.

The reproductive biology of this species is similar to that of *E. hexandra*. Both submerged and terrestrial plants will flower, but the terrestrial individuals flower and fruit more profusely. Seeds are dispersed by water and by birds: Kerner von Marilaun (1894), working in central Europe, found that *E. hydropiper* was one of the species most frequently found as seeds in mud on the feet of waterside birds. Populations fluctuate greatly in numbers from year to year, being largest in summers in which the rainfall is low and much bare mud is exposed.

The history of *E. hydropiper* in Britain is remarkable. It was first recorded in Anglesey in 1830, and subsequently found at a few localities in the Midlands and south-east England. It was, however, last seen in Berkshire in 1917, Staffordshire in 1923, Surrey in

1935 and Sussex in 1944, so that by 1960 it appeared to be restricted to single sites in Worcestershire (Westwood Great Pool) and Anglesey (Llyn Coron). However, the species was found at Loch Lomond in 1968, the first Scottish record (Idle *et al.*, 1970), and it has subsequently been found at a further nine sites in Scotland. *E. hydropiper* has also been discovered since 1960 at additional sites in Worcestershire and North Wales. It is difficult to account for the extinction of this species in southern England, but the survival of *E. hexandra*. The reason for the apparent spread in Scotland is equally unclear: the species may have been overlooked before 1968 or it may have been spread recently by wildfowl, as Mitchell (1981) suggests. In Ireland the species also appears to have increased in the Lough Neagh area since it was first recorded there in 1837 (Harron, 1986).

E. hydropiper has a more continental distribution than *E. hexandra*, being widespread in N. & C. Europe with scattered localities in the mountains further south. It is also recorded in N.W. Africa, and extends eastwards to E. Asia.

Two taxa are recognized within European *E. hydropiper*, which Uotila (1974) treats as distinct species. The British plant is *E. hydropiper* sensu stricto; *E. orthosperma* Düben has an even more continental distribution than *E. hydropiper*.

BRASSICACEAE (CRUCIFERAE)

A large family of some 390 genera and 3000 species. An excellent identification guide to the British and Irish species is available (Rich, 1991). Only *c*.9 species in the entire family are aquatic. Aquatic species of *Rorippa* and *Subularia* occur in our area but the third genus which contains aquatics, *Cardamine*, is represented here only by terrestrial taxa.

Rorippa Scop.

A genus of some 70 species, most of which are native to temperate regions.

The British and Irish species of *Rorippa* fall into two groups. One group consists of the white-flowered species which form the *R. nasturtium-aquaticum* aggregate. These are sometimes segregated in the genus *Nasturtium* R. Br. The aggregate contains two closely related species, *R. microphylla* and *R. nasturtium-aquaticum* sensu stricto, and their hybrid, *R. × sterilis*. The aggregate species is mapped first, and its ecology and distribution described. The segregates are then mapped separately, with a text which deals only with their particular characteristics. Our knowledge of the distribution and ecology of the segregates is limited because they can be identified only from fruiting material, and many records are therefore referable only to the aggregate. In addition to this unavoidable difficulty, many authors of county floras (especially in southern England) appear to combine records of the aggregate with those of *R. nasturtium-aquaticum* sensu stricto, thus obscuring the distribution of the taxa.

The second and larger group of species in our area are the yellow-flowered species in *Rorippa* sensu stricto. The only aquatic species in this group is *R. amphibia*, but the other native plants in this group *(R. islandica, R. palustris* and *R. sylvestris)* are usually found in waterside habitats, and can be particularly frequent in sites which are flooded in winter but dry out in summer. *R. amphibia* and its hybrids with *R. palustris* (*R. × erythrocaulis*) and *R. sylvestris* (*R. × anceps*) are treated in the following text. Much of the information in the accounts of these taxa is taken from Jonsell's (1968) masterly account of the northwest European species; Jonsell (1975) is also a useful source of information on *R. × anceps*.

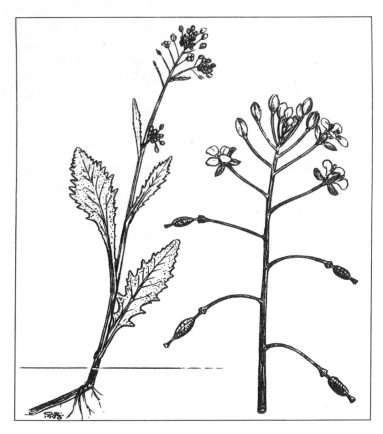

Rorippa amphibia (L.) Besser

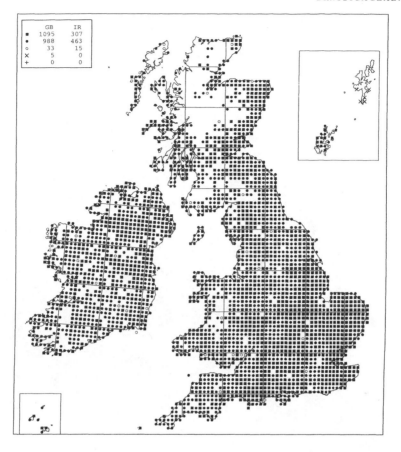

	GB	IR
■	1095	307
●	988	463
○	33	15
×	5	0
+	0	0

Rorippa nasturtium-aquaticum (L.) Hayek
sensu lato   Water-cress

The taxa in this aggregate are plants of wet ground and of shallow water. They are most frequent in or alongside ditches, streams and rivers, occurring in mildly acidic or basic water but in particular abundance in calcareous water (even if this is flowing over base-poor substrates). Their ability to grow terrestrially means that they are well adapted to growth in streams where the water flows only intermittently, including the winterbournes of southern England. Although particularly frequent by running waters, they also grow by lakes, ponds, turloughs, canals and in marshes, where they occur both in open habitats and amongst emergents such as *Glyceria maxima*, *Phalaris arundinacea* and *Sparganium erectum*. They are found over a range of mineral substrates including sands, gravels, silts and clays, but do not grow over peat. *Apium nodiflorum*, *Mentha aquatica*, *Myosotis scorpioides* and *Veronica beccabunga* are characteristic associates. All the taxa in the aggregate are predominantly lowland, being only rarely found above 300m.

These plants are perennials which may flower in their first year. The aggregate contains two fertile species and a sterile hybrid: details of their sexual reproduction are therefore given under the individual taxa. All the taxa

reproduce vegetatively by the growth of creeping stems which root at the nodes, and 'detached stem pieces have a remarkable ability to root and form new plants' (Grime *et al.*, 1988). Plants normally overwinter as small individuals, but they will continue to grow through the winter in water which is issuing from chalk springs at a constant temperature.

This aggregate is frequent in lowland Britain and Ireland, and is absent only from the most acidic and oligotrophic areas. There is documentary evidence for its use as a medicinal plant in Europe in the 1st century AD. The first commercial watercress farm in Britain was opened in 1808 in Gravesend, to supply the London market (Manton, 1935). Although all three taxa in the aggregate are undoubtedly native, the extent to which their natural distribution has been modified by the spread of cultivated stock is unknown. There is no evidence that their national distribution has changed in recent years.

R. nasturtium-aquaticum sensu lato is widespread in Europe, north to Scotland, Denmark and S. Sweden. It extends eastwards through Turkey and Iran to the western edge of the Himalaya. It is widely naturalized in N. America (Green, 1962) and in many parts of the southern hemisphere.

For an account of the ecology of the plants in this aggregate, see Howard & Lyon (1952).

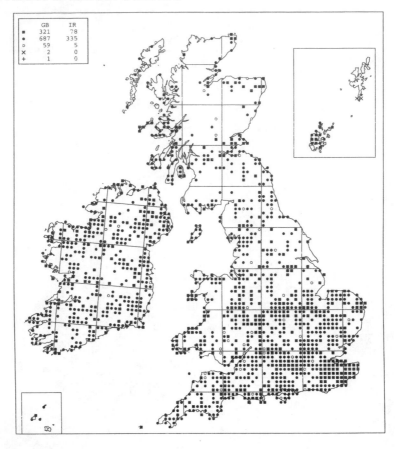

Rorippa nasturtium-aquaticum (L.) Hayek
Water-cress

R. microphylla and *R. nasturtium-aquaticum* grow in similar habitats, and sometimes in mixed populations. However, some differences in their ecological requirements are apparent in areas where there is a range of soil types and relief. In Gloucestershire *R. nasturtium-aquaticum* tends to be restricted to calcareous soils (Airy Shaw, 1949). In the Sheffield area *R. nasturtium-aquaticum* is the usual species of low-lying calcareous springs, being replaced by *R. microphylla* or *R.* × *sterilis* at higher altitudes or on non-calcareous soils (Grime *et al.*, 1988). These preferences have also been noted by Graham (1988) in Co. Durham. *R. nasturtium-aquaticum* is most frequent in the lowlands, but it reaches 550m at Moor House.

This species is a fertile tetraploid (2n=32). Although it is normally perennial, it can behave as an annual in sites where the mature individuals are killed by summer drought (Howard & Lyon, 1952). It is self-compatible and regularly sets fruit by self-pollination, although flowers are visited by small insects and some cross-pollination is thought to occur. Freshly shed seeds will float and are capable of germinating immediately. Those which fail to germinate probably retain viability for a period, as Howard & Lyon (1952) obtained 32% germination after seeds had been stored in packets for five years. Seedlings have been observed on bare soil, but the species is thought to be incapable of reproducing by seed in closed vegetation.

The presence of diploid, triploid and tetraploid plants within the *R. nasturtium-aquaticum* aggregate was discovered by Manton (1932, 1935); Howard & Manton (1946) later described the morphological differences between the cytotypes in a classic cytotaxonomic study. The accounts by Howard & Lyon (1950, 1951) brought the species to the attention of most British and Irish botanists, and provided a provisional assessment of their distribution. The map is based on records of *R. nasturtium-aquaticum* sensu stricto. The species is almost certainly under-recorded. It is known to be more frequent than *R. microphylla* in southern England, but rarer in north-east England and eastern Scotland. There is little information on the relative frequency of the two in Ireland, although Hackney (1992) comments that they seem to be equally common in the north-east.

R. nasturtium-aquaticum is found throughout the European range of the aggregate, and also extends into western Asia (Jalas & Souminen, 1994). It is the most frequent segregate in many of the areas where watercress is naturalized, including N. & S. America (Green, 1962) and Australasia (Aston, 1973).

This is one of the members of the aggregate which is grown as watercress. Howard & Lyon (1952) reported that pot-grown plants were considerably less frost-resistant than those of the other two segregates, and this may be one reason for its more southerly distribution.

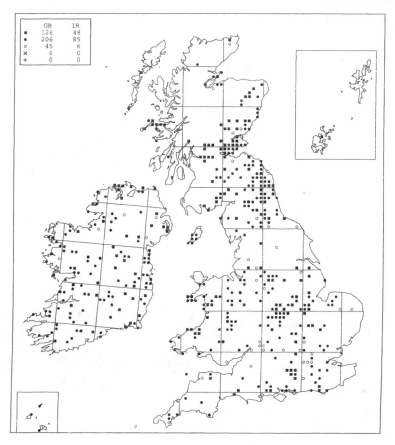

	GB	IR
■	126	48
●	206	85
○	45	6
×	0	0
+	0	0

Rorippa × *sterilis* Airy Shaw (*R. microphylla* (Boenn.) Hyl. ex Å. Löve & D. Löve × *R. nasturtium-aquaticum* (L.) Hayek)   Hybrid Water-cress

R. × *sterilis* can sometimes be found growing with both of its parents, but it also occurs in sites where one or both parents are absent. Howard & Lyon (1952) suggest that it may have pH and calcium requirements which are intermediate between those of its parents, but there has been little subsequent research on this. In the Sheffield area the hybrid tends to replace its parents in the uplands, particularly in limestone areas (Grime, *et al.*, 1988), and it is recorded at 395m in flushes east of Craig Cerrig-gleisiad.

R. × *sterilis* is a sterile hexaploid (2n=48). Most of the pollen is sterile and the fruits contain only the occasional well-formed seed. Effective reproduction is by vegetative spread.

The characteristics of this hybrid were outlined by Howard & Manton (1946) and Howard & Lyon (1950), and the name *R.* × *sterilis* was given to it by Airy Shaw (1951). The hybrid is almost certainly under-recorded in many areas. Plants of either parent sometimes fail to set seed and these non-fruiting individuals (which may have suffered from adverse weather or water-levels) might be misidentified as the sterile hybrid. If possible, identification of the hybrid should be based on the intermediate characters of the occasional seeds produced by *R.* × *sterilis*. Populations of *R.* × *sterilis* which occur in the absence of the parents could have originated by vegetative dispersal from naturally occurring clones, or from cultivated stock.

R. × *sterilis* is 'fairly common' in Europe (Tutin *et al.*, 1993) but its detailed distribution has not been documented. It is also naturalized in north-eastern states of the U.S.A., Japan, New Zealand and almost certainly elsewhere.

Hybrids between *R. microphylla* and *R. nasturtium-aquaticum* can be produced artificially if *R. microphylla* is the female parent. The hybrid is the plant which is most frequently cultivated as watercress. In cultivation it is more frost resistant than *R. nasturtium-aquaticum* (Howard & Lyon, 1952).

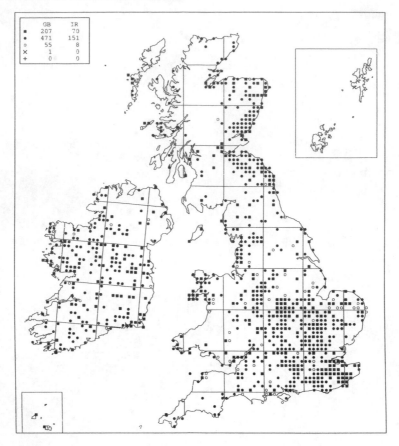

	GB	IR
■	207	70
●	471	151
○	55	8
×	1	0
+	0	0

Rorippa microphylla (Boenn.) Hyl. ex Á. Löve & D. Löve   Narrow-fruited Water-cress

This species extends into somewhat more acidic sites than *R. nasturtium-aquaticum*. In Gloucestershire and the Sheffield area it is the most common segregate on non-calcareous soils (Airy Shaw, 1949; Grime *et al.*, 1988) and in Co. Durham it is more widespread than that species in acidic water and ascends higher up the dales (Graham, 1988). However, it may also grow in highly calcareous sites. It is recorded from sea-level to 365m (Malham Tarn).

R. *microphylla* is an octoploid (2n=64). It is a fertile plant with similar floral biology and germination requirements to *R. nasturtium-aquaticum*.

This species was initially described as *Nasturtium uniseriatum* by Howard & Manton (1946), but Airy Shaw (1947) concluded that the earlier name *Nasturtium microphyllum* was based on the octoploid plant, and therefore had priority. The species is almost certainly under-recorded. However, there is no doubt that it is more frequent than *R. nasturtium-aquaticum* in those areas of northern England and eastern Scotland for which reliable records are available. These areas include Durham

(Graham, 1988), Northumberland (Swan, 1993), Berwickshire (Braithwaite & Long, 1990), Angus (Ingram & Noltie, 1981) and East Ross (Duncan, 1980).

This species is recorded from much of the native range of the *R. nasturtium-aquaticum* aggregate. In Europe it is commoner than *R. nasturtium-aquaticum* in Denmark and the Netherlands, and extends south and east to Italy, Slovenia and Poland, but the details of its distribution are still imperfectly known (Tutin *et al.*, 1993; Jalas & Suominen, 1994). It is known to be naturalized in C. & S. Africa, N. America (especially in south-eastern Canada and north-eastern U.S.A.), Australia and New Zealand.

This is the only member of the aggregate in Britain and Ireland which is not commercially cultivated as watercress. It is believed to have evolved following hybridization between *R. nasturtium-aquaticum* and some other species, the identity of which is not known. Because the seeds of *R. microphylla* tend to occur in one rather than two rows, Howard & Manton (1946) suggested that the other parent is a species of *Cardamine*, but the great similarity of *R. microphylla* to *R. nasturtium-aquaticum* suggests that the missing parent is more likely to be an undiscovered member of the *R. nasturtium-aquaticum* aggregate.

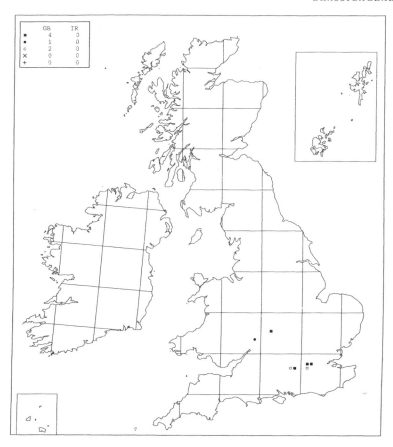

	GB	IR
■	4	0
●	1	0
○	2	0
×	0	0
+	0	0

Rorippa × *erythrocaulis* **Borbás** (*R. amphibia* (L.) **Besser** × *R. palustris* (L.) **Besser**)

Thames Yellow-cress

This hybrid has been recorded from the banks and artificial embankments of large rivers, ditches near rivers and from mud in a reed-bed. It is sometimes found in the absence of one or both parents. It is confined to the lowlands.

This hybrid is a vigorous perennial. The well-known population along the R. Thames is a triploid hybrid between the diploid cytotype of *R. amphibia* (2n=16) and *R. palustris*, which is a tetraploid (2n=32). It is highly sterile, producing no good pollen, but spreads vegetatively. A plant from a ditch near the R. Avon at Tewkesbury has also been investigated cytologically and has the tetraploid cytotype of *R. amphibia* (2n=32) as one parent. This hybrid is much more fertile, with 85% good pollen. As diploid *R. amphibia* is only known from the Thames, hybrids from other localities are likely to involve the tetraploid plant.

This hybrid was first detected in Britain by Britton (1909), who discovered it in several places along the R. Thames between Putney and Richmond, and in particular abundance on the sloping river wall near Hammersmith. Lousley (1976) reported that it was still plentiful by Hammersmith Bridge in 1932, but later became much rarer as a result of the rebuilding of the embankment. Nevertheless, it was still present at Hammersmith Bridge in 1989, growing with *R. amphibia* but not *R. palustris*. There are records of the hybrid from several other localities, but little detailed information on its ecology, distribution or abundance at these sites is available.

According to B. Jonsell (in Rich, 1991), the only confirmed records of this hybrid are from Britain and Sweden, although it has been reported from C. & E. Europe. Tutin *et al.* (1993) state that it is fairly frequent in Europe.

Jonsell's (1968) attempts to synthesize the triploid hybrid between *R. amphibia* and *R. palustris* failed. His artificial tetraploid hybrids were similar to the wild hybrid, but much less fertile.

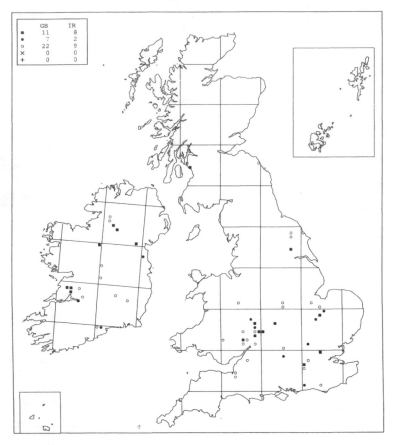

Rorippa × anceps (Wahlenb.) Rchb. (*R. amphibia*
(L.) Besser × *R. sylvestris* (L.) Besser)
Hybrid Yellow-cress

Most of the British records of this hybrid are from river
banks; in Ireland it appears to be at least as frequent on
the shores of lakes as it is by rivers. Other habitats from
which specimens have been collected include ponds, wet
ditches, marshes, damp fields and a rubbish dump. The
hybrid requires more open habitats than those of *R.
amphibia*. All the sites are lowland.

This hybrid is a fertile perennial, and is capable of
reproducing by seed as well as by vegetative means. Like
the parents, the hybrid is self-incompatible. Some populations have rhizomes derived from *R. sylvestris*, and
these assist in local vegetative spread.

Although this hybrid has been recognized at some
British and Irish localities for many years, it is almost
certainly under-recorded and it may be much more frequent than the map suggests. Jonsell (1968) commented
that British records of this hybrid were sparse, but he
concluded as a result of his field and herbarium studies
that extensive hybridization had taken place along some
river systems. There is also evidence that the hybrids

have backcrossed to the parents. All the plants of *R.
sylvestris* from the R. Severn and R. Wye seen by Jonsell
(1968) showed distinct signs of introgression with *R.
amphibia*. Jonsell (1968) also suggests that Irish plants of
R. amphibia have a distinctive leaf shape which may be
attributable to introgression with *R. sylvestris*.

This is a frequent hybrid in Europe, although (like *R.
amphibia*) it is rare in the Mediterranean region. The
European plants include a series of hybrid derivatives
which bridge the morphological divide between the
parents. The hybrid also occurs as an introduction in
N. America (Stuckey, 1972).

Hybrids in our area include both tetraploid plants,
which are derived from crosses between tetraploid *R.
amphibia* (2n=32) and tetraploid *R. sylvestris* (2n=32), and
pentaploids, derived from crosses between tetraploid *R.
amphibia* and hexaploid *R. sylvestris* (2n=48). There is no
cytological evidence that diploid *R. amphibia* (2n=16) has
hybridized with *R. sylvestris*, but records of *R. × anceps*
from the Thames at Chertsey and Richmond suggest that
it may have done so. Most of our knowledge of its distribution comes from herbarium specimens, and many of the
details of its reproductive biology, distribution and habitat
requirements need to be established by fieldwork.

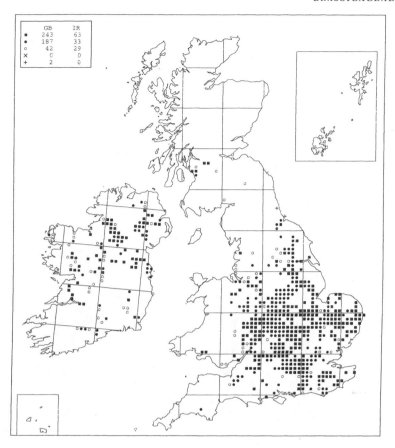

	GB	IR
■	243	63
●	187	33
○	42	29
×	0	0
+	2	0

Rorippa amphibia (L.) Besser   Great Yellow-cress

This species grows in lowland sites which are flooded in winter, and where shallow water may persist through the summer or recede to leave a damp and often muddy substrate. It often grows amongst the fringe of emergents along the edge of streams or rivers, or on riverside banks of muddy shingle or silty gravel. Other habitats include canals, fenland drains, the sheltered shores of lakes and reservoirs, ditches, ponds (especially on flood plains), turloughs and marshes. *R. amphibia* is usually found where the water is both calcareous and eutrophic. In rivers it is most frequent where calcareous water flows over a clay substrate, as the water-level in clay rivers is subject to marked fluctuations, but the substrate is sufficiently deep and stable to allow the development of a fringe of emergents (Holmes, 1983).

R. amphibia is a perennial. Plants are highly self-incompatible, and perhaps for this reason seed-set in the field is often poor. Seeds may become established in areas where the vegetation is open but relatively stable. Seedlings form a vegetative rosette in the first year and flower in subsequent seasons. Established plants may spread to form loose patches by the growth of axillary leaf rosettes which develop at the base of flowering shoots towards the end of the season. Vegetative reproduction by fragmentation of the hollow stems and the

subsequent dispersal by floods of stem fragments with axillary buds is probably the main method of spread along the edges of lakes and rivers.

R. amphibia is locally frequent in England as far north as the R. Tees; there is no evidence that its frequency or abundance have changed appreciably in recent years. Populations in southern Scotland are probably introduced. In Ireland it is now known to be more frequent than the *Atlas of the British Flora* suggests, and it is probably still under-recorded.

This species is native to Europe, N. Africa and Asia. In Europe it is most frequent in the middle latitudes, being rare in the north and in the Mediterranean area (Jalas & Suominen, 1994). In Asia it is most frequent in the west but it is recorded in scattered localities eastwards to Japan. *R. amphibia* is also recorded as an introduction in N. America, where it was first reported in 1930 (Stuckey, 1972), Argentina and New Zealand.

Two cytotypes of *R. amphibia* are present in the British Isles. A diploid plant (2n=16) is only recorded from the Thames valley, where it is apparently the only variant present. The tetraploid (2n=32) is recorded from other localities in England and Ireland, and is the only cytotype recorded in Scandinavia. There is a strong sterility barrier between the cytotypes: attempts to hybridize them have failed and they appear to be genetically isolated.

Subularia L.

A genus of two species, the widespread *S. aquatica* and the rarer *S. monticola* Schweinf. which is endemic to the East African mountains. The species have an isoetid growth form but unlike other isoetids they are annuals or biennials, not perennials. The existence of a genus of isoetids in this predominantly terrestrial family is surprising, and provides a remarkable example of convergent evolution.

Subularia aquatica L.

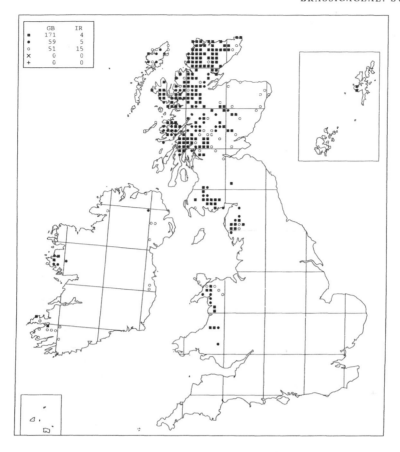

	GB	IR
■	171	4
●	59	5
○	51	15
×	0	0
+	0	0

Sububaria aquatica L.   Awlwort

S. aquatica is most frequent at the edge of acidic, oligo-trophic lakes but it can also grow in outflow streams and, exceptionally, in canals leading from them. It usually grows in water shallower than 1m, although it can occur down to depths of at least 1.5m. Plants occur as scattered individuals in open vegetation, with associates such as *Isoetes echinospora, I. lacustris, Littorella uniflora, Lobelia dortmanna*, the aquatic form of *Juncus bulbosus, Myriophyllum alterniflorum* and *Potamogeton gramineus*. They are also found amongst stems of *Carex rostrata, Eleocharis palustris* or *Equisetum fluviatile* where these emergents form open swamps. Substrates over which *S. aquatica* grows include fine silt, gravel and stones; in some areas it shows a preference for fine gravelly beds. Although it is one of the most reliable indicators of oligo-trophic water, it has been found in more eutrophic lakes (Seddon, 1972) but it has not been recorded over base-rich substrates. Plants are able to tolerate short periods of emersion, although terrestrial forms are smaller than those which grow in water. *S. aquatica* is recorded from sea-level to 825m at Ffynnon Llyffant.

S. aquatica is more ruderal than the other isoetids (Farmer & Spence, 1986). Most plants are annuals. There is some doubt about the maximum possible longevity, but some individuals can overwinter as rosettes, thus behaving as biennials. Plants flower from June to September. Submerged plants may produce cleistogamous flowers or flowers which open fully; in

either case they are self-pollinated while in bud. The flowers of terrestrial plants open fully, and may be insect-pollinated or self-pollinated. Individual plants produce 8–125 seeds, the terrestrial plants producing fewer than the aquatics. The seeds do not float, but germinate readily. *S. aquatica* lacks any means of veget-ative reproduction.

The map of *S. aquatica* in the *Atlas of the British Flora* (Perring & Walters, 1962) grossly underestimates its distribution. It is an inconspicuous species, easily over-looked as it usually grows with other isoetids and rarely occurs in abundance. The species is probably still under-recorded in areas which have not been surveyed inten-sively, but is much less frequent than *Isoetes lacustris, Littorella uniflora* or *Lobelia dortmanna*. Woodhead (1951) considered that it was less abundant at the time he was writing than it had been 90 or even 25 years pre-viously. However, little evidence is available to assess trends in the abundance of *Subularia* in the western parts of Britain or Ireland. Like *Isoetes lacustris* and *Lobelia dortmanna*, it has certainly been lost from some sites at the eastern edge of its range because of eutroph-ication. Its disappearance from Lough Neagh, where it was first recorded in the British Isles in the 1690s and was formerly plentiful, may be attributable to the lower-ing of the water-table in the 19th century (Harron, 1986).

S. aquatica has a more or less circumboreal distribu-tion, although it is rare in Asia. It is widespread in N. Europe, and although absent from the Alps extends south very locally to the Pyrenees and Bulgaria. American plants have been described as *S. aquatica* subsp. *americana* G. A. Mulligan & Calder.

An account of the ecology of *S. aquatica* is given by Woodhead (1951). Few publications since then have dealt with its ecology in any detail, and the species is too scarce to figure prominently in general accounts of aquatic vegetation.

PRIMULACEAE

This is a predominantly terrestrial family of 22 genera and 800 species. The only truly aquatic genus in Europe is *Hottonia*, but species of *Anagallis*, *Lysimachia* and *Samolus* can be found in waterside communities, fens and swamps in Britain and Ireland. Some African *Anagallis* species are submerged aquatics.

Hottonia L.

There are two species of *Hottonia*, the Eurasian *H. palustris* and the American *H. inflata* Elliott. The American species differs from ours in being an annual, with small autogamous flowers, and in having swollen and buoyant flowering stems. For an account of the genus, see Prankerd (1911).

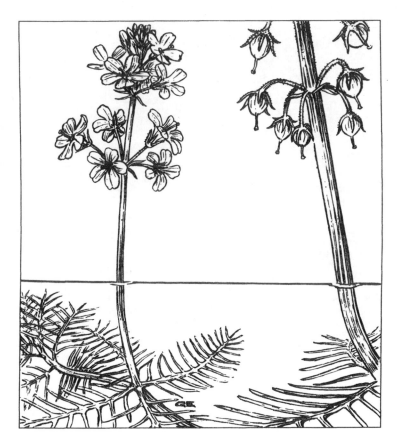

Hottonia palustris L.

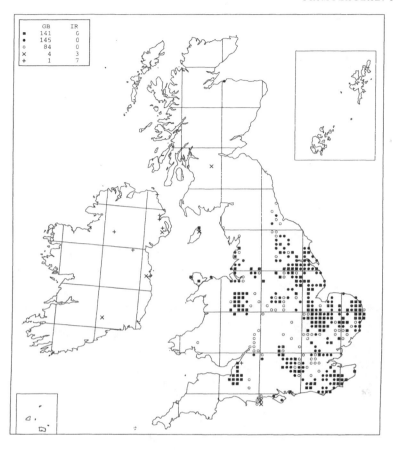

	GB	IR
■	141	0
●	145	0
○	84	0
×	4	3
+	1	7

Hottonia palustris L.   **Water-violet**

This species is confined to small, sheltered, lowland water bodies, where it grows submerged in shallow water or as a terrestrial form at the water's edge. Almost all records are from ponds, ditches and fenland drains. *H. palustris* is notably shade-tolerant and can sometimes be found in ponds in woods and plantations. It is also recorded as a colonist of pools in disused clay-, gravel- and sand-pits, and disused canals. It is a plant of mesotrophic or meso-eutrophic, usually calcareous, water. It grows over both organic and mineral substrates but is often particularly frequent over peat. Characteristic associates include *Callitriche obtusangula, Elodea canadensis, Hippuris vulgaris, Myriophyllum verticillatum* and *Potamogeton lucens.*

This species is a shallow-rooted perennial. It remains green throughout the winter, and can survive periods when it is embedded in ice. It flowers in May; the aerial inflorescence arises from, and is supported by, a whorl of leaves just below the surface of the water. *H. palustris* is heterostylous, with 'pin' and 'thrum' flowers. The inflorescences remain above the water until the seeds mature. After the seeds are shed the flowering stem disintegrates and the fragments can become established as new plants. Plant fragments may be dispersed by birds or by machinery used to clear ponds and ditches. The seeds have no innate dormancy; they can germinate under a wide range of environmental conditions or survive at least 15 months desiccation. The seeds sink as soon as they are shed (Ridley, 1930) but most seedlings

float for a period after germination in water. Seedlings develop much more rapidly on moist soil than in water. If mature plants are stranded above the water-level they persist as a land form with smaller, more rigid leaves.

H. palustris is a native species in England and Wales. Like many plants found in small water bodies it has decreased in many areas because of the infilling of ponds, and the mechanical dredging or elimination of drainage ditches. The conversion of grazing marshes to arable land has also had an adverse effect on this species (Mountford, 1994a; Palmer, 1986). An isolated Scottish population in the Culbin Forest may have been introduced by waterfowl; it persisted from 1923 until the pond in which it grew dried up in 1958 (McCallum Webster, 1978). A colony which was discovered in the Isle of Man in 1988 may also owe its origin to bird-dispersal. In Ireland the species is probably introduced at all its sites. The only area where it might be native is in Co. Down, where it was first recorded in 1810. Here it has declined and is now very rare (Hackney, 1992).

This species is virtually confined to Europe, where it is found from C. Sweden south to S. France, C. Italy and Romania; it is absent from the Mediterranean area. It just extends into W. Asia.

H. palustris is probably favoured by fluctuating water-levels (Fitter & Smith, 1979; Brock *et al.*, 1989). Emersed stands have a higher biomass than submerged stands, and reproduction by seed is more likely if water-levels fall during the late summer, allowing seedlings to grow on newly exposed mud.

CRASSULACEAE

This family is widespread in both temperate and tropical areas. Most of the 33 genera and 1280 species are succulent herbs or shrubs of arid habitats, but the genus *Crassula* includes a few aquatics.

Crassula L.

Only ten of the 300 species of *Crassula* are aquatics. Two of these occur in our area: *C. aquatica* is apparently native whereas *C. helmsii* is a widespread introduction.

Crassula helmsii (Kirk) Cockayne

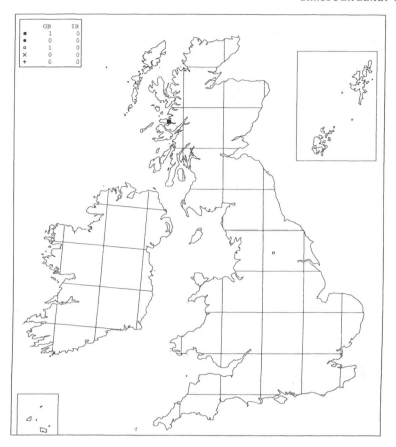

Crassula aquatica (L.) Schönland   Pigmyweed

At its Scottish locality *C. aquatica* grows at the side of the R. Shiel, and in a ditch and a shallow scrape created for ducks nearby. In the sheltered shallows of the river it can grow as scattered plants or in dense swards in water up to 25 cm deep, associated with frequent *Glyceria fluitans* and *Alisma plantago-aquatica*, *Callitriche* sp., *Eleocharis palustris*, *Elodea canadensis*, *Hydrocotyle vulgaris*, the aquatic form of *Juncus bulbosus*, *Littorella uniflora*, *Persicaria hydropiper*, *Ranunculus flammula* and *Fontinalis antipyretica*. It also grows terrestrially as scattered plants on moist and rather bare sandy soil, with *Agrostis stolonifera*, *Carex viridula* subsp. *oedocarpa*, *Galium palustre*, *Gnaphalium uliginosum*, *Juncus bulbosus*, *Ranunculus flammula*, *Viola palustris*, *Fossombronia* sp. and *Pellia epiphylla*. In Yorkshire *C. aquatica* grew in shallow water at the edge of an artificial lake, and with *Callitriche stagnalis*, *Limosella aquatica*, *Persicaria hydropiper*, *P. minor* and *Rorippa palustris* on damp mud. Both sites are below 300m.

C. aquatica is a small annual which presumably germinates in spring and summer. Crassulacean Acid Metabolism has been demonstrated in submerged plants (Keeley & Morton, 1982). Plants flower and fruit from July to September. Both terrestrial and aquatic plants flower freely. The flowers are inconspicuous and are probably automatically self-pollinated. The number of plants in the Scottish population varies from year to year: the plant was present in 1969 and 1971, for example, but could not be found in 1970 (Sowter *et al.*, 1972).

C. aquatica was first discovered in Britain at Adel Dam in 1921 (Butcher, 1921, 1922). It increased in abundance in the years following its discovery, but later declined as the soft mud on which it grew became colonized by dense beds of *Carex rostrata*, *Juncus effusus* and *Typha latifolia* (Sledge, 1945). It was last seen in 1938 and had gone by 1945. The Westerness population was discovered in 1969, when it occupied an area of *c*.6 square metres (Sowter, 1971). In 1990 it was present along a length of river at least 200m long, but the population has not been surveyed sufficiently frequently to establish whether this represents a long-term increase or whether numbers simply fluctuate from season to season.

This species is widespread in the northern hemisphere but has a disjunct distribution. The main concentrations of records are from Europe, E. Asia and both western and eastern N. America, south to Mexico, but it is also recorded at scattered localities elsewhere. In Europe it occurs from Iceland and northern Scandinavia south to Austria.

It is difficult to account for the presence of this species in just two British sites, but their disjunct nature and their northerly location fit in with the European distribution of the species. The tiny seeds may have arrived on the feet of aquatic birds, as Druce (1922) suggests. The Scottish site is a salmon river, and *C. aquatica* could have been brought to this site on the waders of a visiting fisherman.

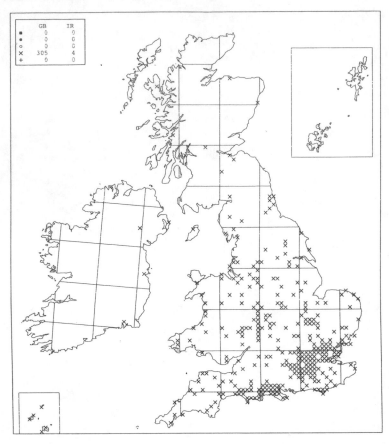

	GB	IR
■	0	0
●	0	0
○	0	0
×	305	4
+	0	0

Crassula helmsii (Kirk) Cockayne
New Zealand Pigmyweed

This introduced species grows on damp ground or in sheltered, standing or very slowly flowing water up to 3m deep. In swamps or shallow water it grows as an emergent, and it can be present in dense masses which may exclude competing species. In deeper water the submerged stems grow in more open stands. *C. helmsii* is usually found in ponds but it is also an increasingly frequent colonist of flooded sand- and gravel-pits; additional habitats include lakes, reservoirs, marl-pits, dune-slacks, canals, ditches and fenland drains. It grows over a range of soft substrates including clay, coarse and fine silt and peat, and tolerates a wide range of water chemistry from nutrient-poor and acidic to eutrophic and calcareous. It is usually absent from exposed sites with wave-washed gravel substrates and from very soft, easily disturbed silts. It is confined to the lowlands.

C. helmsii is a monoecious perennial which remains green throughout the winter months. Flowers are produced only on the emergent stems; submerged populations persist in a vegetative state. British plants set seed but this may not be viable, as Dawson & Warman's (1987) efforts to germinate it failed. Effective reproduction is by vegetative fragmentation: a portion of stem with only a single node can become established as a new plant. Short apical shoots are produced in autumn and may float away to colonise new sites. Plant fragments may be inadvertently introduced into ponds when the

contents of aquaria are emptied out, or spread on footwear or fishing tackle. Spread by animals such as waterfowl or (in some areas) ponies is thought to be a less important means of dispersal, but this is a difficult subject to study and few direct observations on the method of spread are available.

This species has been sold commercially in Britain since 1927 as an aquatic suitable for outdoor ponds. It was first discovered as a naturalized plant in 1956, when it was found in a pond at Greensted (Laundon, 1961). Kirby (1964) predicted that it would gradually invade more sites, and the species began to be reported from an increasing number of scattered localities in southern England in the late 1970s. This spread has continued, and the number of recorded sites is increasing rapidly.

C. helmsii is a native of S.E. Australia, Tasmania and New Zealand. In addition to its localities in our area, it has been reported as an introduction in Belgium, Germany and Russia.

This account is based on the detailed studies of *C. helmsii* by Dawson & Warman (1987) and Dawson (1994). The rapid spread of this species has caused considerable concern because of its ability to outcompete native vegetation in shallow ponds. In areas such as the New Forest, *C. helmsii* may represent a threat to the habitat of several rare or scarce plant species, including *Galium constrictum* and *Ludwigia palustris* (Byfield, 1984). Methods of control are discussed by Dawson & Warman (1987) and Spencer-Jones (1994) and in a leaflet published by the Natural Environment Research Council.

ROSACEAE

There are 107 genera and 3100 species in the Rosaceae. The family includes both shrubby and herbaceous genera, many of which make an important contribution to the vegetation of temperate regions. Cook (1990a) does not classify any members of the family as aquatics, but one species qualifies for inclusion in this work.

Potentilla L.

This is a large genus of 500 species of shrubs and herbs, most of them plants of the north temperate and boreal zones, although a few occur in south temperate regions. The single species considered here is sufficiently different from the others to have been placed by Linnaeus in its own genus, as *Comarum palustre* L. It is now regarded as the sole European representative of subgenus *Comarum* (L.) Syme.

Potentilla palustris (L.) Scop.

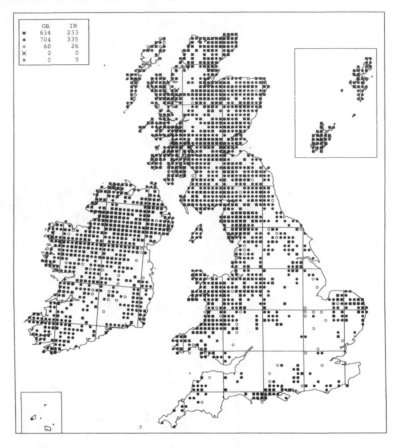

	GB	IR
■	634	233
●	704	335
○	60	26
×	0	0
+	0	0

Potentilla palustris (L.) Scop.   Marsh Cinquefoil

This species grows in permanently flooded swamps, and in mires and wet meadows where the water-table lies below the soil surface during the summer months. In swamps it usually grows in shallow water (less than 20cm deep) at the edge of oligotrophic or mesotrophic lakes. It usually occurs amongst open stands of *Carex rostrata*, but it can also be found with *C. vesicaria*, *Equisetum fluviatile*, *Phragmites australis* and *Schoenoplectus lacustris*. *Menyanthes trifoliata* is a characteristic associate. *P. palustris* is also frequent and locally abundant in mires at the edges of lakes, in natural hollows, bog pools, peat-cuttings and on floating rafts which extend out over the surface of lakes and pools. It favours sites where the water is nutrient-poor but slightly or moderately base-rich; in very acidic and oligotrophic mires it is confined to areas where there is an inflow of base-rich water. It often grows with *Galium palustre* and *Mentha aquatica* and bryophytes such as the more base-tolerant sphagna or, in more calcareous sites, pleurocarpous species such as *Campylium stellatum*, *Calliergon cordifolium*, *C. cuspidatum*, *Plagiomnium* spp. and *Scorpidium scorpidioides*. *P. palustris* grows over a wide range of organic and mineral soils. It will persist in sites which have become partially shaded by the encroachment of willow or alder carr. It ranges from sea-level to 800m at Tom a'Choinnich (McVean & Ratcliffe, 1962); there is an unlocalized record from

850m in the Scottish Highlands (Hooker, 1884).

P. palustris is a rhizomatous perennial. The flowers are protandrous, secrete nectar and are visited by a range of insects. Cross-pollination enhances seed-set but in the absence of insect visitors flowers will self-pollinate and produce some seed (Olesen & Warncke, 1992). The seeds will float for long periods (Guppy, 1906; Praeger, 1913), but little is known about the circumstances under which the species reproduces by seed. Established plants spread by growth of the rhizome, and floating rafts of vegetation are often bound together by the rhizomes of *Menyanthes trifoliata* and *P. palustris*.

This distinctive species is still frequent in northern and western parts of Britain and Ireland. Like many species which grow in mires it has decreased in S.E. England, where it is extinct in a number of vice-counties, including Cambridgeshire (last recorded in 1886), Hertfordshire (1919) and Bedfordshire (1969), and in others is confined to a few populations. Detailed studies in the Dorset heathlands have shown that it is no longer present in 25 of the 37 sites where it was recorded by R. d'O. Good in the 1930s. Twelve of its former sites have been destroyed by agricultural or other development; the remaining sites may now be too dry to support this species (Byfield & Pearman, 1994).

P. palustris is a circumboreal species which is found in arctic and boreal regions throughout the northern hemisphere. In Europe it extends southwards in the mountains to Spain, Italy and Bulgaria.

HALORAGACEAE

This is a small family of nine genera and 120 species. All the native European species are members of the largest genus, *Myriophyllum*. The other genera include both terrestrial and aquatic plants, most of which are found in the tropics and the southern hemisphere. One species, *Haloragis micrantha*, was recently discovered naturalized in western Ireland, where it grew in a very wet, peaty heath (Green, 1989); it is often found in water in its native range (Cook, 1990a).

Myriophyllum L.

This genus includes some 60 species. All are aquatics, but the genus includes both submerged plants and emergents as well as some species which can be found in seasonally flooded habitats. The genus attains its greatest diversity in Australia, where many new species have been discovered in recent years.

All three native species of *Myriophyllum* have inflorescences in the form of terminal spikes which project above the water-level. These bear female flowers in the lower half and male flowers towards the apex; the sexes are often separated by a few hermaphrodite flowers. The flowers are well adapted to wind-pollination (Cook, 1988), with petals which are very small or absent.

Myriophyllum exalbescens Fernald is closely allied to *M. spicatum*, or perhaps a subspecies or variety of that species. The most notable difference between the two taxa is the fact that *M. exalbescens* produces turions whereas *M. spicatum* does not. *M. exalbescens* is widespread in N. America and also occurs in northern Eurasia from Scandinavia to Kamchatka (Faegri, 1982; Ceska & Ceska, 1986). It ought to be looked for in northern Scotland.

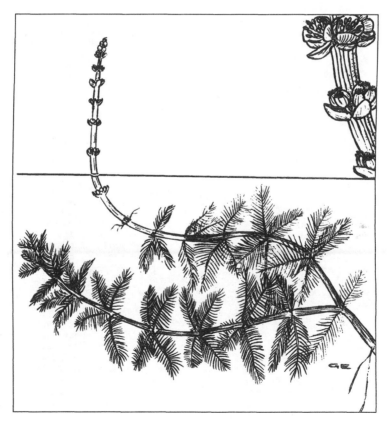

Myriophyllum spicatum L.

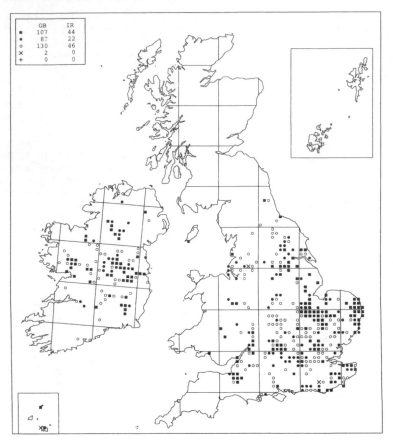

Myriophyllum verticillatum L.
Whorled Water-milfoil

M. verticillatum grows in still or slowly flowing, calcareous water in lowland lakes, ponds, streams, canals, fenland drains and disused clay- and gravel-pits. It is a plant of clear or somewhat turbid water which is moderately but not excessively eutrophic: in the Norfolk broads it increased in abundance when nutrient levels first began to rise but then declined to extinction as eutrophication continued (George, 1992). It can grow in abundance in suitable habitats, where its associates often include *Elodea canadensis*, *Hippuris vulgaris*, *Lemna trisulca*, *Potamogeton crispus*, *P. natans* and *Ranunculus circinatus*. *M. verticillatum* can be found over a range of substrates including both peat and mineral soils. It can persist as a terrestrial form for brief periods if stranded above the water-level.

M. verticillatum is a perennial species. Seed is set in Britain but little is known about its subsequent fate. Vegetative spread is probably more important than sexual reproduction. The species overwinters as turions, which develop on the stems in late August and September. Field studies in Michigan have shown that the turions are initially dormant but germinate in the early months of the year, when the water temperature is still 0–4°C. Under experimental conditions their dormancy is broken by cold treatments of 0–4°C. Detached stems initially float and produce turions much earlier than rooted plants. They eventually sink and develop adventitious roots, and can therefore become established as new plants (Weber & Noodén, 1974, 1976).

Fossil evidence suggests that this species may have been present in the British Isles throughout the Pleistocene (Godwin, 1975). The map indicates that in Britain its distribution has contracted in recent years, and this conclusion is supported by the results of the BSBI Monitoring Scheme (Rich & Woodruff, 1996) and by detailed local studies (Mountford, 1994a). The reasons probably include eutrophication, habitat destruction (the filling in of ponds and ditches) and land-use change (particularly the conversion of pasture to arable). The adverse effects of land-use change on this species in coastal grazing marshes was demonstrated by Driscoll (1983, 1985a), who found that it had decreased markedly in frequency in an area which was converted to arable in the late 1970s, but remained virtually unchanged in a similar area which was still managed as pasture. The apparent decline in Ireland may be due to under-recording.

M. verticillatum is widespread in the temperate regions of the northern hemisphere. In Europe it extends from northern Scandinavia to the Mediterranean area, but it is rare near the northern and southern edges of its range. Records from S. America are erroneous (Orchard, 1981).

Although both the inflorescences and the turions of *M. verticillatum* are distinctive, the identification of vegetative material is not always as simple as most floras imply. Records of *M. verticillatum* from outside its known range in Britain and Ireland should be treated with caution, as it is sometimes confused with *M. alterniflorum* or *M. aquaticum*. Many of the Welsh records listed by Ellis (1983), for example, have proved to be erroneous.

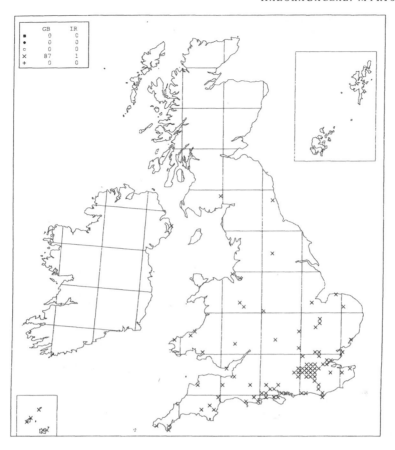

	GB	IR
■	0	0
●	0	0
○	0	0
×	87	1
+	0	0

Myriophyllum aquaticum (Vell.) Verdc.
Parrot's-feather

This introduced species is recorded from shallow, still or very slowly moving water in a range of lowland habitats. Most records are from ponds but *M. aquaticum* has also colonized reservoirs, flooded gravel-pits, streams, canals and ditches. There is a single record from the shallows of the R. Wharfe near Otley, but this population may not have persisted. Almost all the water bodies where *M. aquaticum* has become established are eutrophic, but it is difficult to generalize about the ecological requirements of the species from the scattered localities in which it has become established. It is able to persist as a terrestrial plant when ponds dry out, and has even been recorded on the dry bank of a council tip in West Cornwall (Grenfell, 1984).

M. aquaticum is a perennial which, unlike the native species, produces emergent as well as submerged shoots. The species is dioecious, with flowers which are borne in the axils of the upper emergent leaves. Both sexes occur in the native range of the species, but even there female plants very rarely set seed. Only female plants have become naturalized, and introduced populations spread by asexual means. There are no specialized vegetative propagules, but the stems are brittle and small fragments can become established as new colonies.

This species is widely grown in garden ponds in both Britain and Ireland. The first naturalized plants to be discovered were collected from a pond at Lingfield in Surrey in 1960 but misidentified as *Hottonia palustris* (Leslie, 1987). *M. aquaticum* was recorded in E. Sussex in 1969, W. Cornwall in 1972 and Glamorgan in 1975. It was subsequently found in an increasing number of sites in Britain, and it was discovered in clay-pits in Ireland in 1990 (Hackney, 1992). It has survived the winter in ponds which became covered with ice during cold spells (Milner, 1979), and has persisted at individual sites for at least ten years. Most populations probably originate as plants discarded from aquaria or garden ponds. Dawson (1993) noted that *M. aquaticum* is spreading less rapidly than *Crassula helmsii* because it does not produce the numerous small vegetative fragments which act as propagules in that species.

M. aquaticum is a native species of the lowlands of central S. America; its native range is mapped by Orchard (1981). Female plants are naturalized in warm temperate and tropical areas elsewhere in S. America and in Africa, Asia, Australia, New Zealand, N. & C. America and Hawaii. In mainland Europe it is naturalized in France and it is also recorded from Austria.

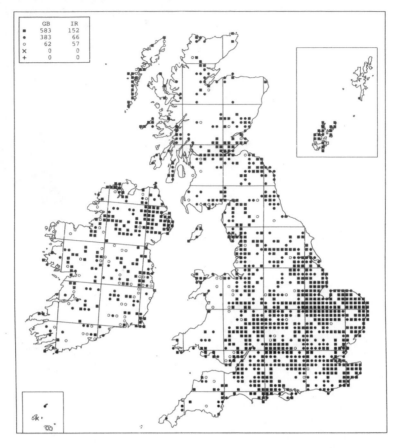

	GB	IR
■	583	152
●	383	66
○	62	57
×	0	0
+	0	0

Myriophyllum spicatum L.   Spiked Water-milfoil

M. spicatum usually grows in water which is both calcareous and meso-eutrophic or eutrophic. It is found at depths of 0.5–3m in lakes, ponds, rivers, streams, canals, major ditches and fenland drains, and in flooded quarries and clay-, sand- and gravel-pits. It grows in both sluggish and rapidly flowing rivers, where it roots in a range of substrates but is perhaps more frequent over sand and gravel rather than over fine clay or silt. In some sites, including mesotrophic rivers and machair lochs, it grows with *M. alterniflorum*. Like *Potamogeton pectinatus* and *Zannichellia palustris*, it often persists in highly eutrophic water at sites where other, less tolerant, macrophytes have disappeared. *M. spicatum* can also be found in brackish water, and it even grows in rocky cliff-top pools in western Scotland. Although normally found in eutrophic sites, it also occurs in abundance in highly calcareous but nutrient-poor limestone lochs. It is a primarily lowland species, which reaches 380m at Malham Tarn (Sinker, 1960). Wilson's (1938, 1956) record at a much higher altitude in Westmorland is an error, as the plant was *M. alterniflorum* (G. Halliday, pers. comm.).

M. spicatum is a perennial which dies back to the roots in winter. Populations usually flower and set seed freely, unless they grow in deep or turbulent water. The seeds can float for a few hours or a day. They exhibit prolonged dormancy, which can be broken by freezing or other mechanical or chemical treatments which rupture or erode the stony endocarp (Patten, 1955). Seeds germinate at temperatures above 10°C; some will germinate in the dark but light significantly enhances the proportion which germinate (Hartleb *et al.*, 1993). Some 30% of seeds dried and stored for seven years by Davis *et al.* (1973) germinated when rewetted. The seeds of *Myriophyllum* are eaten by wildfowl, including pochard and teal (Olney 1963, 1968), and these might act as dispersal agents. Reproduction by seed may be rare: in N. America seedlings have not even been found in some lakes where an estimated 4 million seeds are set per hectare (Aiken *et al.*, 1979). Plants can regenerate vegetatively from axillary buds, which are easily detached, and from fragments of stem, which are shed naturally during the growing season as well as broken off by disturbance (Guppy, 1894; Aiken *et al.*, 1979). Turions have not been seen in British or Irish material, and have only rarely been reported from mainland Europe (Orchard, 1981). Plants survive periods of emersion in a terrestrial form.

Like *M. verticillatum*, this species may have been present in the British Isles throughout the Pleistocene (Godwin, 1975). It is still frequent in suitable habitats in Britain and Ireland. It may have increased in abundance in the last 150 years in some areas of grazing marsh (Mountford, 1994a) and it has also invaded newly available habitats such as sand- and gravel-pits. These gains more than offset losses in other areas caused by the conversion of grazing marshes to arable land (Driscoll, 1983).

M. spicatum sensu lato is widespread in boreal and

temperate regions of the northern hemisphere, extending south to N. Africa, the Himalaya and Japan. The native American populations are sometimes treated as a separate species, *M. exalbescens* Fernald, but Orchard (1981) concluded that they are best recognized as a variety of the Eurasian plant. The Eurasian taxon has been introduced into N. America, where it behaves as an extremely aggressive aquatic weed, 'hampering fishing, boating, and swimming...choking out water plants, providing mosquito breeding habitats, and lowering real estate values' (Blackburn & Weldon, 1967).

For a more detailed account of the biology of this species, based on research in N. America, see Aiken *et al.* (1979). Some of the physiological reasons for its ability to grow rapidly are reviewed by Grace & Wetzel (1978).

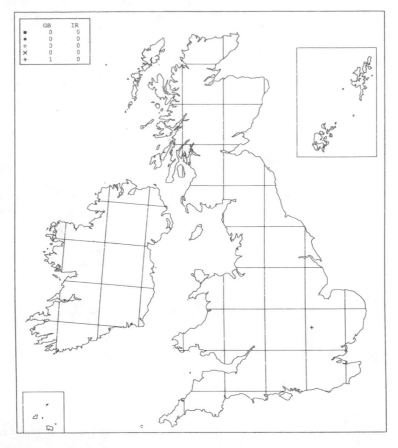

	GB	IR
■	0	0
●	0	0
○	0	0
×	0	0
+	1	0

Myriophyllum verrucosum Lindl.

This introduced species was formerly naturalized in a large, disused gravel-pit at Eaton Socon. Here it grew in deep water in large pools in the gravel-pit, and in a dwarfed terrestrial state on wet gravel. In its native range it is typically found in stagnant or slowly flowing water 30–60cm deep, but it is also found in the relatively deep water of fast-flowing rivers and on damp mud (Orchard, 1986).

M. verrucosum is a perennial species. Brenan & Chapple (1949) described the flowers of the naturalized material as hermaphrodite; Australian plants are also monoecious but usually have male flowers at the top of the spike, female flowers below and hermaphrodites restricted to a zone between them (Orchard, 1986). The British plant flowered only sparingly when growing in water, but both flowers and fruits were produced in abundance in terrestrial habitats. In Australia it is an opportunistic species which can flower at any time of year. Plants in deep water do not flower but flowering is probably triggered by falling water-levels, and terrestrial populations flower and fruit profusely (Orchard, 1986).

M. verrucosum was first collected in Bedfordshire in 1944, but initially identified as *M. verticillatum*. It was thoroughly naturalized by 1946, when its correct identity was established, but it failed to survive the severe winter of 1946–47. It was almost certainly introduced to Britain as seeds in Australian wool-shoddy, which was then extensively used by farmers and market gardeners as a manure and water-retaining agent. A number of other alien species of Australian origin were recorded in the gravel pit at Eaton Socon. For a detailed account of the Bedfordshire plant, see Brenan & Chapple (1949).

This species is a native of Australia, where it is the most widely distributed *Myriophyllum* species. It grows 'almost throughout the mainland wherever there is semi-permanent water available', and has spread into the arid regions by colonizing stock watering facilities (Orchard, 1986). The British records are the only known occurrences outside Australia.

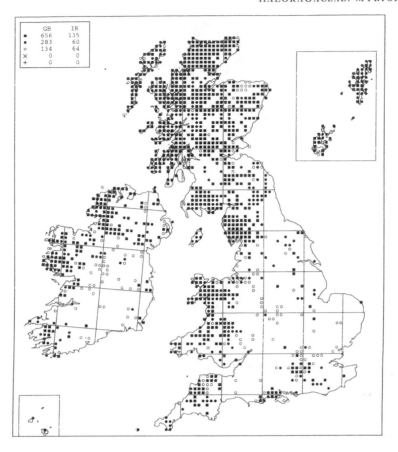

	GB	IR
■	656	135
▪	283	60
○	134	64
×	0	0
+	0	0

Myriophyllum alterniflorum DC.
Alternate Water-milfoil

This species has a wide ecological amplitude, as it grows in standing waters, including lakes and pools, and in slowly or rapidly flowing streams and rivers. In addition to these natural habitats, it also colonizes flooded quarries and gravel-pits. In southern England and Wales it is confined to acidic and mesotrophic or oligotrophic water. In Scotland and Ireland it is most often found in such habitats, and it is one of the few macrophytes which are frequent in the acidic, peat-stained streams and rivers of the north and west. However, it can also grow in highly calcareous water, as in the lochs on the Durness limestone, although it is absent from sites which are both calcareous and eutrophic. It is found over a wide range of nutrient-poor substrates from fine sand, peat and marl to stones and boulders. It is usually found at depths of 0.3–3m, but it can grow in a terrestrial state if exposed above the water-level, and it descends to depths of 4.5m in Loch Ness (West, 1905). It is frequent in both lowland and upland waters, ascending to 780m at Lochan an Tairbh-uisge.

M. alterniflorum is a perennial species. Viable seed is set in Britain but little is known about the frequency with which new plants are established from seed. The species lacks specialized turions, but it can regenerate from short lengths of shoot. In north-west Scotland the species has proved to be very variable genetically, both within and between populations (Harris *et al.*, 1992).

The fossil record of *M. alterniflorum* demonstrates that it has certainly persisted in the British Isles since the middle of the last glacial period, and it may have been continuously present for much longer (Godwin, 1975). It is frequent and locally abundant in the acidic areas of the north and west. Like many calcifuges, it has declined in the south and east where acidic habitats are infrequent and have often been destroyed. It was, for example, last recorded in Essex in the second half of the 19th century (Jermyn, 1974), and became extinct in Cambridgeshire, Hertfordshire and Leicestershire in the first half of the 20th (Perring *et al.*, 1964; Dony, 1967; Primavesi & Evans, 1988).

This species is found in the boreal and temperate zones of the northern hemisphere. In Europe it is most frequent in the north and west but extends south to Sicily. It is also recorded from N. Africa, eastern and western N. America and Greenland.

Pugsley (1938) drew attention to the presence of slender plants with very short leaves in shallow water in Lough Beg, Lough Neagh and Lough Ree. He considered that these were similar to the normal variant of the species in N. America, and therefore named them var. *americanum*. In Lough Beg and Lough Neagh they grew in shallow water over a sandy substrate, in the apparent absence of typical *M. alterniflorum* (Praeger, 1938b). They have now become very rare in these loughs, which have undergone eutrophication (Harron, 1986). Similar plants occur in the Hebrides. They may simply be compact, shallow-water phenotypes, but the problem requires reinvestigation.

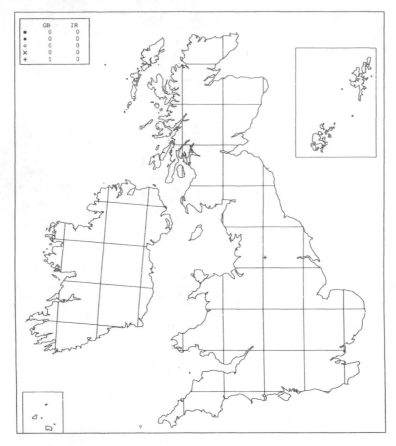

	GB	IR
■	0	0
●	0	0
○	0	0
×	0	0
+	1	0

Myriophyllum heterophyllum Michx.

This introduced species was formerly naturalized in a lowland canal between Halifax and Salterhebble. It grew in some quantity in a stretch of canal which also contained the alien aquatics *Potamogeton epihydrus* and *Vallisneria spiralis*.

M. heterophyllum is a monoecious perennial. The reproductive biology of the naturalized population does not appear to have been studied.

This plant was first discovered in its British locality in 1941 (Walsh, 1944). It persisted there until the drainage of the canal in 1947 or 1948 (Lavin & Wilmore, 1994; G. T. D. Wilmore, pers. comm.).

M. heterophyllum is a native of N. America, where it extends from S.W. Quebec, Ontario and N. Dakota south to Florida and New Mexico. In its native range its habitats include ponds, lakes, streams, ditches, canals, swamps, sloughs and borrow-pits (Godfrey & Wooten, 1981).

LYTHRACEAE

Most of the 30 genera and 580 species in the Lythraceae are herbs, but the family also includes some trees and shrubs. Eight of the genera contain aquatic plants. The distribution of the family is primarily tropical, and only the most widespread genus, *Lythrum*, occurs as a native in western Europe.

Lythrum L.

This genus includes 38 species which range from robust perennial herbs to small annuals. Both these extremes are represented in the British Isles: *L. salicaria* is a perennial with a spectacular inflorescence and *L. portula* is an inconspicuous annual. The contrast is so great that *L. portula* was placed in the separate genus *Peplis* by Linnaeus, where it remained until Webb (1967) pointed out that the two species were linked by intermediates. The only aquatic species in our area is *L. portula*, but *L. salicaria* is frequent in fens and *L. hyssopifolia* is a rare species of seasonally flooded ground.

Two subspecies of *L. portula* are recognized in the British Isles. There is no evidence to suggest that they differ in their ecology, so the text accompanying the maps simply deals with their distribution and taxonomy.

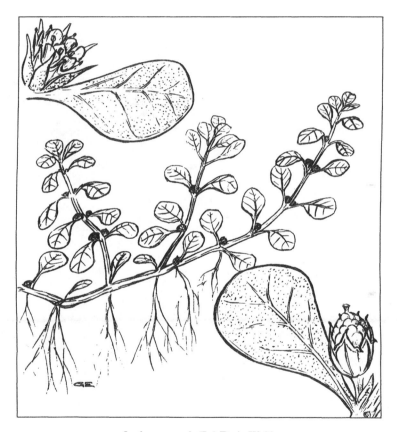

Lythrum portula (L.) D. A. Webb

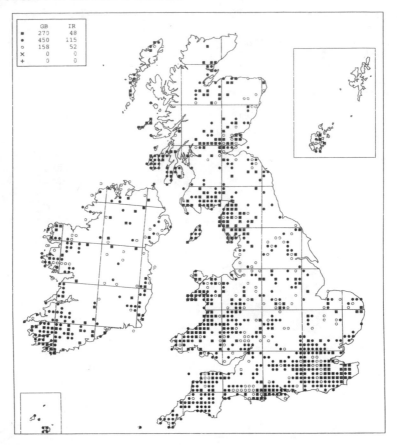

	GB	IR
■	270	48
●	450	115
○	158	52
×	0	0
+	0	0

Lythrum portula (L.) D. A. Webb

Water-purslane

L. portula is found in water and on damp ground by the water's edge. It is sometimes found in dense masses in shallow ditches, pools and sheltered lakes, occasionally occurring in water as deep as 1m (West, 1910). However, it is much more frequently found in habitats which flood in winter but dry out in summer. Typical sites include the stony, gravelly or muddy margins of ponds, lakes and reservoirs, springs, flushes and moist streamsides, dried-up pools and ditches, depressions in heaths and sand dunes, damp areas in sand and gravel workings and rutted tracks and woodland rides. It is a distinct calcifuge, which is typically found over base-poor mineral soils (including sands, gravels and clay) but avoids the most acidic and nutrient-poor soils and is rarely found over peat. The occurrence of a large population on limestone by Coole Lough in the Burren is exceptional, and an example of the anomalous behaviour of normally calcifuge species in western Ireland (Webb & Scannell, 1983). *L. portula* is a primarily lowland species, which ascends to 460m at the source of the R. Teme (Wilson, 1956).

Most plants of *L. portula* are prostrate, and easily root at the nodes, although erect variants sometimes occur (Salisbury, 1970). The species behaves as an annual in most sites but plants can persist as short-lived perennials, regrowing from the rooted stems in spring (Briggs,

1880). The flowers are usually self-pollinated (Knuth, 1906–9). Plants which grow in water fruit less readily than terrestrial individuals (Allen, 1954). The seeds are small and have no obvious adaptation to dispersal, but they somehow reach isolated sites such as disused sand- and gravel-pits. The seeds require light for germination. A substantial proportion of the fresh seeds sown by Salisbury (1970) at the end of August germinated within six days, with further seeds appearing sporadically or in flushes thereafter. If seeds do not germinate they can probably remain dormant for many years, as plants can reappear at a site after long intervals (Croft, 1994). The stems of *L. portula* are brittle and the species may also spread vegetatively within a water body (Arber, 1920).

L. portula has decreased in some areas, especially in the eastern part of its British range. Many of the ponds on commons and heaths where *L. portula* once grew in south-east England are now overgrown or drained, and formerly rutted tracks are now surfaced and quite unsuitable. Nevertheless, the species is able to persist on woodland rides, and occasionally colonizes newly available habitats. This may explain why it has not undergone the severe decline shown by some other species of seasonally flooded habitats.

This is primarily a European species, extending from Scandinavia to southern Europe. It also occurs in N. Africa and W. Asia and, as an introduction, in California, C. & S. America and New Zealand.

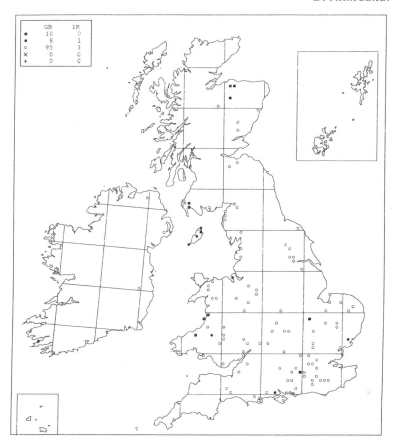

Lythrum portula (L.) D. A. Webb subsp. *portula*

The infraspecific taxa currently recognized in *L. portula* were defined by Allen (1954). The records of subsp. *portula* plotted on the map are based on herbarium specimens determined by Allen and by Perring (Perring & Sell, 1968), with some later additions. Only material with mature fruits can be determined to subspecific level, and this limits the number of records which can be obtained from herbaria. The preponderance of pre 1950 records is a reflection of the period when most specimens were collected, and does not indicate that the subspecies has declined. The subspecies of *L. portula* have not been recorded systematically by field botanists, and their distribution is imperfectly known.

Subsp. *portula* is widespread in Europe, and extends south to Italy, Sardinia and Corsica (Allen, 1954).

L. portula shows considerable variation in the length of the epicalyx segments. In some plants these are almost obsolete, whereas in others they can be up to 2mm long. In the British Isles there is a cline from west to east, the plants with the longer segments growing in the west (Allen, 1954). The extremes of variation are recognized as distinct taxa, which are defined somewhat arbitrarily. Subsp. *portula* includes plants with segments up to 0.5mm long, whereas subsp. *longidentata* has segments 1.5–2.0mm long. The subspecies are linked by a series of intermediates, and some populations of the species include both intermediates and plants referable to subsp. *portula*. The taxa are perhaps best treated as varieties, as Allen (1954) suggests, rather than as subspecies.

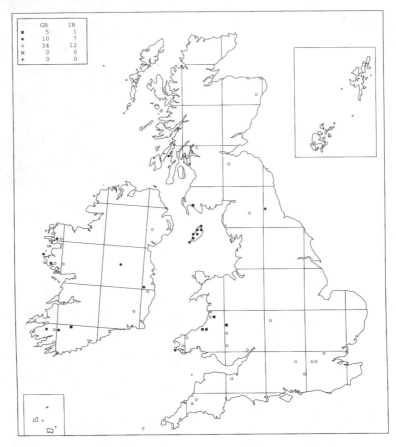

Lythrum portula subsp. *longidentata* (J. Gay) P. D. Sell

The records plotted on the map are based on those listed by Allen (1954), with some later additions. As with subsp. *portula*, the preponderance of pre-1950 records simply reflects the fact that most records are based on herbarium specimens.

Subsp. *longidentata* is found in western Europe and the Mediterranean region, being recorded from France, Portugal, Spain, Algeria and the Azores.

ONAGRACEAE

There are 24 genera and 650 species in the Onagraceae; most of the species are herbs or shrubs but some are trees. The family has a cosmopolitan distribution but the species are most numerous in N. and S. America. The largest genus in Britain and Ireland is *Epilobium*, which contains a number of species that grow in damp habitats (e.g. *E. hirsutum, E. palustre*). Two genera contain aquatics, the widespread *Ludwigia* and the American *Boisduvalia*.

Ludwigia L.

This is a genus of 75 species, including 15 aquatics. Only a single species, *L. palustris*, occurs as a native in Europe but two South American species are naturalized in France and Spain. The species which occur in the Old World have been monographed by Raven (1963).

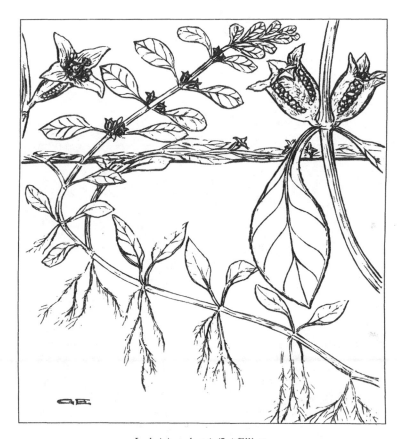

Ludwigia palustris (L.) Elliott

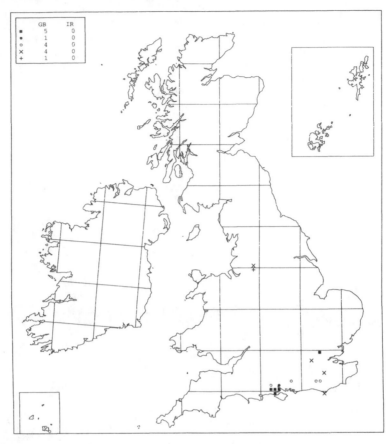

	GB	IR
■	5	0
●	1	0
○	4	0
×	4	0
+	1	0

Ludwigia palustris (L.) Elliott

Hampshire-purslane

In the New Forest this species grows as an emergent in shallow water in ponds, streams, ditches and old gravel-, sand-, clay- and marl-pits, and as a terrestrial plant on mud at the edge of ponds, the dried-up beds of ponds and streams and damp ground in marshy hollows. In shallow water at the edge of ponds it can be the dominant macrophyte, and some stands cover hundreds of square metres. The sites are on grazed common land over mildly acidic soils, and are subject to grazing and trampling by cattle and ponies. Characteristic associates include *Apium inundatum*, *Baldellia ranunculoides*, *Eleogiton fluitans*, *Galium palustre*, *Glyceria* spp., *Hydrocotyle vulgaris*, *Hypericum elodes*, *Juncus bulbosus*, *Lythrum portula*, *Mentha aquatica*, *Pilularia globulifera*, *Potamogeton polygonifolius* and *Ranunculus flammula*; the nationally rare *Galium constrictum* and *Mentha pulegium* also grow with it at some sites. Outside the New Forest, the only recent record of *L. palustris* as a native species is from Epping Forest, where it grows on mud at the edge of a pond over acidic sands and gravels. It appeared in 1989 in a garden pond at Tonbridge, where it was not planted deliberately. In 1991 it was found in a dew pond at Seaford Head, where it also seems likely to be an introduction although one can only speculate about the circumstances in which it arrived.

L. palustris is an annual or a short-lived perennial. Terrestrial plants flower and fruit freely; aquatic plants tend to flower less freely but are more likely to perennate. The fruits are indehiscent and mature fruits float; they are therefore likely to be dispersed locally by water movement. Like other species of similar habitats *L. palustris* seeds require light for germination, but desiccation does not enhance germination. If seeds are removed from the fresh capsules they show 'quasi-simultaneous' germination, with some germinating rapidly but the majority remaining dormant. Seeds can germinate within the fruits, emerging through the fruit wall. For further details of the reproductive biology of this species, see Salisbury (1972). Populations of *L. palustris* can vary in size from year to year: one at Brockenhurst reappeared in the dry autumn of 1969 after a long absence (Salisbury, 1972) and the species was discovered at Epping Forest in the dry summer of 1976, growing on mud which is normally under water.

L. palustris was formerly more widespread as a native plant than it is today, but it has been lost from at least four localities. It became extinct during the 19th century at two outlying sites: Petersfield Heath, where the marshy areas in which it grew dried out (Townsend, 1883), and a pit near Buxted, which became completely overgrown with scrub (Wolley-Dod, 1937). In Jersey *L. palustris* was known from two areas of marshland: it disappeared from one when the area was drained in the 1870s and was last seen in the other in 1920 (Le Sueur, 1984). In Hampshire, it was formerly regarded as an extreme rarity, being known to Townsend (1883) from only one locality, but 'there has been a continuing expansion of its New Forest range' (Brewis *et al.*, 1996). The future of the species appears to be secure, although there has been some concern that it may be threatened by the spread of *Crassula helmsii* (Byfield, 1984). *L. palustris* has not been refound in Essex since the original discovery in 1976.

This is a plant with a rather southerly distribution. It is found in W., C. & S. Europe northwards to England and the Netherlands; it appears to be most frequent in the south-west, and in Portugal it is common near the coast (Raven, 1963). Elsewhere it is found in N. Africa, Asia east to Iran and in temperate N. America. It is established as an introduced species in southern Africa, Hawaii and New Zealand; in the North Island of New Zealand it spread explosively following the first record in 1929 (Raven, 1963).

Salisbury (1972) points out that the indehiscent capsules of *L. palustris* are likely to limit its powers of dispersal, and suggests that this might partly explain its limited distribution in southern England. The recent records of plants which have presumably become established from horticultural sources are therefore particularly noteworthy, and it will be interesting to see if the species becomes more widespread as a result of such introductions.

APIACEAE (UMBELLIFERAE)

A large family of 420 genera and over 3,000 species. The family has a cosmopolitan distribution, but attains its greatest diversity in the temperate regions of the northern hemisphere. Most members of the family are herbs but there are some woody species. The Apiaceae is closely related to the Araliaceae, a smaller and predominantly tropical family in which most species are woody. There is little or no scientific justification for separating these two families (Mabberley, 1987), but they are kept apart by tradition. An illustrated account of the British and Irish members of the Apiaceae is given by Tutin (1980). Our plants include representatives of all three subfamilies which are recognized in the family: Hydrocotyloideae, Saniculoideae and Apioideae.

Cook (1990a) classifies 15 genera and 70 species as aquatics, but many of these are only seasonally submerged. In addition to the species considered below, *Carum verticillatum*, *Cicuta virosa*, *Hydrocotyle vulgaris* and *Sium latifolium* occur in fens or seasonally flooded habitats in our area, and are included within Cook's broader definition of an aquatic.

Hydrocotyle L.

Most of the species in the subfamily Hydrocotyloideae occur in the temperate regions of the southern hemisphere, particularly in S. America. The only genus which reaches our area is *Hydrocotyle*, with one native and three naturalized species. The native *H. vulgaris* grows in a range of moist and seasonally flooded habitats. The alien aquatic *H. ranunculoides* was discovered in Britain only in recent years, and is not listed by Stace (1991) or Kent (1992), but it is now sufficiently well naturalized to be considered here.

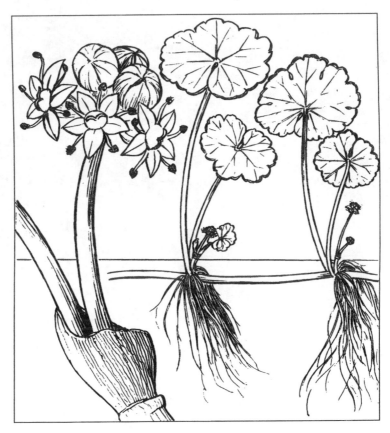

Hydrocotyle ranunculoides L. f.

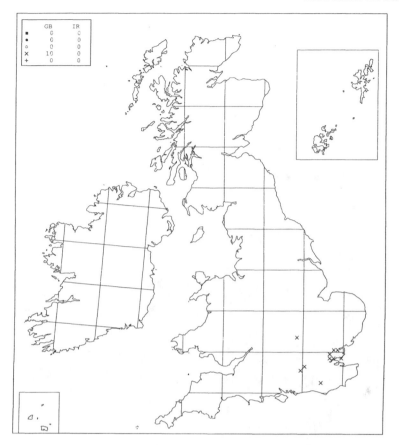

	GB	IR
■	0	0
●	0	0
○	0	0
×	10	0
+	0	0

Hydrocotyle ranunculoides L. f.
Floating Pennywort

This robust species grows as a rooted plant on mud or in free-floating colonies on the surface of still or slowly moving water. The floating colonies can be quite large: the most extensive patch so far recorded covered an area of 24m². *H. ranunculoides* is naturalized in the R. Chelmer and the Chelmer & Blackwater Navigation canal, and has also become established in artificial lakes, ponds, streams, ditches and a flooded gravel-pit.

Plants growing on mud flower and set seed. Floating plants do not flower, but reproduce vegetatively. The stems root at the nodes and are brittle. Fragments are easily broken off and can float to new localities, where they become established.

H. ranunculoides is sold in aquatic plant nurseries, often as 'marsh pennywort', the English name of *H. vulgaris*. It was first recorded as a naturalized alien in 1990, when M. Heywood discovered the population in the R. Chelmer at Chelmsford. By 1992 it had been recorded in the Chelmer & Blackwater Navigation up to 12km

downstream of Chelmsford, and in 1993 it was found in lakes, ponds and streams elsewhere in Essex. It is also known from a lake near Milton Keynes, a pond at Piltdown and a roadside ditch at Hamm Moor. On the basis of its behaviour in Essex, Payne (1994) suggests that it may spread to become a common species.

This species is a native of N. America, where it is widespread from Washington State and Pennsylvania southwards. Godfrey & Wooten (1981) describe it as growing 'in dense mats on mucky shores, in shallow water near shores, or in floating mats'. It is naturalized to the south of its native range in C. & S. America, and in some areas it is a serious weed. In the Old World it is recorded from S. Europe (Italy, Sardinia, Spain) and Africa (Ethiopia). It was discovered as an invasive weed in several localities in the Netherlands in 1995 (Bass & Hoverda, 1996). Tutin *et al.* (1968) consider that it may be a native of Italy and Sardinia, but this seems unlikely.

For further information about this plant in Britain, see Payne (1994). The above notes are based on this account and on information collated and supplied by K. J. Adams.

Berula Besser ex W. D. J. Koch

This is a small genus containing two species, the
widespread *B. erecta* and the African *B. thunbergii*.

Berula erecta (Huds.) Coville

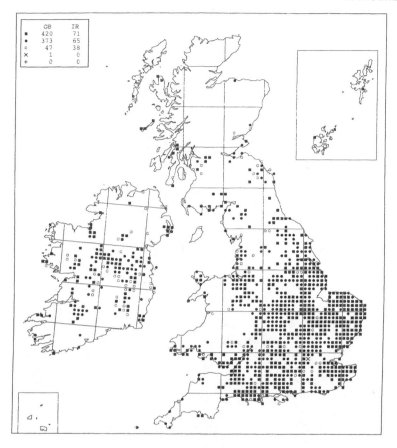

	GB	IR
■	420	71
●	373	65
○	47	38
×	1	0
+	0	0

Berula erecta (Huds.) Coville

Lesser Water-parsnip

B. erecta can grow as a submerged aquatic or as an emergent species in shallow water or seasonally flooded ground. Completely submerged leafy clumps are frequent in clear, calcareous streams and the upper reaches of rivers, where the flow of water is rapid but stable and the substrate is a coarse gravel. In this habitat *B. erecta* is often associated with *Ranunculus penicillatus* subsp. *pseudofluitans*, and it frequently grows immediately upstream of a patch of *Ranunculus*, in the zone where the flow of water is reduced (Haslam, 1978). As an emergent, *B. erecta* grows at the edge of lakes, ponds, streams, rivers, canals, ditches and in marshes. It is also associated with calcareous, mesotrophic water in these habitats, but it usually grows on substrates such as fine silt or mud. *Apium nodiflorum, Mentha aquatica, Rorippa nasturtium-aquaticum, Veronica anagallis-aquatica* and *V. beccabunga* often grow with it. *B. erecta* is an exclusively lowland species, reaching 260m between Croglin and Renwick.

The submerged leaves of this species remain green throughout the winter, and clumps spread in early spring by the growth of short-lived stolons or rhizomes. Sediment often accumulates around the shoots and the clumps become raised above the surface of the substrate, until they become vulnerable to erosion and are washed away. The entire cycle from establishment to erosion can take less than a year (Haslam, 1978). The species can become established from small floating plants which become detached from the parent clone in autumn, and from vegetative fragments. Submerged clumps often fail to flower, but terrestrial plants flower more freely. Little is known about the frequency with which they reproduce by seed.

B. erecta is sometimes overlooked because of the similarity of its leaves to those of *Apium nodiflorum*. The differences were not clearly set out by successive editions of Clapham *et al.* (1952), but they were described by Haslam *et al.* (1975) and Wigginton & Graham (1981) and are now widely known; the mapped distribution should therefore be fairly accurate. The overall distribution of the species appears to be fairly stable, but the plant has declined in frequency in some areas, including Essex (Tarpey & Heath, 1990) and Gloucestershire (Holland, 1986), probably because of habitat destruction. The most northerly population in our area, in a limestone valley in Shetland, was greatly reduced in quantity in 1986, when the stream in which it grew was widened and deepened to facilitate drainage (Scott & Palmer, 1987).

B. erecta is found in Europe from southern Scandinavia southwards, although it is rare in the Mediterranean region. It extends eastwards to C. Asia, and it is also widespread in temperate N. America. It is also recorded as an introduction in Africa.

Oenanthe L.

A genus of 40 species, most of which are found in wetland habitats in the temperate northern hemisphere. The genus is not a critical one, but the species are more difficult to identify than those in other genera of the Apiaceae in our area. Four of the seven British and Irish species are considered below; in addition *O. silaifolia* is a plant of unimproved, seasonally flooded meadows. All our species are poisonous, *O. crocata* being especially toxic (Cooper & Johnson, 1984).

Oenanthe aquatica (L.) Poir.

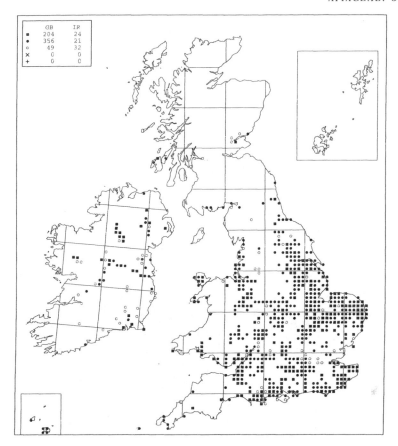

	GB	IR
■	204	24
●	356	21
○	49	32
×	0	0
+	0	0

Oenanthe fistulosa L.　　　Tubular Water-dropwort

This plant occupies a range of moist or wet habitats in the lowlands. It is usually found in places which are flooded for at least part of the winter but where the summer water-table lies at or below the surface. These include meadows and pastures on the flood plain of rivers, marshes and fens with a species-rich sward which is not dominated by tall reed-swamp species and in the fringing vegetation at the edge of lakes, ponds, rivers, streams, canals and ditches. However, plants can also be found growing as emergents in shallow water, particularly in grazing marsh ditches and in dune-slacks.

O. fistulosa is a perennial with tuberous roots. The young leaves differ from the mature leaves in having markedly broader lobes. Both terrestrial and emergent plants flower and fruit freely. Swards of seedlings can be found in moist, open vegetation in the summer and autumn. The species has no specialized means of vegetative reproduction.

In areas such as the Somerset Levels where grazing marshes provide much suitable habitat this is a frequent species, but elsewhere it is rather uncommon. It is able to persist at the edge of ditches or in winter-flooded hollows in fields where the grassland has been improved for agriculture, but it has undoubtedly decreased in abundance in regions where low-lying pastoral land has been converted to arable. The results of the BSBI Monitoring Scheme indicated that it had decreased in frequency in both England and Scotland since 1960 (Rich & Woodruff, 1996) and its decline has been documented in Bedfordshire (Dony, 1976) and Essex (Tarpey & Heath, 1990).

O. fistulosa is a predominantly European species, occurring from Scotland and southern Scandinavia southwards, and in adjacent parts of N. Africa and S.W. Asia.

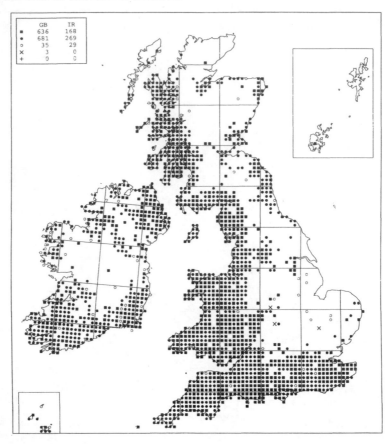

	GB	IR
■	636	168
●	681	269
○	35	29
×	3	0
+	0	0

Oenanthe crocata L.   Hemlock Water-dropwort

This is the most robust member of the genus in our area. It grows in shallow water in streams, on the banks of rivers, streams, lakes, ponds and canals, in marshes and wet woodland, in crevices in waterside masonry, amongst flushed stones and boulders at the top of beaches and on dripping or flushed sea-cliffs. It is one of the few macrophytes which will grow in small, rocky coastal streams in western Britain, where its roots lodge firmly in rock crevices. However, the largest plants are found in muddy swamps, especially where these are enriched by run-off from nearby farmyards. *O. crocata* is most frequent in or by acidic water, or where calcareous waters flow over acidic substrates, but it also occurs in calcareous habitats throughout its British and Irish range. These include wetlands in the Burren (Webb and Scannell, 1983) and the classic chalk rivers of south–east England (Holmes, 1983). It is tolerant of a variable water-level, and is therefore able to grow at the edge of rivers over clay. In the Isles of Scilly it extends into drier places than is usual on the mainland, and it can even be found on wall-tops (Lousley, 1971). It is a distinctly lowland species, which is only rarely recorded above 300m; it occurs up to 320m on The Paps in Co. Kerry (Scully, 1916).

This is a perennial with tuberous roots and leaves which remain green through the winter. The reproductive biology of the species does not appear to have been studied in any detail. Plants flower in early summer and after flowering the stout but hollow flowering stems often collapse. They lack specialized means of vegetative dispersal and spread, although the tuberous roots of riverside plants could conceivably be washed out of the substrate and become established downstream. Reproduction is usually by seed.

O. crocata is frequent and often abundant within its rather sharply delimited British and Irish range. It is highly poisonous to both man and his animals, but it is rarely persecuted by farmers who perhaps consider that the effort to control such a frequent species would be wasted. However, Marquand (1901) suggested that the disappearance of the plant from Alderney may have been caused by deliberate eradication. The only indication that the species is declining comes from the London area, at the very edge of its range; even here it is still frequent by the Thames (Burton, 1983).

This species has an Atlantic distribution, extending from Belgium and Britain south to Spain, Portugal and Morocco; it extends into the western Mediterranean region as far as Italy.

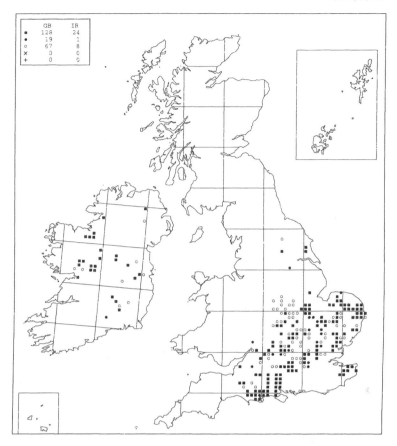

	GB	IR
▪	128	24
●	19	1
○	67	8
×	0	0
+	0	0

Oenanthe fluviatilis (Bab.) Coleman
River Water-dropwort

O. fluviatilis is most frequent in the crystal-clear, meso-eutrophic water of calcareous streams and rivers, where it can often form large beds. It is also found in canals, fenland lodes and ditches, but only rarely grows in ponds. It occurs in both still or sluggish water and in streams where the flow is more rapid, and at a range of water depths, from shallow water in streams where the flow is sufficiently constant to protect the plant from desiccation to water at least 1.5m deep in fenland lodes. It is often found over sands and gravel, or in places where these substrates are mixed with clay; it avoids deep, fine silt. *Ranunculus penicillatus* subsp. *pseudofluitans* is an almost constant associate in streams and rivers; other species which often grow with *O. fluviatilis* include *Callitriche obtusangula*, *Nuphar lutea*, *Sagittaria sagittifolia* and *Zannichellia palustris*. *O. fluviatilis* is confined to the lowlands.

This is a perennial species with leaves which remain green throughout the winter. In rapidly flowing water the prostrate stems root at the nodes and form large vegetative clumps which rarely flower; reproduction of these colonies is by vegetative fragmentation. Plants in still or sluggish water flower more freely. If the flowering stems are not broken off or dragged down below the water-level by the current they produce viable seed, but the frequency with which the species reproduces from seed is unknown.

This species was first described as *O. phellandrium* var. *fluviatilis* by Babington (1843); it had previously been confused with *O. aquatica* (*O. phellandrium*). The differences between the two species were clarified by Coleman (1844, 1849), who recognized that *O. fluviatilis* deserved specific rank. There is little doubt that *O. fluviatilis* has declined since its discovery some 150 years ago, as rivers have become very eutrophic or have suffered from deep dredging and channel 'improvement' or greatly increased boat traffic. Rivers in which it is now much less frequent than it was, or from which it has disappeared, include the Thames in Berkshire (Bowen, 1968) and the London area (Lousley, 1976; Burton, 1983), the Blackwater and the Stour in Essex (Tarpey & Heath, 1990) and the Welland in Leicestershire (Primavesi & Evans, 1988). The species may, however, be somewhat under-recorded. Its vegetative clumps, once known, could scarcely be mistaken for anything else, but a surprising number of botanists manage to overlook them. Flowering plants are sometimes mistaken for *O. aquatica*.

O. fluviatilis is endemic to W. Europe; its world distribution is mapped by Cook (1983). On the European mainland it is known from Denmark and from a rather compact area between the Seine and the Rhine in Belgium, France and Germany. It is declining in N.W. France (Mériaux, 1981) and is apparently extinct in Germany (Haeupler & Schönfelder, 1989).

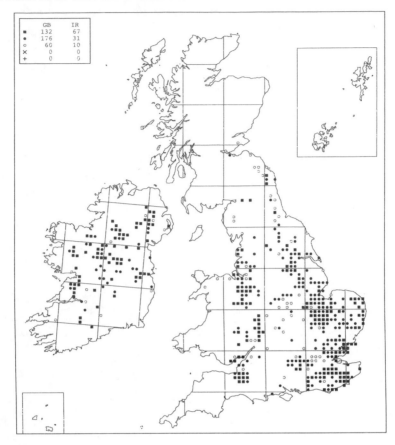

	GB	IR
■	132	67
▪	176	31
○	60	10
×	0	0
+	0	0

Oenanthe aquatica (L.) Poir.
Fine-leaved Water-dropwort

Unlike the closely related *O. fluviatilis*, this is found in lowland sites where the water is still or slowly flowing, and subject to marked fluctuations in depth. It is most frequent in shallow ponds and ditches, where it often grows on deep, silty and often eutrophic substrates in areas where the vegetation is kept open by trampling cattle. It is also found in open vegetation at the edge of sheltered lakes, reservoirs, canals, sluggish streams and rivers, in marshes and in seasonally flooded depressions. It can persist in moderate shade in ponds which are surrounded by trees, but is rarely found in closed vegetation in reed-swamps. In many areas it is particularly frequent on clay soils, but it is not confined to them.

The seeds of *O. aquatica* will germinate soon after they are shed, but germination is enhanced by a period of stratification under water and then favoured by light, aerobic conditions and moderately high temperatures. Swards of seedlings can therefore be found in summer on ground from which water has receded. Once seedlings have become established they continue to grow terrestrially or, if the water-level rises, as aquatics with more finely divided leaves. Plants may behave as winter annuals, germinating in autumn and flowering the next summer, or as summer annuals, flowering in the same

year that they germinate. The winter annuals tend to grow into larger plants which produce more seeds. The seeds can float for some hours after they are shed; seeds which sink may germinate and give rise to seedlings which then float. The number of plants at a site may vary greatly from year to year, depending on the fluctuations in water-level.

This is an uncommon species throughout much of its British and Irish range. It has undoubtedly decreased in some areas of eastern Britain, including Berwickshire (Braithwaite & Long, 1990), Berkshire (Bowen, 1968), Essex (Jermyn, 1974; Tarpey & Heath, 1990) and Leicestershire (Primavesi & Evans, 1988). In Suffolk it was 'formerly frequent and generally distributed, now rare and disappearing, as its habitats have been rapidly destroyed in recent years' (Simpson, 1982). *O. aquatica* may be over-recorded in some areas of S. England as *O. fluviatilis* is sometimes mistaken for it.

O. aquatica is widespread in Europe and western Asia, especially in the middle latitudes; it extends north to southern Scandinavia. It is also known in a few scattered sites in C. Asia and, as an introduction, in New Zealand.

See Hroudová *et al.* (1992) for a detailed account of the reproductive biology of this species in Czechoslovakia; the details of its germination and establishment are taken from this as there has been little research in Britain or Ireland.

Apium L.

A genus of 30 annual or perennial species which
are confined, as natives, to the Old World. Four of
these are aquatic, including two which occur in
our area.

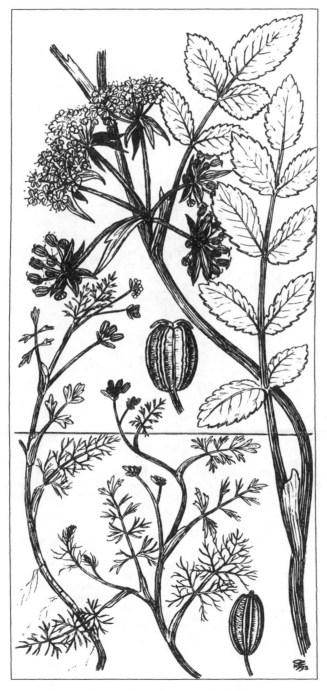

Apium inundatum (L.) Rchb. f.; *A. nodiflorum* (L.) Lag.

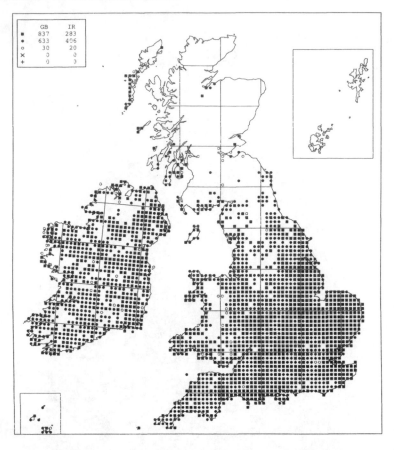

	GB	IR
■	837	283
●	633	496
○	30	20
×	0	0
+	0	0

Apium nodiflorum (L.) Lag.   Fool's Water-cress

A. nodiflorum can be found in shallow water in marshes, springs, streams and ditches, where it may form dense, pure stands, and at the edges of rivers, lakes, ponds and canals. It is usually emergent but can also form submerged patches. It characteristically grows in fertile sites where the growth of taller species is restricted by erosion or by ditch clearance or other disturbance (Grime *et al.*, 1988). Although it can be particularly abundant in rapidly flowing, calcareous streams, including winterbournes which dry out in summer, it is also found in non-calcareous habitats. It is a highly palatable species which can be eliminated by grazing. It has a predominantly lowland distribution, reaching 335m at Harley Dingle.

A. nodiflorum is a perennial with shoots which die back in winter and regrow in spring. Plants flower and fruit freely. Seeds germinate from early spring to autumn on damp substrates or in shallow water. Seedlings which become established in open vegetation may reach maturity but those which arise in or near established stands of the species tend to be outcompeted by mature plants (Thommen & Westlake, 1981). Vegetative reproduction and spread takes place by the

regeneration of detached shoots, which can develop new roots within two days under laboratory conditions (Grime *et al.*, 1988).

This species is still frequent throughout much of its British and Irish range, and no decline is apparent at the scale of the map. Even in the London area it is still frequent outside the most heavily built-up areas (Burton, 1983; Kent, 1975).

W., C. & S. Europe, most frequent in the west and reaching its northern limit in Scotland. It also occurs in S.W. Asia and N. Africa, and as a naturalized and spreading alien in N. & S. America (Cook, 1990a).

A. nodiflorum is ecologically very similar to *Rorippa nasturtium-aquaticum* sensu lato. *A. nodiflorum* is at a selective advantage in sites which dry out in summer, as its roots penetrate more deeply into the substrate. It replaced *R. nasturtium-aquaticum* as the dominant species at some sites in Dorset after the 1976 drought. *R. nasturtium-aquaticum* is favoured where the water is deeper than 40cm, and its superior ability to regenerate from fragments also leads to its dominance in heavily grazed or disturbed situations (Thommen & Westlake, 1981). *Berula erecta* differs from *A. nodiflorum* in being restricted to calcareous sites, and is more frequently found as submerged plants.

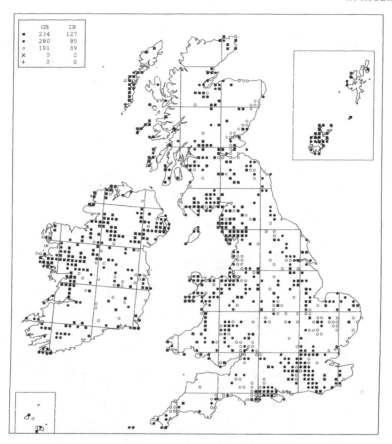

	GB	IR
■	234	127
●	280	85
○	191	39
×	0	0
+	0	0

Apium inundatum (L.) Rchb. f.   **Lesser Marshwort**

A. inundatum is a heterophyllous species which can grow as an aquatic with finely divided submerged leaves or as a terrestrial plant with broader leaves. It grows in sites where the water-level fluctuates and which are subject to periodic desiccation: these include the edges of lakes and reservoirs, pools and dune-slacks. It is able to colonize freshly created habitats, including new ponds and, in one area, scrapes dug for natterjack toads. It can also be found in more permanent shallow water in streams, ditches, canals, the backwaters of rivers and marshes. It is confined to oligotrophic or mesotrophic habitats, occurring over a wide range of substrates including sand, silt, gravelly stones and peat, but it appears to avoid the least fertile sites. Most of its sites are base-poor, but it can also grow in highly calcareous water where the nutrient levels are sufficiently low. Characteristic associates include *Baldellia ranunculoides*, *Callitriche hamulata*, *Eleogiton fluitans*, the aquatic form of *Juncus bulbosus*, *Littorella uniflora*, *Myriophyllum alterniflorum* and, in western Ireland, *Eriocaulon aquaticum*. *A. inundatum* is a primarily lowland species, which reaches 450m at Rhyd Galed and 500m on Cronkley Fell.

A. inundatum flowers and fruits freely in both terrestrial habitats and in shallow water. The small flowers are probably self-pollinated (Knuth, 1908). Reproduction is by seed. It fluctuates in abundance in response to changes in water-level, and it can be the most abundant species in shallow ponds. It sometimes reappears in pools and ditches following clearance, apparently from buried seed, but it has suffered a massive decline in the Basingstoke Canal recently because of dredging (Brewis *et al.*, 1996).

This species appears to be declining throughout eastern Britain: the decrease has been documented in numerous areas including S. England (Bowen, 1968), East Anglia (Perring *et al.*, 1964; Simpson, 1982), N. England (Graham, 1988; Swan, 1993) and Scotland (Ingram & Noltie, 1981). Like many species of shallow water, it has suffered from the destruction of its habitats by drainage. It has also been lost from sites which are now overgrown following the cessation of grazing (Byfield & Pearman, 1994), and perhaps from some which have become eutrophic. By contrast, Allen (1984) draws attention to the numerous recent records from the Isle of Man and Galloway, and suggests that they may represent an increase in the plant rather than more adequate recording of this inconspicuous species. There are insufficient data to assess trends in Ireland.

A suboceanic species, found in W. & C. Europe from Britain, S. Scandinavia and Poland south to the Iberian peninsula and Sicily. It is listed as 'vulnerable' in the Netherlands (Weeda *et al.*, 1990), and appears to be declining in Germany (Haeupler & Schönfelder, 1988). It also occurs in N. Africa.

The sterile hybrid between *A. inundatum* and *A. nodiflorum*, *A.* × *moorei*, is a very local plant in Britain, but is widespread in Ireland; it is not known on the European mainland (Perring & Sell, 1968; Tutin, 1975). It grows in marshes and at the side of lakes, rivers, streams and canals.

MENYANTHACEAE

This small family contains five genera and about 40 species. All the species are submerged, floating or emergent aquatics. The family was once thought to be closely allied to the Gentianaceae, but it is now placed in the Solanales rather than the Gentianales.

Menyanthes L.

This is a monotypic genus which contains the familiar bogbean, *Menyanthes trifoliata*. This differs from other members of the family in having trifoliate rather than entire leaves.

Menyanthes trifoliata L.

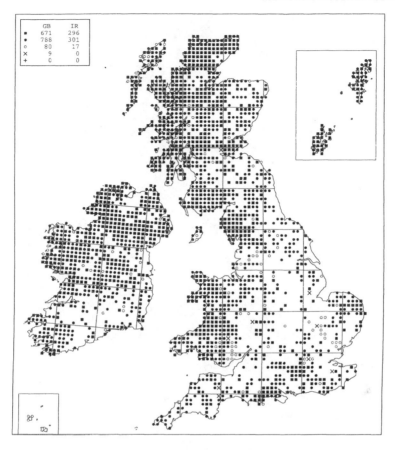

	GB	IR
■	671	296
●	788	301
○	80	17
×	9	0
+	0	0

Menyanthes trifoliata L.   Bogbean

M. trifoliata grows at the fringe of lakes or slowly flow-ing rivers, and in pools and wet flushes. At the edge of lakes, pure stands of this species are often found where the emergent zone gives way to open water. At their landward edge, pure stands usually grade into mixed communities of *Carex rostrata*, *M. trifoliata* and *Potentilla palustris* in shallower water, and then to car-pets of *Calliergon cuspidatum* or *Sphagnum* spp. where the water-table lies near the surface of the soil. In these fen communities *M. trifoliata* can be accompanied by a wide range of other herbs. It may also grow in bog pools or flooded peat-cuttings, where *Eriophorum angusti-folium*, *Sphagnum auriculatum* and *Utricularia minor* are frequent associates. It is also found, often as dwarf shoots, in the wettest parts of flushes dominated by small sedge species and in dune-slacks. It occurs over both organic and inorganic substrates, in calcareous and acidic waters and over a wide range of nutrient levels. It is recorded from sea-level to over 1000m near the summit of Beinn Heasgarnich.

M. trifoliata is a rhizomatous, deciduous perennial. Established colonies spread by growth of the rhizome, and in open water they can occupy many square metres. The flowers are heterostylous and self-incompatible, and 'pin' and 'thrum' clones tend not to grow intermixed. They are pollinated by insects. The pollen of thrum plants is less viable than that of pin plants, and the num-ber of seeds set per flower is often low (Hewett, 1964; Nic Lughadha & Parnell, 1989). The seeds are buoyant and under experimental conditions may remain floating for 15 months. The seed coat is hard and germination will not take place until it is ruptured. Reproduction by seed appears to be rare (Hewett, 1964). *M. trifoliata* grows in isolated ponds and Ridley (1930) suggested that its fruits were eaten and dispersed by duck.

The fossil record of *M. trifoliata* in our area is remark-ably complete: it has been found in deposits from the last four interglacials. It may have persisted through the last glacial period, and Godwin (1975) suggests that it was a characteristic species of the aquatic communities which developed after the melting of the ice-sheets. It is still a frequent plant in northern and western Britain and Ireland, especially in peatland areas. However, it has decreased in almost all the counties of southern and east-ern England because of the drainage of wetland habitats in historic and recent times. It is an ornamental species which is sometimes planted out in the wild, and several of the extant populations in the London area, for ex-ample, are believed to originate from such introductions (Burton, 1983).

M. trifoliata occurs in boreal and temperate regions throughout the northern hemisphere.

The rhizomes of *M. trifoliata* are still used medicin-ally in Donegal. They are collected in March and boiled to produce a liquid which is taken to cure eczema and to purify the blood.

Nymphoides Ség.

There are 20 species in this genus, all of which have the 'water-lily' growth form with floating leaves and aerial flowers; the flowers of many species are insect-pollinated. The genus includes both annual and perennial species. Although the genus has a world-wide distribution, only a single species, *N. peltata*, is native to Europe.

Some species of *Nymphoides*, including *N. peltata*, are heterostylous, with a weak incompat-

ibility mechanism between the 'pin' and 'thrum' forms. Other species in the genus have a strong incompatibility mechanism (e.g. *N. indica* (L.) Kuntze) or are dioecious (e.g. *N. cordata* (Elliott) Fernald). Ornduff (1966) suggests that both the strongly incompatible and the dioecious species have evolved from weakly incompatible ancestors, under selection pressures that favoured outbreeding.

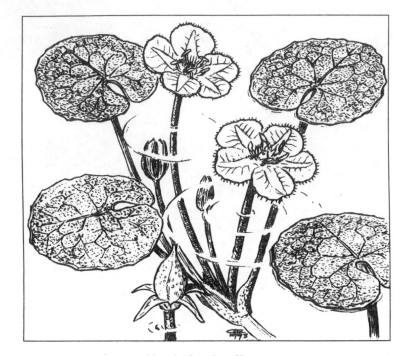

Nymphoides peltata Kuntze

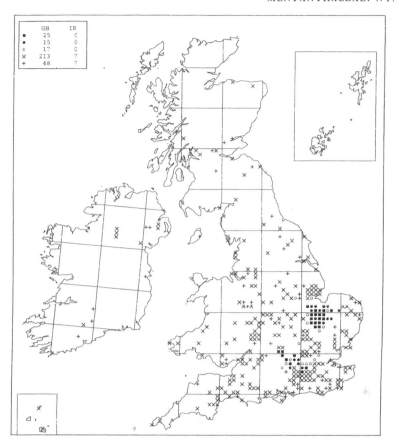

	GB	IR
■	25	0
●	15	0
○	17	0
×	213	7
+	48	7

Nymphoides peltata Kuntze Fringed Water-lily

Within the area in which it is thought to be native, *N. peltata* is a plant of lowland, calcareous and eutrophic water over inorganic substrates at the edge of lakes, slowly flowing rivers and fenland lodes. In these sites it often grows in dense masses in water 0.5–2.0m deep. The species is also grown as an ornamental plant, and it is thoroughly naturalized in ponds, natural and ornamental lakes, flooded gravel-pits, clay-pits and quarries and in rivers, canals and ditches.

N. peltata is a rhizomatous perennial. It is heterostylous, individual plants having 'pin' or 'thrum' flowers. Each flower lasts for only one day, but plants flower over a long period. The flowers appear patterned to organisms with vision which extends into the ultraviolet wavelengths, and are visited by a range of insects (Velde & Heijden, 1981). The fruits develop under water. More seed is set from crosses between pin and thrum plants than if both parents have the same floral morphology, and the viability of seeds and vigour of the seedlings is greater (Ornduff, 1966). The seeds float and may be dispersed by water. They are completely digested when eaten by waterfowl, but they are tolerant of desiccation and may be carried on the feathers of birds (Cook, 1990b; Smits *et al.*, 1989). The seeds require a short cold treatment before germination. In the Netherlands they will germinate in quantity in spring in aerobic conditions on exposed mudflats or in very shallow water; in anaerobic situations they remain dormant in a persistent seed-bank (Smits *et al.*, 1990). Plants may reproduce vegetatively from rhizomatous spread or fragmentation of the stems.

N. peltata is thought to be native in the Thames valley, where it was recorded as frequent in the R. Thames about London in 1570 (Kent, 1975), and in East Anglia, where Ray (1660) found it 'in many rivers about the Fens in great plenty'. The native range has, however, been obscured by the spread of plants which have been deliberately introduced into the wild or have spread from cultivated stock, and it cannot now be delimited with certainty. The species has decreased in its native sites in the Thames valley (Bowen, 1968; Burton, 1983), but it continues to be introduced and to spread both within the area in which it also occurs as a native and elsewhere.

As a native plant *N. peltata* is widespread in Europe from England and the Baltic southwards, and extends eastwards through Asia to Japan. It is naturalized in N. America (Stuckey, 1974).

All the native plants in East Anglia appear to be the pin morphotype; in the Thames valley both forms occur. Many introduced populations consist of either pin or thrum plants, suggesting that they may be spreading clonally.

BORAGINACEAE

This is a large but predominantly terrestrial family, which contains 156 genera and 2,500 species of herbs, shrubs or trees. All the European representatives are herbs or dwarf shrubs. In the British Isles, eight genera and 23 species occur as natives, and there are numerous introduced taxa. Cook (1990a) does not list any aquatic Boraginaceae, but a single species is sufficiently aquatic in our area to be mapped in this book.

Myosotis L.

Five of the ten native species of *Myosotis* in the British Isles belong to series *Palustres* Popov.

They are closely related ecologically as well as taxonomically, as they all grow in seasonally flooded or permanently damp or wet habitats. Only *M. scorpioides* can regularly be found growing in water, but *M. laxa* subsp. *caespitosa* and *M. secunda* often occur at the water's edge. *M. stolonifera* is a scarce species of upland springs, flushes and other waterside habitats (Stewart *et al.*, 1994), and *M. sicula* is confined to a single poolside in Jersey (Le Sueur, 1984). A taxonomic account of the European species in series *Palustres* has been given by Schuster (1967), but some of his conclusions have not been adopted by later authors. Useful notes on the identification of our species are given by Welch (1961).

Myosotis scorpioides L.

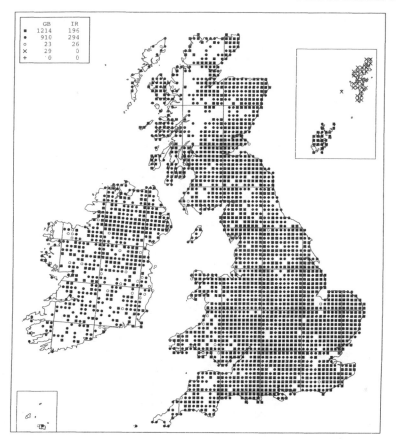

Myosotis scorpioides L.   **Water Forget-me-not**

M. scorpioides grows in a range of open or shaded, wet or damp habitats. It can be found as an aquatic, forming submerged patches or floating rafts at the edge of lakes, rivers, streams, canals and ditches. It is, however, more frequent above the water-level, usually growing at the edge of still and flowing water in situations where the growth of more robust species is restricted by the disturbance associated with winter flooding (Grime *et al.*, 1988). It is usually found in fertile sites over calcareous or mildly acidic soils. It is a predominantly lowland plant, ascending to at least 600m at Moor House.

M. scorpioides is a rhizomatous perennial which over-winters as leafy shoots. Submerged patches may not flower, but emergent and terrestrial stands flower freely. The incompatibility system of *M. scorpioides* is controlled by several genes and the degree of self-incompatibilty differs from plant to plant: some hermaphrodite individuals are completely self-incompatible whereas in others 95% of self-fertilized ovules may develop into seeds (Varopoulos, 1979). The seeds sink or float for only a short period after they are released (Guppy, 1906; Praeger, 1913). Plants also reproduce vegetatively: stems

are broken off by flowing water and the resulting fragments can regenerate (Ridley, 1930). Vegetative reproduction is probably more important than reproduction by seed (Grime *et al.*, 1988).

This species is frequent in suitable habitats throughout Britain and Ireland. There is no evidence that it is declining, except in totally urban areas where its habitat has been eliminated. In Shetland this species is often grown in gardens and it has become naturalized near houses and crofts; in Orkney it is doubtfully native (Scott & Palmer, 1987).

M. scorpioides is a variable species which is widely distributed in temperate Eurasia. It also occurs as an alien in N. & S. America and New Zealand.

Botanists sometimes confuse this species with *M. secunda*, which replaces it in areas where the soils are infertile and base-poor. The third frequent taxon in series *Palustres*, *M. laxa* subsp. *caespitosa*, is an annual of disturbed, waterside habitats. The hybrid between *M. laxa* and *M. scorpioides*, *M.* × *suzae*, is a vigorous and partially fertile perennial which is known from a few localities in England and Wales, growing in sites where both parents occur (Benoit, 1975). It is probably under-recorded.

HIPPURIDACEAE

This family contains a single genus, *Hippuris*. The relationships of the family, like those of many families with highly reduced flowers, are uncertain.

Hippuris L.

This is a small genus, treated by some authors as monotypic although others recognize two or more closely related species. The genus has given its name to the '*Hippuris* syndrome', used to describe plants with erect, unbranched stems with simple, elongate leaves borne in symmetrical whorls. This character-complex has evolved in several unrelated flowering–plant families, all of them dicotyledons, including Elatinaceae (*Elatine alsinastrum* L.), Haloragaceae (*Myriophyllum elatinoides* Gaudin, *M. hippuroides* Nutt. ex Torr. & A. Gray), Lythraceae (*Rotala hippuris* Makino), Ranunculaceae (*Ranunculus polyphyllus* Waldst. & Kit. ex Willd.) and Scrophulariaceae (*Hydrotriche hottoniiflora* Zucc., *Limnophila hippuridoides* Philcox). It is also found in some cryptogams, including *Chara* and *Equisetum*. The syndrome was described by Cook (1978), who suggested that it represents one way in which aquatic plants can increase the surface area of their assimilating organs.

Hippuris vulgaris L.

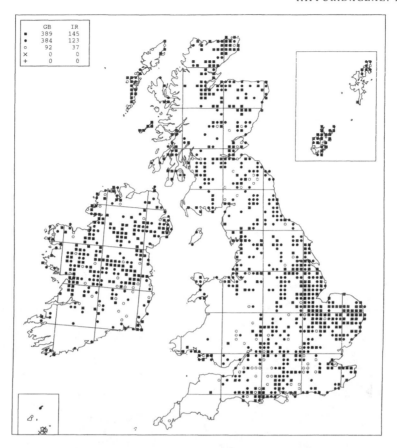

	GB	IR
■	389	145
●	384	123
○	92	37
×	0	0
+	0	0

Hippuris vulgaris L.   Mare's-tail

This is a heterophyllous species which may occur as an aquatic with submerged or emergent shoots or as a terrestrial plant. As an aquatic it is found in still or sluggish water of lakes, ponds, flooded gravel-pits, rivers, streams, canals and ditches. Aquatic plants are usually found in clear, base-rich water, and are abundant in the limestone lakes of Lismore and the Durness area of Sutherland. As an emergent or terrestrial plant *Hippuris* grows at the edge of standing waters, and in swamps, marshes and upland flushes. Terrestrial plants are most vigorous on soft, eutrophic substrates, and *H. vulgaris* is one of the few species to persist in the nutrient-enriched Norfolk broads. However, small scattered shoots of *H. vulgaris* can be found over mesotrophic or meso-oligotrophic substrates, including sphagnum lawns. It is also tolerant of saline conditions, and it may grow with *Bolboschoenus maritimus*, *Ranunculus baudotii* and *Schoenoplectus tabernaemontani* in coastal habitats. *H. vulgaris* is a predominantly lowland species, rarely found above 400m but ascending to 900m in bryophyte-dominated springs on Moine Mhor.

H. vulgaris is a rhizomatous perennial which can form large stands in favourable habitats. It has bisexual or unisexual flowers. There appears to be little information on its reproductive biology. It is able to colonize newly available habitats such as freshly dug ponds and bird scrapes and abandoned clay-, gravel- and sand-pits.

Palynological studies suggest that *H. vulgaris* may have been present in our area throughout the Pleistocene period (Godwin, 1975). Its fruits are abundant in deposits of Early, Middle and Late Weichselian age, suggesting that it was a frequent plant in periglacial vegetation. Currently, there do not appear to be any marked trends in its distribution. It is reported to be declining in some areas, including Essex (Tarpey & Heath, 1990), Suffolk (Simpson, 1982) and two of the three English grazing marshes studied by Mountford (1994a), but increasing in others, such as Gloucestershire (Holland, 1986). It may be spread by deliberate introduction to some sites: Sykes (1993) comments that in the North York Moors it is 'often planted in new fish-ponds and ornamental lakes'.

H. vulgaris is widespread in arctic, boreal and temperate regions of the northern hemisphere, and also occurs as a native species in southern S. America.

The submerged-type of shoot differs greatly from the aerial-type, and intermediates do not occur. The switch from submerged to aerial shoots is triggered by high light intensities, and takes place only at temperatures above 10°C. Although the transition normally takes place as shoots emerge from the water, aerial-type shoots can also be formed under very clear water (Bodkin *et al.*, 1980).

CALLITRICHACEAE

A small family which contains a single genus, *Callitriche*.

Callitriche L.

This is a genus of at least 25 species, most of which are aquatics with submerged or submerged and floating leaves. Some species grow in damp, terrestrial habitats.

The genus is taxonomically difficult, as the species are very variable in vegetative morphology, often having both aquatic and terrestrial forms, and the flowers are very reduced. Ripe fruits are often needed for identification. The species have been studied intensively by H. D. Schotsman. Her account in *Flora Europaea* (Tutin *et al.*, 1972), and more detailed studies of the species in Bavaria, Belgium, France, the Netherlands, Portugal and Switzerland (Schotsman, 1954, 1958a, 1961a,b, 1967, 1972; Duvigneaud & Schotsman, 1977) are particularly useful to British and Irish botanists. For southern European taxa the *Flora Europaea* account is now rather outdated, and it should be read in conjunction with Schotsman's later papers (e.g. 1977, 1982).

Schotsman's early account of the Dutch species was summarized by David (1958). Unfortunately British and Irish botanists have continued to neglect the genus, which in our area still retains its 'somewhat shady reputation' (Schotsman, 1954). The few critical studies of our taxa include Savidge's (1960) outline of his work on the experimental taxonomy of the European species, and Lewis-Jones & Kay's (1977) survey of the genus in W. Glamorgan. Most field botanists do not attempt critical identifications of the plants they find, and there are no experts in our area who will identify material. As a result of this neglect the distribution of our more difficult species is still imperfectly known, and the distribution maps of these taxa remain 'patchy and discontinuous' (Lewis-Jones & Kay, 1977).

C. hermaphroditica and *C. truncata* differ from the other species in our area in having translucent submerged leaves; they both lack floating leaves and never produce modified terrestrial variants. One would not expect them to be confused with other species, but examination of the voucher specimens has revealed that many of the records of *C. hermaphroditica* reported by McCallum Webster (1978) are erroneous, and other botanists in Scotland and Ireland have also mistakenly

Callitriche hermaphroditica L.; *C. obtusangula* Le Gall

recorded this species. We have attempted to eliminate dubious records of this species from the map. Although *C. hermaphroditica* is similar to *C. truncata*, their British and Irish distributions are virtually vicarious and confusion between them is therefore unlikely. Of the other species, *C. obtusangula* is a distinct species, identifiable in the field or (in cases of doubt) by microscopic examination of the pollen grains. This leaves two pairs of similar species: *C. platycarpa* and *C. stagnalis* are very difficult to separate, as are *C. brutia* and *C. hamulata*. The degree to which these species are under-recorded is discussed in the accounts of the species.

Plants of *Callitriche* are monoecious, with male and female flowers borne on the same stem. There are three pollination syndromes in the British and Irish species (Schotsman, 1982). The pollination of the totally submerged *C. hermaphroditica* and *C. truncata* occurs under water. Male and female flowers are initially enveloped in several pairs of imbricate young leaves. The pollen germinates within the anthers after they dehisce (Philbrick & Anderson, 1992). It is released into the closed space formed by the imbricate leaves and the pollen tubes make contact with the stigma. In three species (*C. obtusangula*, *C. platycarpa*, *C. stagnalis*) the flowers develop only in the axils of the floating rosette leaves. The stamens project above the surface of the water, and the emergent stigmas are fertilized by wind-blown pollen, or perhaps by pollen which falls on to the surface of the water and floats amongst the rosette leaves. As the stem grows and new rosette leaves develop, the fruit gradually becomes submerged. The male flowers of the terrestrial forms also release their pollen into the air. Self-pollination of these species is usual, but cross-pollination may occur. Finally, *C. brutia* and *C. hamulata* have flowers which (in aquatic forms) are borne under water in the axils of the submerged linear leaves or the rosette leaves. There is usually a male flower in one axil of a pair of leaves and a female flower in the other. The single stamen of the male flower usually bends towards the two stigmas of the female. The pollen germinates in the anther and the pollen tubes grow towards the stigma. In terrestrial forms pollination is similar, but the anthers only dehisce when moistened by a drop of rainwater or dew in the leaf axils. These species have fewer pollen grains in the anther than the wind-pollinated taxa. Their mechanism of self-pollination is very effective and results in good fruit set.

Most species of *Callitriche* are pioneers which colonize disturbed substrates (Schotsman, 1972). They lack specialized means of vegetative reproduction, but detached fragments root freely at the nodes and may become established as new plants in water and (if the species grows terrestrially) on moist soil.

No *Callitriche* hybrids are known in our area. In mainland Europe sterile hybrids are known from several sites: all are probably crosses between *C. platycarpa* and *C. cophocarpa* Sendtn.

In addition to the species treated below, Stace (1991) lists *C. palustris* L. as a British plant, apparently on the basis of a single specimen in the Herbarium Delessert, Geneva, which was collected at Petersham, Surrey, in 1877 and listed by Schotsman (1954). R. D. Meikle, N. Y. Sandwith and J. P. Savidge examined the only specimen from Petersham in this herbarium, collected by C. B. Clarke on 2 September 1877, and were in no doubt that it was a terrestrial form of *C. stagnalis* (Meikle & Sandwith, 1956). Unless further evidence comes to light, it seems best to disregard this record.

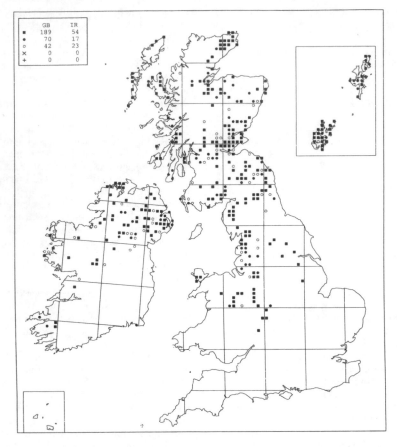

	GB	IR
■	189	54
●	70	17
○	42	23
×	0	0
+	0	0

Callitriche hermaphroditica L.
Autumnal Water-starwort

This species typically grows submerged on silty substrates in mesotrophic or eutrophic lakes. It can be found in shallow water at the edge of sheltered lakes, but in more exposed sites it is confined to deeper water, and sometimes descends to depths of 5–6m. In addition to lakes, it occasionally grows in deep, eutrophic streams and also occurs as a colonist of canals and flooded gravel-pits. In mesotrophic waters characteristic associates include *Myriophyllum alterniflorum*, *Potamogeton gramineus*, *P. × nitens* and *Nitella flexilis*, and it often grows with the scarce *Najas flexilis*. In more eutrophic sites it can be found in abundance, growing with *Potamogeton crispus*, *P. pusillus* and *Zannichellia palustris*. It is a predominantly lowland species, which is found up to 350m at Kingside Loch and was formerly recorded at 380m in Malham Tarn.

C. *hermaphroditica* is a shallow-rooted, self-pollinated annual. Plants usually fruit prolifically, and plants laden with fruit are often uprooted and dispersed by autumn gales. Little is known about its germination requirements.

The presence of fossil fruits of *C. hermaphroditica* in late glacial deposits in Cornwall, Co. Wexford and Co. Louth suggests that it was widespread at that period but has since contracted to a predominantly northerly distribution. Its current distribution was under-recorded in both Britain and Ireland in the *Atlas of the British Flora* (Perring & Walters, 1962). It is therefore difficult to assess any recent changes. There is no evidence to suggest that it is declining, and the recent discovery of plants in gravel-pits in Lincolnshire suggests that it may even be extending its British range.

C. *hermaphroditica* has a circumboreal distribution. In Europe it has a predominantly northern distribution, extending south to the British Isles, northern Germany and Romania. It is absent from the Alps.

Populations of *C. hermaphroditica* differ in fruit morphology. In Britain two variants have been reported by Savidge (1958) and Schotsman (1958b); one has fruits *c*.1.5mm in diameter and is known from canals in England and Wales, whereas the other has more broadly winged fruits *c*.3mm in diameter and is more common in Scotland and Ireland, although it also occurs in Anglesey. Plants from Anglesey were described as *C. autumnalis* var. *macrocarpa* by Hegelmaier (1867). A recent study of Swedish material has confirmed the existence of two variants there, a small-fruited plant in the south and a large-fruited plant in the north (Martinsson, 1991). More detailed studies of the variation of this species in Britain and Ireland are required.

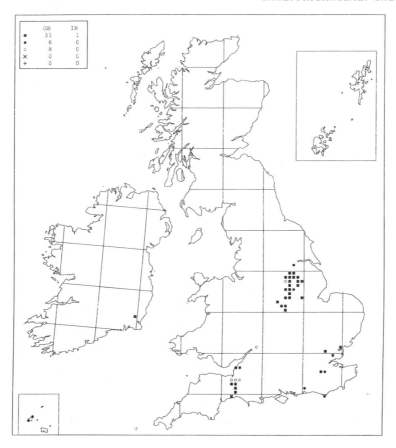

	GB	IR
■	31	1
●	6	0
○	8	0
×	0	0
+	0	0

Callitriche truncata Guss.
Short-leaved Water-starwort

C. truncata is a pioneer species which colonizes newly-created lowland habitats, or areas which have been disturbed by major management operations. It can be found in abundance in lakes, reservoirs, ponds, rivers, canals, ditches and flooded gravel-pits. It is usually found in base-rich, mesotrophic or eutrophic waters; in Ireland it is found in the Slaney River and in nearby brackish ditches. It is found in both shaded and unshaded sites, and grows over a wide range of substrates from pebbles and stones to soft organic sediments. It usually grows in shallow water, but can descend to depths of 1.5m. Plants are most vigorous in clear, still or very slowly flowing water but are occasionally found in sites where the water flows steadily but rather more rapidly, or in turbid conditions. *Elodea canadensis* is a frequent associate.

C. truncata is an annual, although plants which break up in autumn and float to the surface can overwinter as vegetative fragments. Recent observations indicate that it fruits freely at most of its British sites (Barry & Wade, 1986), these contrast with the traditional view (e.g. Pearsall, 1935; David, 1958) that the British populations rarely set seed. Dispersal of seed is by water or possibly by wildfowl: seed fed to ducks has germinated when recovered from the faeces. The seeds have no innate dormancy, but in the wild seeds shed in autumn do not germinate until spring. The species may reappear in a stretch of water after a long absence (Lousley, 1937), presumably germinating from dormant seed.

It is difficult to account for the very restricted distribution of *C. truncata*. In recent years it has apparently become extinct in Sussex (last seen in 1966), Guernsey (1968) and in Kent (1971). Some of its sites have been destroyed in Guernsey, but there is no obvious reason for its disappearance from the other localities. It has, however, been recorded in new localities in recent years: Barry & Wade (1986) had records from 12 water bodies since 1980 whereas it is now known from 28. The recent discovery of plants in a fishing lake and a reservoir in S. Essex in 1988 (Adams, 1988), and a reservoir in N. Essex in 1991, represent a significant extension to the species' range. *C. truncata* is an inconspicuous plant which may be refound at some of its old localities, or discovered elsewhere.

C. truncata has a Mediterranean-Atlantic distribution. The western European plant is subsp. *occidentalis*, which occurs in Malta, Spain, Portugal, France, Belgium, Britain and Ireland. It is replaced by subsp. *truncata* in the central and eastern Mediterranean region, and in N Africa, and by subsp. *fimbriata* Schotsman in S. Russia.

Much of the information in the above account is taken from Barry & Wade's (1986) detailed account of the ecology of this species.

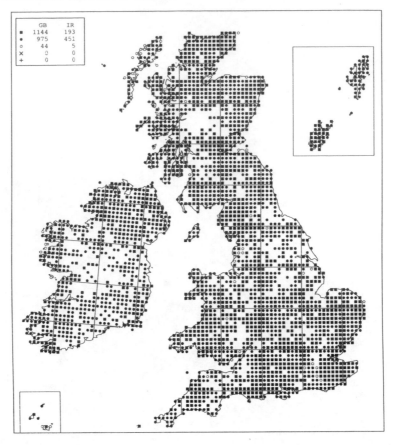

	GB	IR
■	1144	193
●	975	451
○	44	5
×	0	0
+	0	0

Callitriche stagnalis Scop.
Common Water-starwort

C. stagnalis grows as an aquatic in shallow water or as a terrestrial species on open ground. As an aquatic it is found in a wide range of oligotrophic, mesotrophic or eutrophic waters. It is most frequent on non-calcareous soils (Schotsman, 1972; Grime *et al.*, 1988), but it is absent from the most acidic and nutrient-poor localities. It usually grows over fine organic or inorganic substrates and occurs in both shaded and open situations. Typical habitats include the sheltered bays of lakes, the quiet edges or backwaters of rivers, and in pools, streams or ditches. In some western areas (e.g. Donegal) *C. stagnalis* is much more frequent as a terrestrial annual than as an aquatic. Terrestrial plants grow on seasonally flooded or moist, disturbed ground and can typically be found at the edge of lakes or reservoirs, in ditches and dune-slacks which dry up in summer, on tracks and woodland rides and occasionally in arable fields. Unlike the species with linear submerged leaves (*C. hamulata*, *C. obtus-angula*, *C. platycarpa*), it is rarely found in deep water. There are confirmed records of *C. stagnalis* up to altitudes of 610m (Great Dun Fell), and probable although unconfirmed records up to 890m (Foel Grach).

Aquatic plants of *C. stagnalis* are winter-green perennials. The submerged leaves are broadly elliptical rather than linear. Terrestrial plants are annuals with short internodes: flowers are borne at most nodes and such plants usually fruit copiously. Little information is available about seed dormancy and the conditions required for germination, but Grime *et al.* (1988) suggest that the species may have a persistent seed-bank and probably germinates in spring.

C. stagnalis is widespread in Britain and Ireland, and its distribution appears to be stable. All records attributed to *C. stagnalis* sensu lato (an aggregate comprising *C. platycarpa* and *C. stagnalis*) are shown on the map. This probably gives a reasonable picture of the distribution of *C. stagnalis* sensu stricto, but may overestimate its occurrence in south-east England where some records may be referable to *C. platycarpa* or even *C. obtusangula*.

C. stagnalis is found throughout Europe, where it is represented by a number of different chromosome races (Schotsman, 1961c, 1967). It is so frequent in the western Mediterranean that it is sometimes eaten (like *C. brutia*) as a salad (Schotsman, 1961a). It is also found in Macaronesia, N. Africa and W. Asia, and occurs as an introduction in N. America, Australasia and perhaps elsewhere.

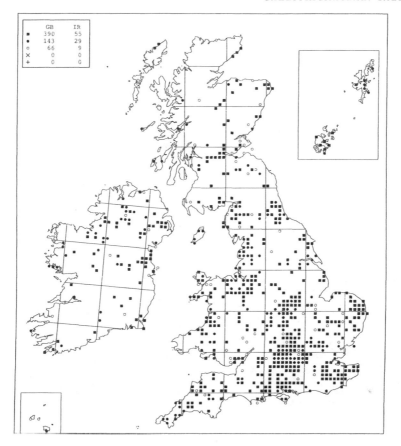

	GB	IR
■	390	55
●	143	29
○	66	9
×	0	0
+	0	0

Callitriche platycarpa Kütz.
Various-leaved Water-starwort

This species is recorded from a wide range of aquatic habitats, including slow-flowing rivers and streams, canals, ditches, lakes, ponds, turloughs, dune-slacks and flooded gravel-pits. It is also found as a terrestrial form on damp peat or mud. It has a wide ecological tolerance, growing in both open and partially shaded sites and across a range of water chemistries from acidic and oligotrophic to calcareous and eutrophic. In lowland, basic rivers it often grows with *C. obtusangula*. The ecological distinction between *C. platycarpa* and *C. stagnalis* is not clear-cut, or it has been obscured by the lack of systematic recording of these species. However, *C. platycarpa* appears to favour deeper or more rapidly flowing water. It is a predominantly lowland species, although it ascends to 520m at Llynoedd Ieuan.

Like *C. stagnalis*, aquatic forms of *C. platycarpa* are perennial. They overwinter as plants with submerged, linear leaves, and plants growing in deep water may persist in this state throughout the summer. However, floating rosettes develop if the stems reach the surface. Although plants of *C. platycarpa* with floating leaves flower freely, the pollen is variable and characteristically contains a proportion of sterile grains. There is some evidence that this may be the result of irregular meiosis (Schotsman & Haldimann, 1981). Plants of *C. platycarpa* set fruit much less freely than those of *C. stagnalis*, and certain populations appear to be virtually sterile (David,

1958; Savidge, 1967). Terrestrial forms are annual and are also wind-pollinated (Schotsman, 1982); they may fruit somewhat more freely.

This species is grossly under-recorded. Even those botanists who attempt to identify *Callitriche* species are frequently unable to find the fruiting plants which they need to confirm the identification of *C. platycarpa*. Lewis-Jones & Kay (1977) concluded that the two species could not be separated unequivocally even on the basis of fruit morphology, and found that a chromosome count was the most rapid and reliable way to distinguish them. There is little or no evidence about trends in the distribution or abundance of this species. However, its tolerance of eutrophic water suggests that it is unlikely to be decreasing in south-east England.

C. platycarpa is widespread in temperate parts of Europe, and it is also recorded from a few localities in the Mediterranean region. Its distribution extends into S.W. Asia. It is particularly frequent in western Europe; towards the east it tends to be replaced by *C. cophocarpa* Sendtn. (Schotsman, 1967; Cook, 1983). In Belgium, as in Britain and Ireland, it has a very wide habitat range (Duvigneaud & Schotsman, 1977).

Savidge (1960) suggests that *C. platycarpa* (2n=20) is an allopolyploid derived from a hybrid between *C. stagnalis* (2n=10) and *C. cophocarpa* (2n=10), a European species which is not known from Britain or Ireland. Schotsman (1967) regards his evidence as far from conclusive. Further cytological work is required to establish the origin of the species and investigate its apparent partial sterility.

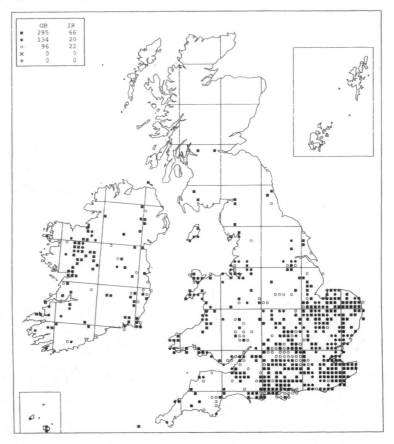

	GB	IR
■	295	66
●	134	20
○	96	22
×	0	0
+	0	0

Callitriche obtusangula Le Gall
Blunt-fruited Water-starwort

C. obtusangula usually grows in still or slowly flowing water in lowland rivers, streams, canals, ditches, ponds and turloughs. It is found in both mesotrophic and eutrophic water, and in open habitats and partial shade. It is most frequent in calcareous waters over both peat and inorganic substrates. In calcareous ditches it can be present in large masses in the summer months, often associated with *Elodea canadensis, Hippuris vulgaris, Lemna minor, L. trisulca, Myriophyllum verticillatum, Potamogeton crispus, P. natans, Ranunculus circinatus* and *Zannichellia palustris*. In chalk streams and rivers it avoids the central channel where the water velocity is greatest, but grows on unstable silt in quieter reaches (Holmes, 1983). *C. obtusangula* is also found in brackish water, typically in ditches in coastal grazing marshes where it may be the commonest member of the genus. Terrestrial forms can be found on damp mud at the water's edge. It is a strictly lowland species.

Aquatic forms of *C. obtusangula* are perennial, with a life-form similar to that of *C. platycarpa*: plants over-winter as shoots with linear leaves and in spring and summer the stems which reach the surface of the water develop floating rosettes. *C. obtusangula* may be more sensitive to winter cold than either *C. platycarpa* or *C. stagnalis* (Schotsman, 1954). Plants in flowing water may fail to fruit, but those in still water or on land fruit freely (David, 1958). Seeds of *C. obtusangula* from coastal populations will germinate in water with a greater salt content than will those of other *Callitriche* species, or those of inland populations of *C. obtusangula* (Schotsman, 1954).

Although *C. obtusangula* is a distinctive species, it is almost certainly under-recorded. There is no evidence to suggest that the species has undergone any major changes in its distribution or abundance, although assessing such changes is hampered by the lack of reliable historical records and the patchiness of the modern survey data.

C. obtusangula is a species of the Atlantic and Mediterranean regions of Europe, found from the British Isles, Spain and Portugal eastwards to Greece; there are also outlying localities in C. Europe (Schotsman, 1961a; Cook, 1983). It is also known from N. Africa. Although all populations studied have the same chromosome number (2n=10), the species consists of a number of cytological races with different caryo-types (Schotsman, 1961c, 1977).

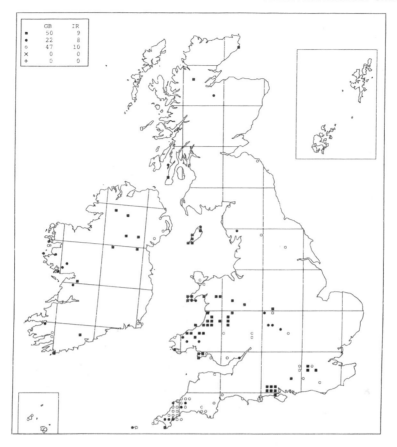

Callitriche brutia Petagna
Pedunculate Water-starwort

This species is usually found growing terrestrially on the bed of dried-up ponds or on exposed mud above the water-level; if conditions are suitable it can be present in abundance. It has also been recorded from damp hollows in sand-dunes, sand- and gravel-pits and woodland rides, and from ditches, the edge of reservoirs, dried-up river meanders, rutted tracks and the moat of Caerphilly Castle. It also grows in shallow water. It is a calcifuge, which is restricted to base-poor and rather infertile sites. Associated species include *Apium inundatum*, *Baldellia ranunculoides*, *Persicaria minor* and *Ranunculus tricho-phyllus*. Most records are from the lowlands, but there are confirmed records up to 425m (Llyn Egnant).

C. brutia is a self-pollinated annual. Fruiting plants have been recorded from mid April to early September. The ripening fruits sometimes become buried in the substrate as their stalks lengthen (Schotsman, 1961a). Like many self-pollinated annuals, individual populations may be characterized by minor morphological traits, which are maintained in cultivation (Schotsman, 1977).

Although variants of *Callitriche hamulata* with pedunculate fruits have been known to British and Irish botanists for well over a century, they have until recently been treated as an environmental modification (David, 1958), a variety (Pearsall, 1935) or a subspecies (Clapham *et al.*, 1952) of that species, or have been ignored completely (Savidge, 1967). It is only in recent years that they have been recognized at specific rank, as *C. brutia* (Clapham *et al.*, 1987; Stace, 1991), and recorded systematically. The map almost certainly seriously underestimates its distribution.

C. brutia is known from southern and western Europe, and reaches its northern limit in southern Scandinavia; it is frequent in the western Mediterranean region where it is eaten as a salad (Schotsman, 1961a). It is also known from N. Africa and extends eastwards in Asia to Iran. It grows in brackish water in Norway, and Naustdal (1974) suggests that it might have been transported to these sites on the sinking-stones, anchors, grapnels, ropes and hauling-lines of Vikings returning from expeditions to Britain!

C. brutia (2n=28) differs from *C. hamulata* (2n=38) in chromosome number. Terrestrial plants of these species are not difficult to separate, as the fruits of *C. brutia* are pedunculate. However, Schotsman (1967) comments, rather disconcertingly, that terrestrial forms of *C. hamulata* may sometimes be found with pedunculate fruits, and that these are nearly impossible to determine. Aquatic forms of both species can have more or less sessile fruits (Schotsman, 1961a), and can be identified only by using much more subtle characters (Schotsman, 1967). Few, if any, British or Irish botanists would be prepared to name aquatic material, and this might not only provide a further reason why the species is under-recorded but also bias the map towards areas where it grows terrestrially. Schotsman (1967) regards *C. brutia* as a plant of shallow water; populations of the aggregate in deeper water are likely to be referable to *C. hamulata*.

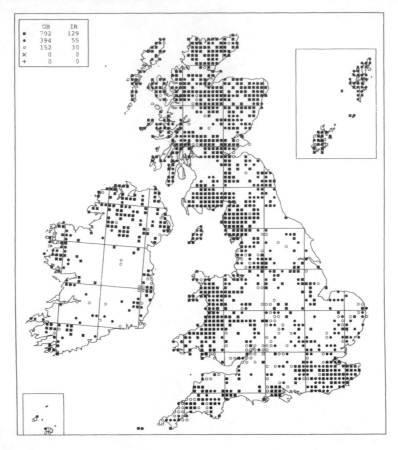

	GB	IR
■	702	129
●	394	55
○	152	30
×	0	0
+	0	0

Callitriche hamulata Kütz. ex W. D. J. Koch
Intermediate Water-starwort

This species is found in a wide range of acidic waters, including lakes, reservoirs, pools, rivers, streams, canals, ditches, and flooded clay-, sand- and gravel-pits. It is found in both shaded and unshaded sites, and from shallow water to depths of over 3m in clear, oligotrophic water. It also grows in upland flushes and, as a terrestrial form, on damp mud. It is the most frequent species of *Callitriche* in the acidic areas of the north and west, where it not only occurs in lakes and pools but also in rocky streams and rivers where few other macrophytes except *Myriophyllum alterniflorum* and the aquatic form of *Juncus bulbosus* will grow. Although it normally behaves as a calcifuge, it sometimes appears to benefit from nutrient-enrichment of acidic waters (e.g. by nesting gulls) and it can occasionally be found in base-rich localities (Seddon, 1972). It is frequent in both lowland and upland areas, ascending to 825m (Carnedd Llewelyn); a plant reported as *C. hermaphroditica* from 915m (Sgurr na Lapaich) by McCallum Webster (1978) is probably referable to this species.

Aquatic plants of *C. hamulata* are perennial. They have a similar life-form to *C. obtusangula* and *C. platycarpa*, with linear lower leaves and a floating rosette of leaves on the surface of the water. Plants in deep or rapidly flowing water may persist throughout the year in the linear-leaved state. The species is self-pollinated and fruits freely. Little is known of its germination requirements.

The map shows all records attributed to *C. hamulata* sensu lato, an aggregate which includes *C. brutia*. Like all the species in this genus, *C. hamulata* is under-recorded in some areas but the map accurately reflects the abundance of the species in areas where acidic waters predominate, and its scarcity elsewhere. Many calcifuge species have declined in south-east England, but few local floras suggest that this is true of *C. hamulata*, perhaps because authors do not have sufficient reliable data to assess historical trends in the distribution of the species.

C. hamulata is a species of W., N. & C. Europe, extending from Iceland and Scandinavia south to the Pyrenees, N. Italy and Romania. It is also recorded from Greenland and plants referable to *C. hamulata* sensu lato have been introduced to Australia and New Zealand (cf. Orchard, 1980).

PLANTAGINACEAE

This is a small family of only three genera, one of which is aquatic.

Littorella P. J. Bergius

Littorella is a genus of three species, all of which are aquatic. Only one species is known in Europe, *L. uniflora*. The closely related *L. americana* Fernald is found in eastern N. America, and is sometimes regarded as a variety of *L. uniflora*. The third species, *L. australis* Griseb., is found in southern S. America. The distribution of all three taxa is mapped by Dietrich (1971).

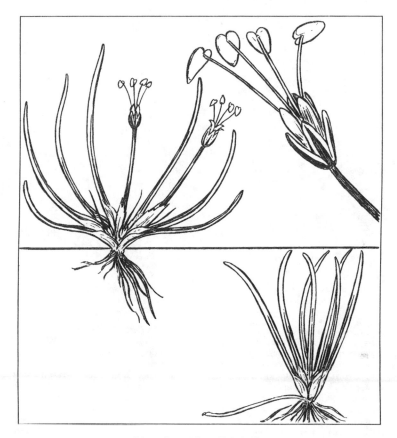

Littorella uniflora (L.) Asch.

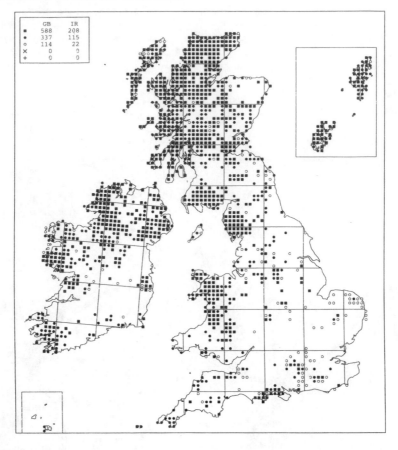

		GB	IR
■		588	208
●		337	115
○		114	22
×		0	0
+		0	0

Littorella uniflora (L.) Asch.   Shoreweed

L. uniflora is a plant of isoetid growth habit, found in shallow, oligotrophic or mesotrophic waters. It grows in lakes, reservoirs, rivers, streams, ponds and dune-slacks, occurring over a wide range of substrates including stones, gravel, sand, peat, marl and soft mud. Although it usually grows in oligotrophic or mesotrophic waters, it can be abundant in lakes such as Esthwaite Water which have become eutrophic. It tolerates emersion and can form a dense band in the draw-down zone around lakes and reservoirs. Some of its sites, such as winter-flooded depressions on heathland and sea-cliffs, dry out completely in summer. Sheep will graze on exposed plants (West, 1910). In deeper water *L. uniflora* grows in open *Eleocharis palustris* and *Equisetum fluviatile* swamps on exposed shores, and below these zones in a carpet on the bottom of lakes, where it is often associated with *Isoetes lacustris* and *Lobelia dortmanna*. It descends to a depth of 3–4m (West, 1905, 1910). In Scotland and Ireland it is most frequent in areas of base-poor rocks and waters, but it can grow in base-rich habitats such as the clear limestone lakes of Lismore, streams and turloughs in the Burren and even in pools on limestone pavement on Inishmore (Webb & Scannell, 1983). In S.E. England, however, it is found only in acidic habitats. It is recorded from sea-level to 825m at Ffynnon Llyffant.

Littorella uniflora usually flowers when emersed; the male flowers are stalked and the pollen is dispersed by wind whereas the female flowers are subsessile. It will also flower in water up to 50cm deep, but it fails to set seed under water. The germination ecology has recently been studied experimentally by Arts & Heijden (1990). Seeds are produced in small quantity, each plant producing a maximum of *c*.20 per annum. They require light in order to germinate. Desiccation for two or four weeks stimulates germination, not only increasing the final number of seeds which germinate once they are rewetted but also causing a rapid and nearly synchronous germination. *Littorella* has a large seed-bank, and the seeds remain viable for decades. In addition to sexual reproduction, *Littorella* also has very effective vegetative propagation which allows colonies to persist even if permanently submerged. Established plants send out stolons bearing new rosettes, and this lateral spread can lead to the development of a dense carpet of plants. In conditions of heavy silt deposition the stolons grow upwards, and by this means plants have survived deposition rates of 2cm of sand per month under experimental conditions (Spence, 1982).

In historical times *L. uniflora* has always been scarce in S.E. England. The map shows that it has decreased, both as a result of the drainage of the damp, acidic habitats it favours and because some of its remaining sites have become overgrown by rank vegetation. In the north and west it is still a frequent plant, and its tolerance of mesotrophic water has enabled it to survive in areas where other isoetids have declined. Experimental studies show that it is able to exploit increases in nutrients to greater effect than *Lobelia dortmanna*, but much less effectively than *Mentha aquatica* (Farmer & Spence, 1986). At

Esthwaite Water, Robe & Griffiths (1992) concluded that *L. uniflora* can adjust well to eutrophic conditions, despite the fact that plants become shaded in summer and autumn by algal blooms and a layer of epiphytes on their leaves. In the Netherlands *Littorella uniflora* has decreased markedly since 1950, being replaced in many sites by the aquatic form of *Juncus bulbosus* and *Sphagnum* spp. These changes are attributed to acidification following increased inputs of ammonium ions (Roelofs, 1983; Roelofs & Schuurkes, 1983; Roelofs *et al.*, 1984).

L. uniflora is widespread in W., C. and N. Europe, becoming rare in the south but extending to Spain, Portugal and Sardinia. It is also recorded from N. Africa (Cook, 1983).

L. uniflora differs from the other isoetids in Britain and Ireland in its wider habitat range; it is the only species which grows on calcareous substrates. It is less exclusively aquatic than the other species, often being found above the water-level and setting seed only when it grows terrestrially.

SCROPHULARIACEAE

A large family of 222 genera and some 4,500
species, although Mabberley (1987) comments
that despite its size, the family is 'of little consequence to mankind'. Most of the species are herbs
but there are also some trees and shrubs. Some 18
genera and perhaps 200 species are aquatics,
although it is even more difficult than usual in this
family to draw the line between aquatic and terrestrial species (Cook, 1990a). The only genus
with aquatic species in Britain and Ireland is
Veronica, although other genera occur in waterside
habitats (*Mimulus*, *Scrophularia*) or on damp mud
in seasonally flooded sites (*Limosella*).

Veronica L.

Most of the 250 species of *Veronica* are found in
the temperate regions of the northern hemisphere,
although some extend southwards to mountains in
the tropics or to the southern hemisphere. Closely
allied shrubby plants in the southern hemisphere
are placed in the genus *Hebe*. All the aquatic

species of *Veronica* are in section *Beccabunga* (Hill)
Dumort., which contains twelve species, although
V. scutellata of section *Veronica* can be found in
fens.

Two subsections of *Veronica* section *Beccabunga*
are recognized, both of which are represented in
Britain and Ireland. Subsection *Beccabunga*
includes *V. beccabunga* and the non-European *V.
americana* (Raf.) Schwein. ex Benth. and poses no
difficult taxonomic problems. Subsection
Anagallides J. Keller comprises the species closely
related to *V. anagallis-aquatica*, which at a world
scale 'is an extremely confusing complex of several
species and subspecies intimately connected to
each other, with abundant hybridisation' (Öztürk
& Fischer, 1982). Even in our area the two species
in this subsection, *V. anagallis-aquatica* and *V.
catenata*, are very similar morphologically and
hybridize to give *V. × lackschewitzii*. The experimental taxonomy of these taxa has been studied by
N. G. Marchant (Marchant, 1970; Walters, 1975).

In the British Isles *V. beccabunga* is diploid
(2n=18) whereas *V. anagallis-aquatica* and *V. catenata* are both tetraploid (2n=36).

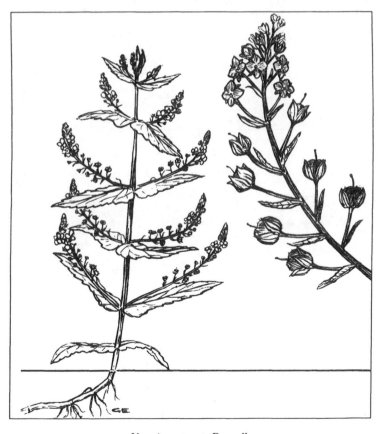

Veronica catenata Pennell

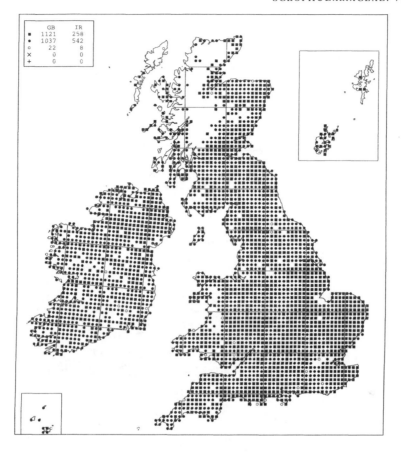

	GB	IR
■	1121	258
●	1037	542
○	22	8
×	0	0
+	0	0

Veronica beccabunga L.   Brooklime

This low-growing species is frequent in very shallow water or on damp ground at the edges of streams and rivers. It is confined to sites where the growth of more vigorous plants is restricted by erosion during floods or by other disturbances. It can usually be found, for example, on muddy ground at those places along a riverbank where cattle come to drink, but is often absent from the intervening stretches where emergents form a continuous sward. In shallow chalk streams it may grow across the entire channel (Haslam, 1978). Other habitats include shallow ditches, marshy hollows in pastures, damp woodland rides, open flushes and on disturbed areas by lakes, ponds and canals or in fens. It is found over a range of substrates from fine silt to stones and rocks, and in both calcareous and non-calcareous sites, but it is absent from very infertile substrates such as acidic peat. It often grows with *Apium nodiflorum*, *Glyceria fluitans*, *Ranunculus flammula*, *R. sceleratus* and *Rorippa nasturtium-aquaticum* sensu lato. It is frequent in the lowlands, and ascends to 730m (Knock Fell); Wilson (1956) cites an unlocalized record from 850m in the Scottish Highlands.

V. beccabunga is a winter-green perennial which flowers freely. The flowers are protogynous and are often cross-pollinated by insects; in unfavourable weather they remain half-closed and are automatically self-pollinated

(Knuth, 1909). Plants set many seeds which germinate readily; there is also a persistent seed bank in the soil (Grime *et al.*, 1988). The seeds are not buoyant, and therefore rarely occur in floating plant debris. However, they produce a mucilaginous coat when exposed to water and might therefore be dispersed by animals (Guppy, 1906). Fragments of stems root readily, and the plant frequently spreads vegetatively along streams (Ridley, 1930; Grime *et al.*, 1988). The species is shallow rooted and in floods entire plants may be swept away (Haslam, 1978).

This species is still frequent throughout much of Britain and Ireland, and there is no evidence for any decline at the scale of the map. Few authors of county floras regard it as a decreasing species, except in the most heavily built-up areas (Kent, 1975; Graham, 1988).

V. beccabunga is widespread in temperate regions of Eurasia, and also occurs in N. Africa. It is naturalized in N. America, where it was probably introduced initially with ships' ballast. Its spread in N. America has been less rapid than that of some other alien aquatics, perhaps because many suitable habitats there are already occupied by the closely allied *V. americana* (Raf.) Schwein. ex Benth. (Les & Stuckey, 1985).

The unusual epithet *beccabunga* is derived from the Low German 'beckbunge': 'beck-' is cognate with our word beck and means stream but the meaning of '-bunge' is disputed (Gilbert-Carter, 1964). *V. beccabunga* was formerly eaten as a salad plant in N. Europe.

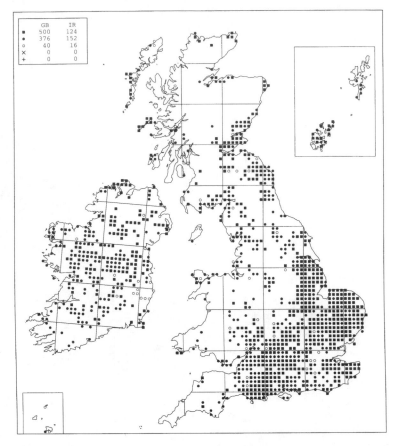

	GB	IR
■	500	124
●	376	152
○	40	16
×	0	0
+	0	0

Veronica anagallis-aquatica L.
Blue Water-speedwell

This species is found as submerged stands in shallow water, and as an emergent or terrestrial species in swamps and on disturbed ground at the water's edge. It grows by a wide range of water bodies, including rivers, streams, ditches, lakes, ponds and flooded clay- and gravel-pits. It is found on fertile substrates, whether acidic or calcareous. However, it is replaced by *V. catenata* in some habitats, such as coastal grazing marshes (Mountford, 1994b). It is primarily a lowland plant, which reaches 380m at Malham Tarn (Sinker, 1960).

Young plants of *V. anagallis-aquatica* soon develop stolon-like lateral branches from their base, which send out numerous adventitious roots (Marchant, 1970). The vegetative vigour and reproductive performance of the species varies greatly with habitat conditions. In flowing water it may persist as submerged vegetative stands. Plants on fertile silt by streams and rivers are much larger than those which grow in disturbed and rather dry terrestrial habitats. Individual plants are self-compatible. Although their inflorescences are usually aerial, plants which are made to flower under water in experimental conditions produce flowers which do not open but which nevertheless set seed (Combes, 1965). The fruits of *V. anagallis-aquatica*, like those of *V. catenata* and

their hybrid, are often galled by the weevil *Gymnetron villosulum* Gyllenhal. Submerged stands fragment when streams are in spate (Haslam, 1978), and the stems can root and become established as new plants.

The two related species *V. anagallis-aquatica* and *V. catenata* were not separated by British botanists until the differences were set out by Druce (1911, 1912); a more detailed account was provided by Britton (1928) and the correct nomenclature established by Burnett (1950). Historical records cannot be attributed to one or other species unless they are supported by herbarium specimens. Both species may still be somewhat under-recorded in some areas as they can be identified only in flower or fruit; in other areas where the hybrid *V.* × *lackschewitzii* is frequent, *V. anagallis-aquatica* may be over-recorded (Crackles, 1990). There is no evidence to suggest that the distribution or abundance of *V. anagallis-aquatica* sensu stricto has changed markedly since it was first recognized in our area.

V. anagallis-aquatica is a native of the temperate and Mediterranean regions of Eurasia and N. Africa, which extends north in Europe to C. Scandinavia. However, it has been introduced to C. & S. Africa, N. & S. America and New Zealand, and so now has an almost cosmopolitan distribution. The species is taxonomically complex, with both diploid and tetraploid subspecies (Öztürk & Fischer, 1982).

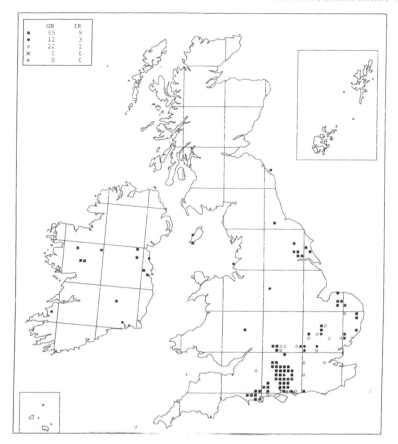

Veronica × *lackschewitzii* J. Keller (*V. anagallis-aquatica* L. × *V. catenata* Pennell)

This hybrid is found in a similar range of habitats to *V. anagallis-aquatica*. Some populations are completely submerged in flowing water over 50cm deep; these rarely if ever flower in the wild. The most conspicuous plants grow in shallow water in lakeside swamps and ditches or at the edge of streams and rivers. Some riverside plants accumulate mounds of silt around their roots and therefore rise above the water-level. In terrestrial habitats *V.* × *lackschewitzii* can be found on muddy cattle-trampled ground at the edge of streams and rivers, on crumbling riverbanks, on moist mud at the edge of reservoirs and on damp sand or gravel in disused mineral workings. The hybrid can often be found with both parents, but it can also occur in the absence of one or both parents. In Hampshire it is more frequent than either parent by large rivers and their tributaries, replacing both parents over large areas (Brewis *et al.*, 1996). The hybrid is confined to the lowlands.

In favourable habitats *V.* × *lackschewitzii* can be a very robust perennial: clumps with 70 flowering stems 1.2m high have been recorded. F1 hybrids are normally highly sterile, and continue flowering long after the parents have set fruit, the racemes becoming longer and longer. Such hybrids are thought to be at a selective advantage in flow-ing water because of their vegetative vigour (Walters, 1975). However, F2 and later generations can be more fertile. Many populations of the hybrid consist of highly sterile plants, and may represent a single clone, but some wild populations with a range of fertility are known (Walters, 1975). The stems root readily at the nodes, even high up on the inflorescence (Wells, 1968) and the hybrid reproduces and spreads vegetatively after fragmentation in floods or as a result of other disturbance. Large populations can develop in rivers which are regularly cleared out. The hybrid can be very persistent: it was still present in 1991, for example, by the R. Piddle at Chamberlayne's Mill, a site where it was first collected in 1921.

The first British record of this hybrid was made by Williams (1929), soon after Britton (1928) clarified the distinction between the parental species. Earlier specimens have since been discovered in herbaria. The highly floriferous but sterile F1 hybrids are distinctive. Nevertheless, many botanists are still unaware of the plant and it is almost certainly under-recorded, especially in Ireland. Submerged plants which persist in the vegetative state can be identified only by bringing them into cultivation.

This hybrid is widespread in mainland Europe. It is also recorded from N. America, where *V. catenata* is native but *V. anagallis-aquatica* is a naturalized alien (Brooks, 1976; Heckard & Rubtzoff, 1977).

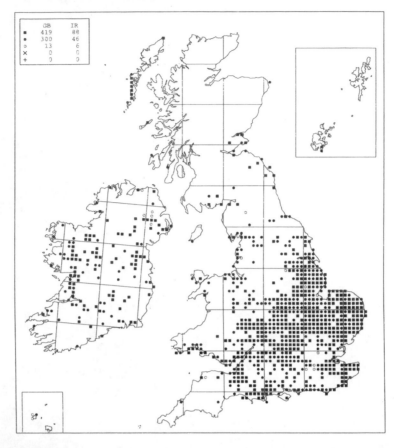

Veronica catenata Pennell

Pink Water-speedwell

This species grows in water up to 1m deep in streams and ditches, in flushes, on mud at the edge of lakes, reservoirs, ponds, rivers and streams and on moist ground in sand-dunes or disused sand-, gravel-, clay- and chalk-pits. It has also been recorded from riverside rocks. Although it can be found on fertile, acidic soils, it tends to be much more frequent in calcareous regions. It also occurs in coastal areas which are occasionally flooded by brackish water. Although it often grows with *V. anagallis-aquatica*, it has a more restricted habitat range than that species, occurring more frequently on exposed mud at the edge of standing waters. Characteristic associates in such habitats include *Alisma plantago-aquatica*, *Lycopus europaeus*, *Ranunculus sceleratus*, *Rorippa nasturtium-aquaticum* sensu lato and *Rumex maritimus* (Sinker *et al.*, 1985). *V. catenata* is confined to the lowlands.

Unlike *V. anagallis-aquatica*, young plants of *V. caten-ata* develop basal branches which grow into flowering stems rather than stolon-like stems which spread vegetatively (Marchant, 1970). The species is a perennial, but tiny plants in disturbed terrestrial habitats may be annuals. There is little detailed information on its reproductive biology, and the extent to which it spreads vegetatively.

This species was not recognized in Britain and Ireland until 1911, having previously been confused with *V. anagallis-aquatica*. Even after Druce (1911, 1912) had outlined the differences, the aggregate continued to be recorded for some time. Botanists in Somerset, for example, began to distinguish the segregates in the 1930s (Roe, 1981) and in Gloucestershire the distribution of the two taxa has not yet been worked out (Holland, 1986). *V. catenata* may therefore be somewhat underrecorded.

V. catenata is widespread in the temperate regions of the northern hemisphere, and in Europe extends north to southern Scandinavia. It occurs as an introduction in C. Africa and Australia.

LENTIBULARIACEAE

This small family contains three genera and 275 species of carnivorous plants. Members of the genus *Pinguicula* are rooted herbs which trap animals on their viscid leaves. All the species are terrestrial. The other two genera, *Genlisea* and *Utricularia*, are rootless, although the plants may be anchored by buried stolons or rhizoids. They possess specialized animal–catching traps. *Genlisia* is a small tropical genus of mostly terrestrial species. *Utricularia* is a much larger genus with a cosmopolitan distribution, although most species are found in the tropics.

Utricularia L.

There are 214 species of *Utricularia*, most of which are terrestrial; three of them are sometimes segregated in the genus *Polypompholyx* Lehm. They have recently been the subject of a superb monograph (Taylor, 1989). The morphology of this genus is unlike that of other flowering plants. Stems are absent or very reduced, and are replaced by stolons. There is much variation within the genus (Sculthorpe, 1967). The largest

species is *U. humboldtii* R. H. Schomb., which grows in pools of water in the leaf axils of South American bromeliads; the most spectacular are perhaps the species in section *Orchidioides* A. DC., which have large white, lilac or orange-red flowers and grow on mossy banks or tree branches in the montane rain forests of Central and South America. At the other extreme are plants such as *U. neottioides* A. St.-Hil. & Girard which form low mats on rocks in swiftly flowing water, to which they are affixed by numerous rhizoids. Excluding these rheophytes, 35 species of *Utricularia* are aquatics. The aquatic species may be free-floating plants with green stolons or may be anchored by additional colourless stolons buried in the substrate. The eight native European species, all of which are aquatics, are closely related members of section *Utricularia*.

Utricularia is a taxonomically troublesome genus on a world scale, partly because it is difficult to collect adequate specimens. Even in Europe there are still unresolved taxonomic problems: Taylor (1989) refers to the 'extraordinary amount of often contradictory written matter' devoted to these few species. One of the major problems in our area is that floral characters are

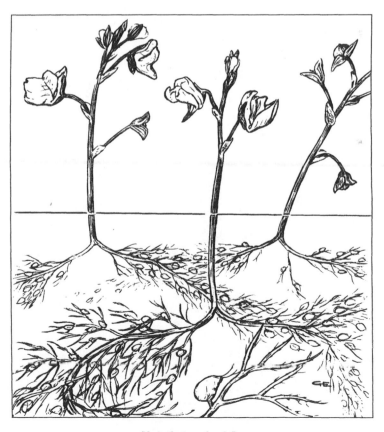

Utricularia vulgaris L.

important for identification, but many populations persist and spread vegetatively, and rarely flower. All the European species reproduce by turions, which are found only in the aquatic species of section *Utricularia* which grow outside the tropics. The turions, like those of *Myriophyllum verticill-atum*, are made up of morphologically reduced, concave, overlapping leaves and they are borne at the apices of the stems and the main lateral branches. They are usually formed in autumn, although they may be produced earlier on plants which are droughted or starved of nutrients (Sculthorpe, 1967).

Flowering plants of *U. australis* and *U. vulgaris* are easily recognized, but vegetative distinctions between the two species are slight and perhaps too variable to give an absolutely certain identi-fication. A map of the aggregate is provided here, followed by maps of the two segregates.

There are more difficult taxonomic problems concerning *U. intermedia* sensu lato. Taylor (1989) suggests that most British material is *U. ochro-leuca*, and that the evidence for the presence of *U. intermedia* sensu stricto is not conclusive. Stace (1991), basing his taxonomy on work by Thor (1988) which was not published in time for Taylor to evaluate, recognized three species in our area, *U. intermedia*, *U. ochroleuca* and the newly

described *U. stygia*. Only the aggregate is mapped, as data on the distribution of the segre-gates are very incomplete.

The final species, *U. minor*, is the only one which is relatively free of taxonomic complica-tions. We may, however, be overlooking *U. bremii* Heer ex Koell., a C. European species which closely resembles *U. minor*. An old herbarium specimen from Morayshire 'could well be this species' (Taylor, 1989).

The 'bladders' of *Utricularia* were initially thought to be flotation devices, or reservoirs of air. It was not until the middle of the 19th century that they were recognized as animal–catching traps. These reports attracted the attention of Darwin (1875), who made the first thorough study of their morphology. He accurately described the traps, and understood the way in which the bris-tles around the entrance provide a funnel leading to the trapdoor. However, he thought that small invertebrates pushed their way past the trapdoor, which then prevented their escape. It was not until 1911 that the true mechanism was dis-covered. There is a negative pressure inside the bladders. The trapdoor opens when animals touch the bristles at the lower margin of the trap-door, and the victim is sucked inside (Lloyd, 1942; Meyers, 1982).

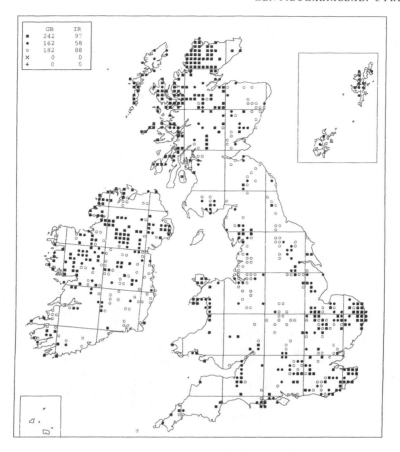

	GB	IR
■	242	97
●	162	58
○	182	88
×	0	0
+	0	0

Utricularia vulgaris L. sensu lato

The floating stolons of *U. vulgaris* sensu lato grow in nutrient-poor, still or very slowly flowing water. They can be found both in acidic and in basic, highly calcareous water over a range of inorganic or peaty substrates, but do not grow in eutrophic conditions. Characteristic habitats include the sheltered bays of lakes, pools and marl-pits, swamps, ditches in bogs, fens and grazing marshes, disused peat-diggings and flooded clay-pits. Plants in this aggregate are shade-tolerant, and can be found in stands of lakeside emergents such as *Carex rostrata* and *Phragmites australis*. Although they are often found in shallow water, they can grow at greater depths in lakes with clear water, and descend to 4m in the limestone lakes of Lismore (West, 1905). Although predominantly lowland, the aggregate is recorded at 500m in Llyn Anafon; there is an unlocalized record from 640m in Perthshire (White, 1898).

For details of the flowering and fruiting behaviour of the species in this aggregate, see under *U. australis* and *U. vulgaris*. Effective reproduction of both species is by turions, which develop on the apex of the shoots from July onwards.

As the map shows, these species have declined in lowland Britain. The marked decline in S. Lancashire was caused by the drainage of the extensive 'mosses' in the early 19th century, and many smaller sites have been lost elsewhere as lowland peatlands have been drained and taken into agricultural use. Other sites have been adversely affected by eutrophication, and *U. vulgaris* sensu lato is now a local plant in most parts of Britain. It remains plentiful only in the north and west, where acid soils predominate, and in a few areas where calcareous waters are still unpolluted. The map suggests that the plant has undergone a similar decline in Ireland, but this needs to be confirmed by further fieldwork.

U. vulgaris sensu lato is widespread in Europe, extending north to scattered localities in northern Scandinavia. The aggregate has a wide distribution outside Europe, occurring in temperate and tropical Asia, throughout Africa and in Australia and New Zealand.

For individual treatments of the two species in this aggregate, *U. australis* and *U. vulgaris*, see the following accounts.

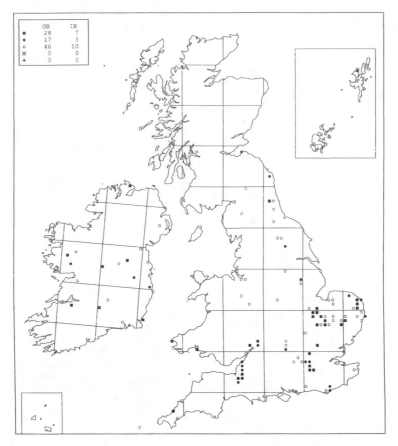

Utricularia vulgaris L.   Greater Bladderwort

This is a species of base-rich waters. Typical habitats include shallow, sheltered bays in limestone lakes, pools and ditches in calcareous fens, ditches in grazing marshes, fenland lodes and flooded clay- and marl-pits. It has also been recorded as a colonist of flooded gravel-pits. It can be found in oligotrophic and mesotrophic waters: it increased in abundance in Broadland as the input of nutrients increased in the first half of this century, only to decline again as the levels of nutrients continued to increase (George, 1992). *U. vulgaris* is apparently confined to the lowlands.

It is not unusual to find flowering plants of *U. vulgaris* in the southern part of its British and Irish range, and at some sites flowers can be found every year. There is considerable variation in the number of flowers from year to year, and flowering may be particularly prolific in hot summers. Gurney (1922) suggests that the temperature when flower buds are initiated in spring may be as important as the temperature at the time of flowering, but the precise factors which govern flowering have not been studied critically. In northern England and Scotland plants flower much less regularly. The pollen of *U. vulgaris* develops normally, and all plants studied by Caspar & Manitz (1975) were euploid (2n=22). Unlike many aquatics, the fruits mature in air. Fruiting plants have been found north to Lincoln, although the frequency with which seed is set is unknown. The usual form of reproduction is by turions, which sink to the bottom of the water in autumn. They may survive periods of freezing (Sculthorpe, 1967), and in East Anglia they begin to grow in late March or early April (Clarke & Gurney, 1921; Friday, 1989).

The map is based on records of flowering plants. Both the historical and recent range of the species must be seriously underestimated. This under-recording makes it almost impossible to demonstrate trends in the distribution of the species, but there is little doubt that its range has retracted in southern and eastern England because of habitat destruction and eutrophication.

U. vulgaris is widespread in Europe, although it is rare both in the Mediterranean region and towards its northern limit in northern Scandinavia. It is also known from N. Africa and temperate Asia east to Siberia and Tibet.

Detailed studies of a population of *U. vulgaris* at Wicken Fen have shown that the traps account for about half the biomass of a *Utricularia* plant. An individual plant has a mixture of large traps on the mid-line of each leaf lobe, and smaller traps on the finer segments. The youngest traps are the most effective: after six days their ability to catch prey declines rapidly, and most traps are lost from the leaves within 32 days. New leaves bearing young traps develop continually. In a single season an average plant may produce a total of 15,000 traps, and catch 230,000 crustaceans, midge larvae and oligochaete worms (Friday, 1988, 1989, 1991, 1992).

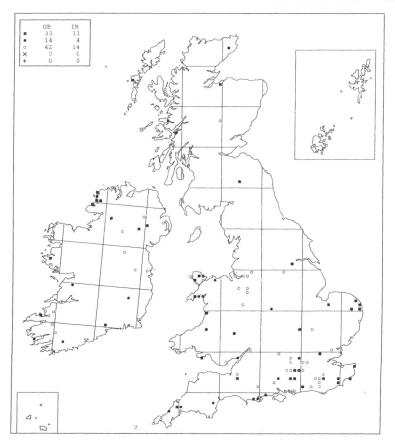

Utricularia australis R. Br.   Bladderwort

It is difficult to obtain precise information about the ecological requirements of *U. australis*, and the extent to which they differ from those of *U. vulgaris,* because the confirmed records of *U. australis* are so few, and are not necessarily representative of the majority of its populations. It is generally agreed that *U. australis* differs from *U. vulgaris* in its preference for base-poor water. It is found in acidic water in lakes, reservoirs, ponds, canals, ditches, and swamps over inorganic or peaty substrates. The two British sites from which both *U. australis* and *U. vulgaris* have been recorded were acid bogs at the foot of calcareous hills, and there is no evidence that the two species actually grew together (Taylor, 1989). However, *U. australis* is recorded from calcareous sites in both England and Ireland, such as a pond on sand dunes on Cruit Island where it grows with abundant *Chara hispida*. *U. australis* is recorded at 335m on Lambrigg Fell, and plants growing at 500m in Llyn Anafon are probably referable to this segregate.

   U. australis appears to flower much more frequently in the southern part of its British range than in more northerly localities, although flowering plants have been found as far north as the Culbin Forest. Meiosis in *U. australis* is abnormal and the pollen of this species,

unlike that of *U. vulgaris*, is often abortive (Caspar & Manitz, 1975). There is no evidence that the British and Irish populations set seed, and reproduction appears to be entirely by turions. Despite its extensive world range, mature fruit of *U. australis* is known only from China and Japan (Taylor, 1989) and the species apparently reproduces vegetatively almost everywhere. In addition to euploid plants, ancuploids with 2n=18, 19 or 20 were recorded by Caspar & Manitz (1975).

   The map is based on records of flowering plants of *U. australis*, plus a few records of vegetative plants determined by experts. It must greatly underestimate both the historical and the current distribution of this segregate, and the fact that bladderworts flower more frequently in the southern part of their range must give it an unduly southern bias. The records are also inadequate to provide an accurate estimate of changes in its distribution. However, *U. australis* has almost certainly decreased in southern and eastern England, but is probably still frequent in those areas of Wales, northern Scotland and western Ireland where the aggregate is recorded.

   U. australis is widespread in Europe, north to C. Scandinavia. It is found in temperate and tropical regions throughout the Old World, and also in Australia and New Zealand.

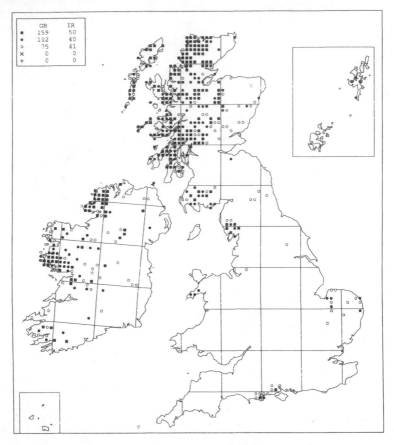

Utricularia intermedia Hayne sensu lato
Intermediate Bladderwort

U. intermedia usually grows in shallow, oligotrophic water, where buried, colourless and trap-bearing stolons anchor it to the substrate. It is most frequent in acidic and peaty sites, including the shallow edges of lakes (where it sometimes grows in stands of emergents such as *Carex rostrata*, *Menyanthes trifoliata* and *Phragmites australis*), pools, wet flushes, slow-flowing streams and ditches. Characteristic associates in these habitats include *Carex nigra*, *Drosera intermedia*, *Juncus bulbosus*, *Potamogeton polygonifolius*, *Ranunculus flammula*, *Chara virgata*, *Sphagnum* spp. and, in western sites, *Hypericum elodes*. *U. intermedia* also grows in highly calcareous but nutrient-poor lakes, fens and flushes over peaty substrates. It was, for example, one of the species which grew in the clear water of the broads in their pristine state (George, 1992). It is recorded from sea-level to 650m at Lochan Achlarich.

Members of the *U. intermedia* aggregate flower much less frequently than either *U. vulgaris* sensu lato or *U. minor*. The populations which flower most frequently are those in Dorset (Hall, 1939), and even here Taylor's (1989) attempts to see flowering plants were 'thwarted by a succession of cool summers'. Reproduction is by turions, which usually begin to develop on the green stolons in late July or August. They can, however,

develop earlier, sometimes on plants in pools which are drying out (Gurney, 1922). They begin to grow when temperatures rise in the spring: in Norfolk, Clarke & Gurney (1921) found that they could be grown in an indoor aquarium in December, but in the wild they began to elongate in late April.

This species had a more restricted distribution in eastern Britain than the other bladderworts. It has decreased in East Anglia and eastern Scotland for the same reasons that *U. vulgaris* sensu lato has declined: the destruction of many lowland peatlands and the eutrophication of the remaining lakes, pools and ditches. Unlike many species of damp heathland it has not shown a marked decline in Dorset, probably because its very wet habitats are not particularly vulnerable to successional changes (Byfield & Pearman, 1994). It remains frequent in its strongholds in the north and west.

U. intermedia sensu lato has a circumboreal distribution. In Europe it is frequent in the boreal zone and extends south in scattered localities to S. France and N. Italy.

Three segregates of *U. intermedia* are thought to occur in Britain: *U. intermedia* sensu stricto (Intermediate Bladderwort), *U. ochroleuca* (Pale Bladderwort) and *U. stygia* (Nordic Bladderwort). The differences between them are outlined by Thor (1979, 1987, 1988). As yet only a few records of these segregates are available, and their distribution in Britain and Ireland needs to be worked out.

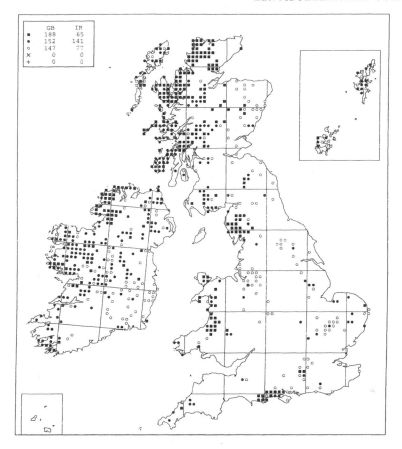

	GB	IR
■	188	65
●	152	141
○	147	77
×	0	0
+	0	0

Utricularia minor L.   **Lesser Bladderwort**

U. minor is most frequently found in peat-stained, acidic and oligotrophic water, where it is often anchored to the substrate by colourless stolons. Its habitats include bog pools and old peat-cuttings, where it can sometimes occur in abundance, shallow water at the edge of lakes, ditches and wet flushes. It often grows in shallow water in stands of emergent *Menyanthes trifoliata*. Other associates include *Carex nigra, Drosera intermedia, Juncus bulbosus, Menyanthes trifoliata, Potamogeton polygonifolius, Utricularia intermedia* sensu lato, *Sphagnum auriculatum* and *S. cuspidatum*. Although *U. minor* is found in base-poor habitats to a greater degree than our other bladderworts, it can also grow in nutrient-poor but base-rich water, as it does in fens and ditches on the Burren limestone near Mullaghmore. It is recorded from sea-level to 600m (Haystacks Tarn), and Wilson (1956) cites an unlocalized record from 690m in Scotland.

U. minor flowers occasionally throughout much of its British and Irish range. In some years plants flower freely in S.E. England: 1910 was a good flowering year in Norfolk, for example, and populations also flowered freely in the hot summer of 1921 in those sites where its pools did not dry out (Clarke & Gurney, 1921; Gurney, 1922). *U. minor* flowers much more frequently in Scotland than the other members of the genus.

Nevertheless, reproduction is usually by turions, which are formed at the end of the green stolons from late July onwards. These will germinate in winter if kept at room temperature; in the wild, growth begins in Norfolk in early April (Clarke & Gurney, 1921).

This species has suffered a marked decline in the eastern part of its British range. The major cause is the destruction of lowland peat-bogs, but shading and scrub encroachment on remaining bogs may also have contributed, and perhaps eutrophication in those sites (e.g. Wicken Fen) where it appears to have grown in calcareous habitats. *U. minor* is apparently extinct in numerous counties or vice-counties including S.E. Yorkshire (last recorded before 1902), Cambridgeshire (last recorded 1951), Northumberland (1959), Lincolnshire (1961), Sussex (1963) and Berwickshire (1974). The map suggests a similar decline in S.E. Ireland, but detailed local studies are needed to confirm this. *U. minor* remains frequent in the peatlands of the north and west, and as the smallest member of the genus it may be somewhat under-recorded in these areas.

U. minor has a circumboreal distribution, extending south to the Himalaya, Burma, Japan and California and also found at high altitudes in Papua New Guinea. In Europe it occurs from Iceland and northern Scandinavia southwards, but it is rare in the Mediterranean region.

CAMPANULACEAE

The only native European aquatic genus in this family is *Lobelia*, which is often placed in the segregate family Lobeliaceae.

Lobelia L.

Only two native species are known in our area, the aquatic *L. dortmanna* and the terrestrial *L. urens*, a rare species of S. England. The familiar garden 'Lobelia', *L. erinus*, is also recorded as an escape from cultivation.

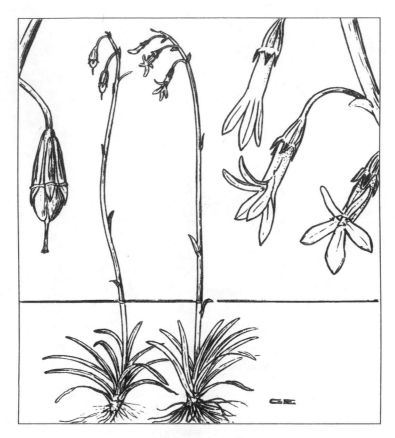

Lobelia dortmanna L.

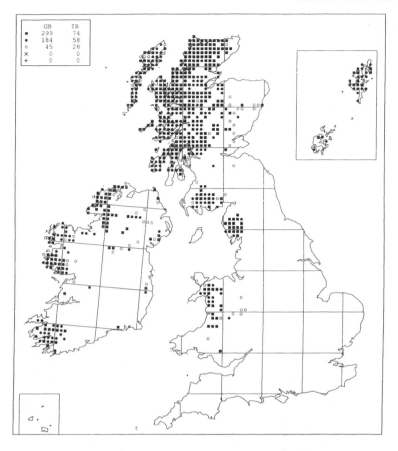

Lobelia dortmanna L.   Water Lobelia

Lobelia dortmanna is one of the most characteristic, and one of the most attractive, species of oligotrophic lakes in Britain and Ireland. Like other isoetids, it is a slow-growing perennial of low competitive ability. It is virtually confined to lakes, but it can sometimes be found in their outflow streams. It grows in shallow water less than 2m deep over a wide range of substrates from coarse sand to highly organic silts. In exposed sites it is found in open communities, where it is often accompanied by *Eleocharis palustris*, the aquatic form of *Juncus bulbosus*, *Littorella uniflora*, *Myriophyllum alterniflorum* and *Ranunculus flammula*. As the vegetation cover increases with depth additional associates include species such as *Callitriche hermaphroditica*, *Potamogeton gramineus* and *P. natans*. *L. dortmanna* also grows in open stands of emergents such as *Carex rostrata*, *Equisetum fluviatile* and *Schoenoplectus lacustris*. Although it is tolerant of exposure, the most luxuriant plants are found in more sheltered sites. It is confined to shallow water because it is intolerant of shade. It is found only over acidic substrates, but it can be found in base-rich but nutrient-poor water where the substrate is siliceous (Webb & Scannell, 1983). *L. dortmanna* may be excluded from marl lochs by an inability to tolerate the deposition of marl on the leaves, and from mesotrophic water by the shading effect of epiphytes (Farmer & Spence, 1986; Farmer, 1989). Small colonies of compact plants can

very occasionally be found growing terrestrially (Farmer, 1987). The species grows from sea-level to 745m at Llyn Bach.

L. dortmanna flowers from May to October. Pollination occurs before the buds open; the aerial flowers do not appear to be visited by insects and almost all set seed (Faegri & Pijl, 1971). Flowers which develop under water also set seed. After liberation from the capsule the seeds sink rapidly. They require a cold pre-treatment of 1–3°C for one month before they will germinate, and they germinate only in the light. Seeds will, however, germinate at light levels which are too low to sustain seedling growth. Populations have a persistent seed bank, and seeds can remain viable for 30 years (Arts & Heijden, 1990). New rosettes are formed vegetatively at the base of the old flower-stalk by the growth of axillary buds. There are no specialized vegetative propagules, and long-distance dispersal is dependent on the spread of seed. A detailed account of the reproduction of *L. dortmanna* is given by Farmer & Spence (1987).

This 'singular and elegant plant' (Wade, 1802) is distinctive when in flower or fruit, and the vegetative rosettes would be overlooked only by inexperienced recorders. The map in the *Atlas of the British Flora* (Perring & Walters, 1962) consequently gave a much truer picture of its distribution than those of other isoetids such as *Isoetes lacustris* and *Subularia aquatica*. Although *L. dortmanna* is still frequent in much of western Britain and Ireland it is vulnerable to eutrophication and the map clearly indicates that sites have been lost at the eastern edge of its range, where human population densities are greatest and agriculture more intensive. It is, for example, extinct in the Shropshire meres (Sinker et al., 1985), in several lochs in S.E. Scotland including Loch Leven (Farmer & Spence, 1986) and at Lough Neagh, where it was first recorded in Ireland by W.

Sherard in 1694 (Harron, 1986; Hackney, 1992).

L. dortmanna is an amphi-atlantic species, widespread in northern Europe, extending south very locally to S.W. France, and in N. America.

For a detailed account of the ecology of L. dortmanna, from which much of the above information is drawn, see Farmer (1989).

BUTOMACEAE

The monotypic genus *Butomus* is the only member of this family, which is allied to the Alismataceae and the non-European Limnocharitaceae.

Butomus L.

The distinctive and highly attractive *Butomus umbellatus* is the only species in this genus.

Butomus umbellatus L.

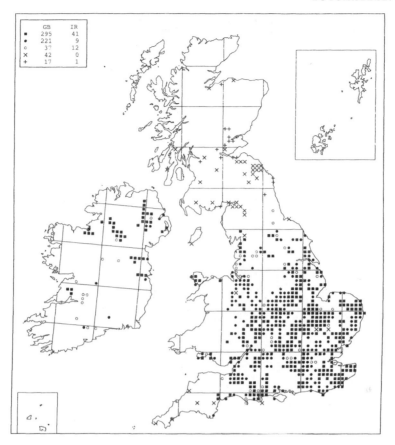

Butomus umbellatus L.   Flowering-rush

B. umbellatus grows as a submerged or emergent aquatic,
or as a terrestrial species near the water's edge. It is usually found where the water is calcareous and eutrophic,
and it can persist, even as a submerged aquatic, in highly
managed and very eutrophic rivers. Submerged plants
are usually found in sluggish rivers, streams and fenland
lodes, where they often grow with *Sagittaria sagittifolia*
and *Sparganium emersum*, two other species with strap-
like leaves. Emergent or terrestrial plants are found at
the edge of lakes, reservoirs, rivers, streams, canals and
ditches. In grazing marshes the species favours ditches
where regular management reduces the competition
from more vigorous species. *B. umbellatus* often occurs
as individual clumps, but extensive stands sometimes
develop in swamps, notably in the Lough Neagh basin.
The species is also planted by ornamental lakes, village
ponds and disused gravel-pits. It is confined to the lowlands.

This is a rhizomatous perennial. The flowers are
insect-pollinated. Both diploids (2n=26) and triploids
(2n=39) are known in Europe (Krahulcová & Jarolímová,
1993). The diploids are self-compatible and usually set
seed, whereas the triploids are self-incompatible and
usually fail to set seed in the wild, although they produce
viable pollen and can set seed if cross-pollinated. The
seeds float and may be dispersed by water. Seedlings
grow slowly and only become established where a moist,
unshaded substrate persists for at least two months

(Hroudová & Zákravský, 1993a). Plants also spread by
lateral buds which develop on the rhizome (Hroudová,
1989). These are easily detached when plants are
uprooted by floods, disturbed by waterfowl or damaged
in dredging operations. They float and readily give rise
to new plants. Triploids are more vigorous vegetatively
and produce more numerous lateral buds on the rhizome; they also produce bulbils in the inflorescence. The
bulbils do not drop off individually, but remain attached
to the inflorescence (Lohammar, 1954; Jans, 1989).
Triploids are more numerous than diploids in the Czech
and Slovak Republics (Hroudová & Zákravský, 1993b).
Unfortunately, little is known about the cytology and
reproductive biology of the British and Irish plants.

B. umbellatus is almost certainly native to Britain
(Godwin, 1975). In England, losses caused by habitat
destruction have been partially offset by its spread along
canals and its deliberate introduction as an ornamental
plant to new sites. The species is believed to reach its
northern limit as a native British plant in Co. Durham,
where it is now extinct. However, there are thriving
populations in natural habitats further north, which are
apparently the result of recent spread. It is locally frequent by the Tweed and its tributaries, for example, but
was not recorded there until 1956 (Braithwaite & Long,
1990; Swan, 1993). It may have spread from planted
stock. In Ireland it is also thought to be native in some
areas, including Co. Clare (Webb & Scannell, 1983).
However, the large populations round L. Neagh and L.
Beg were not known to Stewart & Corry (1888) but the

plant subsequently spread rapidly, perhaps originating from cultivated material (Harron, 1986; Hackney, 1992). As it is especially difficult to distinguish native and naturalized populations in Ireland, all records except those of obviously planted stock are mapped as if they were native.

B. umbellatus is widespread in temperate Eurasia, extending north in Europe to northern Scandinavia; it also occurs in N. Africa. It has spread as a naturalized alien in N. America, particularly in the Great Lakes and St Lawrence River regions (Anderson *et al.*, 1974).

The southerly distribution of *B. umbellatus* in Britain is surprising in view of its presence here in the last full glacial period, and its occurrence in northern Scandinavia (Godwin, 1975). Its apparent ability to spread from introduced plants north of its native range suggests that it might have been limited by its powers of dispersal.

ALISMATACEAE

This is a family of eleven genera and some 95 species, all of which are aquatic. The family has a cosmopolitan distribution.

The Alismataceae and the Butomaceae are members of the Order Alismatales. It has often been suggested that these are the most primitive monocotyledons. Their similarity to dicotyledon families such as the Nymphaeaceae has led to the suggestion that the ancestral monocotyledons arose from ancestors which were not unlike the modern Nymphaeales (Cronquist, 1981). Recent authors have viewed this theory rather sceptically, or rejected it completely (Tomlinson, 1982; Mabberley, 1987), and Dahlgren *et al.* (1985) suggest that the similarity between the two orders is probably the result of convergent evolution in aquatic habitats. However, support for the traditional view has come from studies of chloroplast DNA, which have revealed similarities between the monocotyledons and a group of 'palaeoherbs', including the Nymphaeaceae (Chase *et al.*, 1993). Further work along these lines ought to throw new light on the vexed question of the origin of the monocotyledons.

Many important aquatic families are included with the Alismatales in the Subclass Alismatidae, one of the major subdivisions of the monocotyledons. Those with representatives in our area include the Hydrocharitaceae, Aponogetonaceae, Potamogetonaceae, Ruppiaceae, Najadaceae and Zannichelliaceae, and the marine Zosteraceae. Many of these families appear to have been derived from ancestors similar to the modern Alismatales, but often have very reduced vegetative and floral parts.

In addition to the plants treated below, the aquatic *Caldesia parnassifolia* (L.) Parl. is established in woods at Kinfauns (Clements & Foster, 1994).

Sagittaria L.

Sagittaria is one of the larger genera in the Alismataceae, with 20 species. There are two relatively recent monographs of the genus (Bogin, 1955; Rataj, 1972a,b) but, as Cook (1990a) comments, 'neither of these revisions have found general acceptance'. Two species are native to Europe, one of which occurs in Britain and Ireland. Most of the species are found in the New

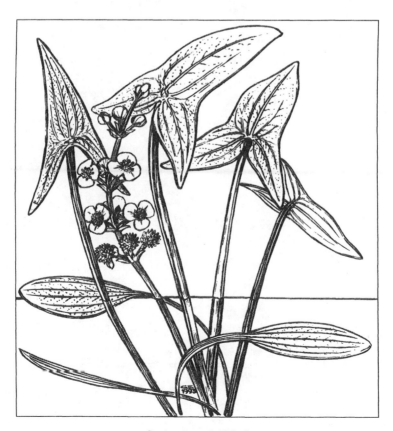

Sagittaria sagittifolia L.

World, of which three are established aliens in our area although none of them is widely naturalized. In addition to these species, which are dealt with in the following accounts, *S. graminea* Michx. is recorded from a gravel-pit near Sandhurst (Clement & Foster, 1994) and has recently been discovered by N. F. Stewart in abandoned peat-cuttings between Ballydehob and Skull, where it may have been planted.

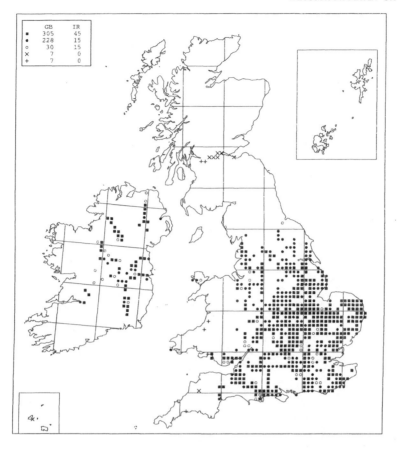

	GB	IR
■	305	45
●	228	15
○	30	15
×	7	0
+	7	0

Sagittaria sagittifolia L. Arrowhead

This is a plant of lowland, shallow, still or slowly flowing water in sheltered lakes and ponds, rivers, streams, ditches and canals. In many areas it is restricted to the major water courses and their associated streams and ditches. It grows in calcareous and eutrophic (but not brackish) water over a range of fine substrates, including clay, sand and silt. *Nuphar lutea* and *Sparganium emersum* are frequent associates in managed rivers; in grazing marsh ditches there is often a richer flora which may include *Callitriche obtusangula*, *Hottonia palustris*, *Hydrocharis morsus-ranae* and *Myriophyllum verticillatum*. *S. sagittifolia* very rarely colonizes newly available aquatic habitats in sand-, gravel- and clay-pits. It is confined to the lowlands.

S. sagittifolia has broad, strap-shaped submerged leaves as well as the familiar arrow–shaped emergent leaves. In the central channel of streams and rivers only submerged leaves may develop. Elsewhere, plants produce emergent leaves and flower freely. Female flowers are borne below male flowers on the same peduncle, and are insect-pollinated. The peduncle with its globose fruiting heads falls into the water. The seeds have a waxy coat and may float until it is damaged by frost. Little is known about the extent to which plants in our

area become established from seed: in Central Europe seeds germinate in spring at temperatures above 13°C, often in anaerobic, submerged sediments (Hroudová *et al.*, 1988). In autumn mature plants die down, perennating by tubers which develop on stolons in the leaf axils. These also provide a means of vegetative dispersal.

Fossil evidence suggests that *S. sagittifolia* is a native British plant: it is known even from late glacial sediments in Cumberland and Skye, north of its current native range. It colonized the canal network in both Britain and Ireland, becoming established as an introduction as far north as the Forth & Clyde Canal, where it was first recorded in 1889 (Hennedy, 1891). In recent years it has decreased in ditches in river valleys, sometimes becoming restricted to the rivers themselves (Tarpey & Heath, 1990). It has also decreased in some grazing marshes (Mountford, 1994a). Its ability to persist in eutrophic, canalized and highly managed rivers has probably prevented a more serious decline.

S. sagittifolia is a variable species which, if broadly interpreted to include *S. trifolia* L., is widespread in boreal, temperate and tropical regions of Eurasia. In Europe it is rare in the Mediterranean region, and extends north to northern Scandinavia. It is replaced by the closely related *S. latifolia* in N. America.

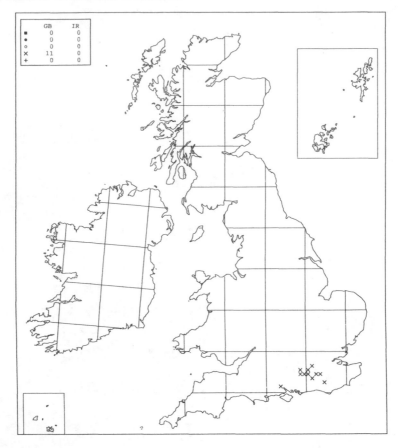

	GB	IR
■	0	0
●	0	0
○	0	0
×	11	0
+	0	0

Sagittaria latifolia Willd.   **Duck-potato**

This alien species is recorded from shallow water in low-land lakes and ponds, roadside ditches and at the edge of canals, streams and rivers. In a pond on Epsom Common it grows with *Menyanthes trifoliata* and in Ashdown Forest it is found in a pond with *Glyceria maxima*, *Lagarosiphon major*, *Myriophyllum aquaticum* and *Nymphaea alba*. In most cases it has become established in the area where it was originally planted, but it may be spreading along the R. Wey, where isolated patches up to 3m in diameter have been recorded in recent years.

Like *S. sagittifolia*, *S. latifolia* dies down in the autumn to tubers which are borne on stolons in the leaf axils. They act as a means of perennation and vegetative spread. The species flowers in Britain, but there are no published observations on the occurrence of ripe seed or seedlings. In America some populations consist of monoecious plants and others of dioecious individuals; the dioecious plants usually reproduce vegetatively (Wooten, 1971). Exposed mud is required for successful seedling establishment. Germination occurs once the seed coat has been eroded or ruptured, which usually happens during the winter. In suitable conditions

seedlings may be very abundant in spring and early summer (Kaul, 1985).

This species was first recorded in 1941 in a pond on Epsom Common, where it survived until the pond was infilled in 1956. It is now known from several other localities in Surrey (Lousley, 1976). It was found in an overgrown pool in Jersey in 1961, and it still persists there, having survived the clearing out of the pool some years after its discovery (Le Sueur, 1984). It was discovered in Hampshire in 1962 and in Sussex in 1976. It may be overlooked elsewhere because of its similarity to *S. sagittifolia*.

S. latifolia is an American species which is naturalized in several European countries. It is the commonest and most widespread species of *Sagittaria* in N. America, found in temperate and subtropical wetlands throughout the continent and extending south to C. America, Colombia, Ecuador and Venezuela. Its native distribution is mapped by Bogin (1955). It is also naturalized in the West Indies and Hawaii.

The starchy tubers of this species were formerly used as food by the N. American Indians, and by Chinese and Japanese immigrants in California (Bogin, 1955). They are, however, too large and too deeply buried to be of much use to wildfowl (Martin *et al.*, 1961).

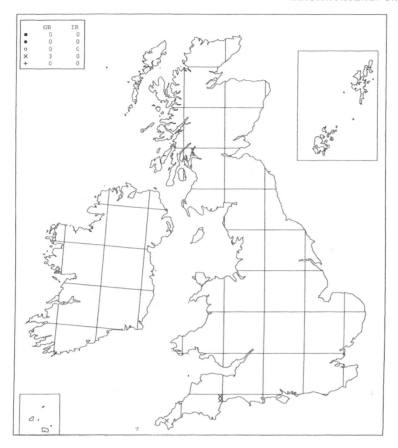

	GB	IR
■	0	0
●	0	0
○	0	0
×	3	0
+	0	0

Sagittaria rigida Pursh   Canadian Arrowhead

S. rigida is recorded as a naturalized alien in the R. Exe in the Exeter area, and from the nearby Exeter Canal. In the Exe it grew along the muddy margin of the river, and plants were exposed above the water when water-levels dropped in hot weather. In the canal it is also found close to the bank as an emergent, associated with *Carex acutiformis*, *C. riparia*, *Lycopus europaeus*, *Oenanthe aquatica*, *Schoenoplectus lacustris* and *Scrophularia auriculata*. It is also known from a pond near Camborne, where it was originally planted.

This is a stoloniferous perennial. Both monoecious and dioecious plants occur in Britain (Hiern, 1908). The inflorescences of the female plants are shorter than those of the male, and have very few whorls of flowers. The monoecious plants have male flowers in the upper whorls of the inflorescence and female flowers below. Hiern (1908) noted that the early flowering plants tended to be dioecious, and the later ones monoecious. Ripe fruit is set in Britain, and the species also reproduces by tubers produced in autumn on the stolons.

S. rigida was first recorded from the R. Exe by Hiern (1908), who discovered it growing in considerable quantity in and near Exeter in July 1908. It was later found in the adjacent Exeter Canal. Keble Martin & Fraser (1939) described it as well established in the river and canal, but it was last recorded in the river in 1952. It was still present in the canal in 1995. The only other British record is from Clowance Lake near Camborne, where it was discovered by H. B. Sargent in 1965 (Margetts & David, 1981).

This species is a native of S.E. Canada and the eastern United States, where it grows in inland, calcareous waters and fresh or brackish tidal creeks. Its native distribution is mapped by Bogin (1955). There are no confirmed records of the species as an established alien elsewhere in Europe (Tutin *et al.*, 1980).

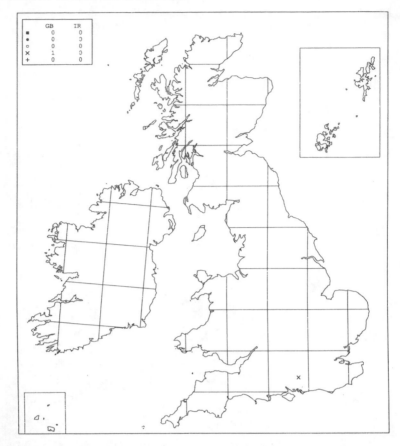

	GB	IR
■	0	0
●	0	0
○	0	0
×	1	0
+	0	0

Sagittaria subulata (L.) Buchenau
Narrow-leaved Arrowhead

This species is naturalized in a single lowland pond. It grows in abundance in shallow, acidic water, accompanied by *Utricularia australis.*

S. subulata has submerged and floating but no emergent leaves. The unisexual flowers open at the surface of the water, as in *Luronium natans,* and only a single flower on each inflorescence opens at any one time. It has a long flowering period in Britain, from June to October. The British population sets ripe fruit, and also reproduces vegetatively by tubers which are produced on stolons in the mud.

This species was discovered at Shortheath Pond by Mr & Mrs D. N. Turner in 1962; it was 'still abundant and obviously established' in 1983 (Grenfell, 1984) and 1995. It persists despite the fact that it is frequently dredged out of the pond by anglers. It is occasionally sold as an aquarium plant, and was probably introduced into Shortheath Pond with discarded material from an aquarium.

S. subulata is a native of N. & S. America, where it is found in the coastal plain from Massachusetts south to Alabama, and in Venezuela and Colombia. The N. American distribution is mapped by Bogin (1955). In this area it grows in tidal waters, small, slow-flowing streams and occasionally in ponds. It is a variable plant, divided by Bogin (1955) into three varieties. Plants of var. *kurziana* (Glück) Bogin in warm streams in Florida may have leaves that reach 15m in length, and flowering scapes up to 90cm. *S. subulata* is recorded as a probable introduction from the highlands of New Guinea (Rataj, 1972a), but the British station is the only one known in Europe (Tutin *et al.*, 1980).

The information on the habitat and reproductive biology of this species in Britain is taken from Brewis (1975).

Baldellia Parl.

There are two species of *Baldellia*: *B. ranunculoides* is widespread in Europe whereas *B. alpestris* (Coss.) Vasc. is endemic to the mountains of the Iberian peninsula. *Baldellia* is one of several small genera in the Alismataceae related to the larger and predominantly tropical genus *Echinodorus*. Their generic limits are not very well defined, and Cook (1990a) suggests that *Baldellia* 'is very close to and should perhaps be united with *Echinodorus*'.

Although Tutin *et al.* (1980) do not subdivide European *B. ranunculoides*, some authors recognize two subspecies, subsp. *ranunculoides* and subsp. *repens* (Lam.) Á. Löve & D. Löve (e.g. Vuille, 1988) and others even treat subsp. *repens* at specific rank (e.g. Lindblad & Ståhl, 1990).

Subsp. *repens* differs from subsp. *ranunculoides* 'by having runners, fewer flowers per umbel, slender and curved pedicels, larger petals, fewer carpels, smaller anthers, a poorer fruit-set, and smaller, papillose fruits' (Lindblad & Ståhl, 1990). Vuille (1988) found that populations of subsp. *ranunculoides* were self-compatible whereas subsp. *repens* was strongly self-incompatible and did not set seed when artificially self-pollinated. The British and Irish populations are treated below as a single taxon, but both subspecies may be present. Scully (1916) reports 'a profusely stoloniferous form' as 'rather common on the shores of some of the larger Kerry lakes'. A study of the variation within our *B. ranunculoides* would be worthwhile.

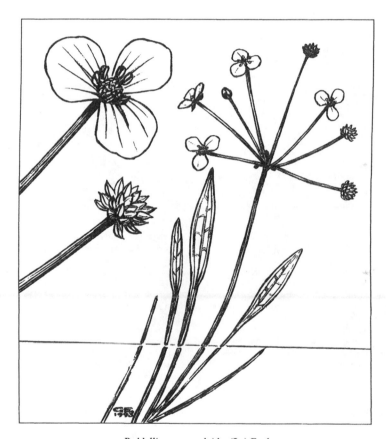

Baldellia ranunculoides (L.) Parl.

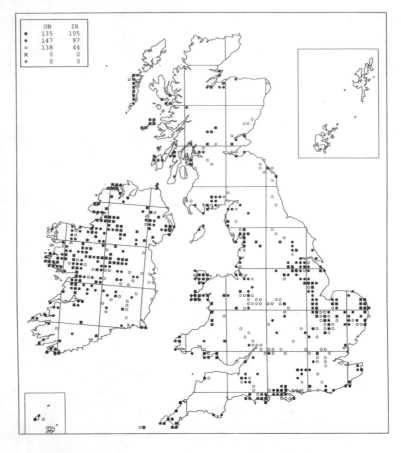

	GB	IR
■	135	105
●	147	97
○	138	44
×	0	0
+	0	0

Baldellia ranunculoides (L.) Parl.
Lesser Water-plantain

B. ranunculoides is found in habitats where the competition from more robust species is restricted by a variable water-level and a low level of nutrients, or by cattle poaching or other disturbances. It is found at the edges of lakes and reservoirs, and in pools, streams and ditches, canals, flooded quarries, abandoned peat-cuttings, marshes and flushes on sea-cliffs and newly constructed bird scrapes. Although it is sometimes found in acidic waters, it is more frequent in sites which are at least slightly basic. These range from those which receive some enrichment from nearby calcareous or ultrabasic rocks, or from dune sand, to highly calcareous waters over chalk or limestone. It can also occur in brackish sites. It grows over a range of organic and inorganic substrates including peat, sand and fine calcareous marl. It is a lowland species which ascends to 320m W. of Libanus.

This is a phenotypically variable species which may persist in relatively deep water as vegetative plants with submerged leaves. These spread by stems which root at the nodes, producing new plants. Plants in shallow water or on mud exposed above the water-level produce aerial leaves and flower freely. Their flowers are pollinated by insects and usually give rise to numerous fruits. Field observations indicate that plants which flower under water may also be capable of fruiting, sug-

gesting that the flowers are self-compatible and capable of self-fertilization. Plants sometimes reappear in abundance in newly cleared ditches, indicating the presence of a persistent seed bank. Established individuals may reproduce vegetatively by division of the basal rootstock, and Vuille (1988) reports pseudovivipary, with plantlets developing from vegetative buds within the inflorescence.

B. ranunculoides has decreased greatly in England and E. Scotland in the historical period. It is now extinct in several counties, and has decreased in many others. Many sites were ponds on commons or heaths which were formerly important for watering stock but have now been filled in or become overgrown. Others were ditches which have disappeared with agricultural improvement, or are now managed unsympathetically, or canals which have become disused and dried out. Some sites have been lost through eutrophication. The species no longer grows at 12 of the 15 sites in Dorset where it was recorded by R. d'O. Good in the 1930s; three of these have been destroyed by agricultural or other development but the remaining pools are still present, but have become overgrown (Byfield & Pearman, 1994). Although *B. ranunculoides* may reappear in areas where it was thought extinct, as it did in Middlesex in 1988, there is no doubt that it continues to decrease in many areas. There is less evidence of a decline in W. Scotland and in Ireland, where it is still frequent in shallow, calcareous water in many areas.

B. ranunculoides is found in W., C. & S. Europe, north to S. Norway and E. to Greece; it also occurs in N. Africa. It is most frequent in the western part of its range (Cook, 1983).

Luronium Raf.

This is a monotypic genus which, like *Baldellia*, is of rather doubtful validity. The only species, *L. natans*, is found in both Britain and Ireland. For a detailed study of the British plant see Willby & Eaton (1993). Rich *et al.* (1995) have recently reviewed the Irish records and decided that there is little doubt that the plant is native in Ireland. This conclusion, however, is not accepted by all Irish botanists (T. G. F. Curtis, pers.comm.).

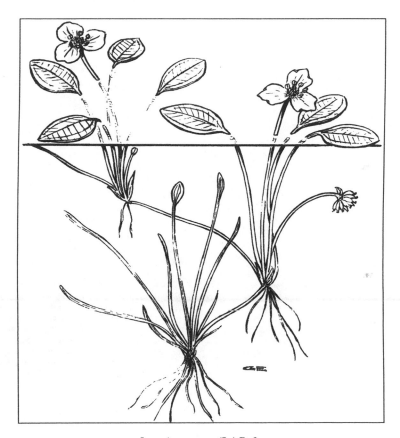

Luronium natans (L.) Raf.

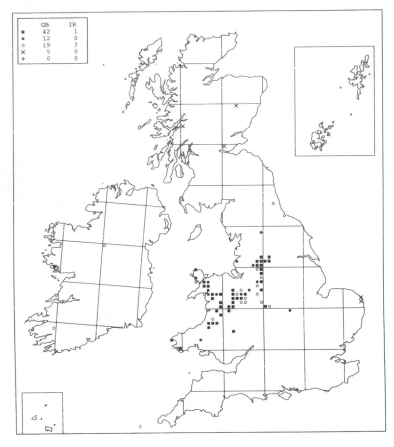

Luronium natans (L.) Raf.
Floating Water-plantain

The most frequent natural habitat for *L. natans* is acidic and oligotrophic lakes, where it grows in water as deep as 2m, with species such as *Callitriche hamulata*, *Isoetes lacustris*, *Littorella uniflora* and *Lobelia dortmanna*, or on bare mud exposed by falling water-levels. It also grows in a slow-flowing, mesotrophic stretch of the Afon Teifi, and it was found in the past in a variety of other habitats including streams, ditches, lowland lakes and pools. At least one of the lakes in which it has been recorded, Lough Leane, was calcareous but nutrient-poor. *L. natans* is also established in the canal system, where it grows in circumneutral or slightly basic, mesotrophic water, often with *Elodea nuttallii*, *Lemna minor*, *Sparganium emersum* and *Potamogeton* spp. In canals it apparently relies for its survival on periodic disturbance, which can be provided by light boat traffic in navigable waters or by periodic dredging or the control of marginal vegetation elsewhere (Willby & Eaton, 1993). The introduced populations in Broadland grow in ditches (Driscoll, 1985b), and the Argyll plants are at the edge of a loch (Slack, 1964). *L. natans* is found from near sea-level to 450m at Bugeilyn.

L. natans is a stoloniferous perennial. In flowing or deep water, or in sites that are shaded or turbid, it can persist as rosettes of submerged leaves, which spread vegetatively. In shallow water or on wet mud it produces floating or terrestrial leaves and flowers. The flowers are entomophilous and fruit freely. Populations in natural habitats appear to be greatest when water-levels are low

and much bare mud is exposed, but when levels are high many plants may be present as inconspicuous, submerged rosettes. In canals, plants are most numerous after disturbance.

Although flowering *L. natans* is very distinct, the vegetative plant can be easily confused with other members of the Alismataceae and has probably been overlooked in some sites and recorded in error at others. A number of outlying records mapped by Perring & Walters (1962) are now known or suspected to be errors. In Britain the populations in oligotrophic, upland lakes appear to be stable, but the plant has disappeared from many of its lowland, mesotrophic localities because of habitat destruction or eutrophication. During the 19th century the species colonized the canal system, but the canal populations have decreased since 1970 as recreational boat traffic has increased. Although *L. natans* was first recorded in Ireland in 1801, the records have long been dismissed as errors and have only recently been reinstated following the discovery of the species in Connemara (Rich *et al.*, 1995). All Irish records are mapped as native although there is doubt about the status of the species in Ireland.

L. natans is endemic to W. and C. Europe, where it extends from S. Norway to N. Spain and eastwards to Bulgaria. It is most frequent in the western part of its range, in Britain, France and N. Germany; in N. Spain, where it has been confused with *Baldellia alpestris* (Coss.) Vasc., it is rare (Rodriguez-Oubiña & Ortiz, 1991). According to Cook (1983), 'in continental Europe it is declining rapidly and has become extinct in many localities'.

Alisma L.

This genus of nine species is found throughout the northern hemisphere. Detailed studies of the taxonomy, cytology, reproductive biology and ecology of the European species have been published by Björkqvist (1967, 1968), from which much of the information below has been taken.

There is a wide range of leaf form in the European species of *Alisma*. Plants may have strap–shaped submerged leaves, floating, emergent or terrestrial leaves; the emergent and terrestrial leaves are very similar. The type of leaf produced varies from species to species and, within a species, with environmental conditions. This plasticity enables plants to adapt to changes in water-level in the marginal habitats in which they grow. Seedlings of *A. plantago-aquatica* with submerged leaves or immature plants with floating leaves often puzzle botanists who are unfamiliar with the variability of this common species.

Alisma plants are perennials which die down in winter, regrowing from a terminal bud on the rootstock. The individual flowers are ephemeral, but the more robust species have large, branched inflorescences and a long flowering period.

Although the flowers are often cross-pollinated by insects, all species are self-compatible and are capable of self-fertilization, and fruit set is usually very high. The seeds will float for long periods, and may also be dispersed on the plumage and feet of waterfowl (Ridley, 1930). They have a hard seed coat and may still be viable after passing through the intestines of birds. The seeds may remain dormant for long periods, as germination occurs once the seed coat has been broken. Seeds will germinate on land or in shallow water. A limited amount of asexual reproduction may occur by the growth of side shoots on the rootstock or by the uprooting of plants in floods. All the species are poor competitors, usually found in open vegetation or on bare mud. They favour water which is at least moderately eutrophic.

There are confirmed records of three of the four European species in our area, *A. gramineum*, *A. lanceolatum* and *A. plantago-aquatica*. The fourth species, *A. wahlenbergii* (Holmb.) Juz., is a Baltic endemic closely related to *A. gramineum* and sometimes treated as a subspecies of that species. *A. wahlenbergii* was reported as occurring with *A. gramineum* in Norfolk (Swann, 1975) but the identity of these plants requires reassessment.

Alisma plantago-aquatica L.

Alisma gramineum (2n=14) and *A. plantago-aquatica* (2n=14) are diploids, whereas *A. lanceolatum* is a tetraploid with two morphologically indistinguishable cytotypes (2n=26, 28). *A. lanceolatum* seems likely to have evolved by polyploidy from *A. plantago-aquatica* or an extinct species with a similar karyotype. It is apparently an ancient species; tetraploid plants of *A. plantago-aquatica* which were artificially synthesized by Björkqvist (1968) did not have the morphology of *A. lanceolatum*. Hybrids between *A. lanceolatum* and *A. plantago-aquatica* are known from the Netherlands, Poland and Sweden; they were described as *A.* × *rhicnocarpum* by Schotsman (1949). Björkqvist (1968) and Pogan (1971) confirmed the identity of the hybrid cytologically and Björkqvist synthesized the hybrids artificially without difficulty. Although the hybrids are intermediate between the parents, they tend to have a rather greater resemblance to *A. lanceolatum* than to *A. plantago-aquatica*. The hybrids are partially sterile, but the degree of sterility varies. Both Polish and Swedish hybrids grew in disturbed habitats, and the hybrid was replaced by its parents in one Swedish site once the disturbance ceased. Hybrids between *A. gramineum* and both the other species have been obtained artificially, but are unknown in the wild. They set very little seed.

There is some doubt whether the hybrid *A.* × *rhicnocarpum* occurs in Britain and Ireland. It is recorded from both countries, and Allen (1984) considers that in the Isle of Man it is commoner than its parents, and more tolerant of the repeated clearing out of drainage ditches. However, Stace (1991) suggests that all records require confirmation. This could be obtained by a study of the microscopic characters which separate the parents (e.g. stomatal length) or by chromosome counts.

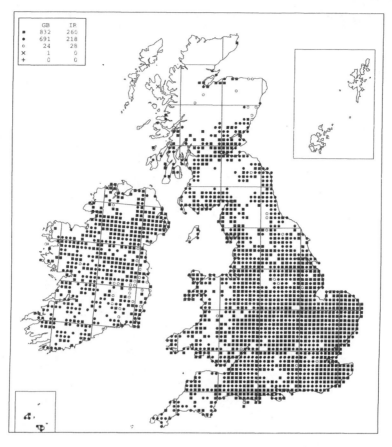

Alisma plantago-aquatica L. Water-plantain

This species is found in shallow water or exposed mud at the edges of a wide range of water bodies, including lakes, reservoirs, pools, slow flowing rivers, streams and ditches, and in marshes and swamps. It usually grows in open habitats, including shallow water where cattle congregate, but it can be found amongst open stands of emergents such as *Glyceria maxima*, *Phragmites australis*, *Sparganium erectum* and *Typha latifolia*. It is found in both mesotrophic and eutrophic water and usually grows on fine-grained, nutrient-rich substrates; plants in shallow water over eutrophic mud can be very robust. It often appears in abundance in newly cleared ditches and is a frequent colonist of newly available habitats such as gravel-pits, and even deep wheel-ruts where water lies in winter (Tarpey & Heath, 1990). It may also be planted around new ponds and lakes (Swan, 1993; Sykes, 1993). It is virtually confined to the lowlands, but reaches 403m at Dock Tarn (Stokoe, 1983).

Plants of *A. plantago-aquatica* in water deeper than *c*.65cm persist as plants with submerged leaves but fail to flower. In shallower water the plants flower and fruit freely. Clapham *et al.* (1952) suggest that *A. lanceolatum* flowers in the morning whereas *A. plantago-aquatica* flowers in the afternoon. This view is supported by Perring *et al.* (1964), who describe the differences as sur-

prisingly constant, but has been questioned by other observers (e.g. Percy, 1964; Morton, 1966). Plants of *A. plantago-aquatica* cultivated by Björkqvist (1967) opened their flowers later than those of *A. lanceolatum*, but there was then a broad overlap in flowering time. In addition to insect pollination, plants may be wind-pollinated (Daumann, 1965). The seed may remain viable for at least ten years if stored at room temperature. *A. plantago-aquatica* reproduces freely by seed, and swards of seedlings can sometimes be seen in shallow water.

Fossil evidence suggests that this species is a long-established native which has persisted in our area through both glacial and interglacial stages. It is still frequent in suitable habitats in both Britain and Ireland, although it is apparently extinct in N. Aberdeenshire, near the northern edge of its British range (Welch, 1993). Its ability to exploit eutrophic waters and colonize new sites doubtless explains why it has survived in areas where some other members of the family have declined.

This species occurs as a native throughout most of Europe (north to the Arctic Circle) and Asia, and all round the Mediterranean; it also occurs in the mountains of E. Africa. It has been introduced to S. Africa, S. America, Australia and New Zealand. Despite its large range, there is little evidence of geographical variation within the species.

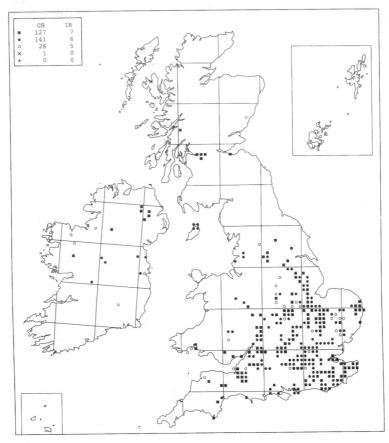

	GB	IR
■	127	7
●	141	6
○	28	5
×	1	0
+	0	0

Alisma lanceolatum With.
Narrow-leaved Water-plantain

This species occurs in shallow water or on damp mud in a similar range of habitats to *A. plantago-aquatica*, with which it often grows. It is found at the edges of lakes, reservoirs, ponds, slowly flowing rivers, canals, fenland lodes and ditches, and in marshes and swamps. It is also found as a colonist of flooded quarries and sand-, gravel- and clay-pits, and even occasionally of seasonally flooded hollows in arable fields. It usually grows in eutrophic water which is often, though not invariably, calcareous. It is most frequent over fine-grained substrates, although it will grow on sand. It is confined to the lowlands.

A. lanceolatum usually has emergent or terrestrial leaves; submerged and floating leaves rarely, if ever, develop. Plants in shallow water flower and fruit freely. The seeds have a less resistant coat than those of *A. plantago-aquatica*: in experiments a greater proportion of untreated seeds germinate and relatively mild pre-treatments such as a short period of frost will enhance their germination to a much greater degree.

This is a rarer species than *A. plantago-aquatica* throughout its British and Irish range. In many counties the largest or even the only populations are found in canals. For many years British and Irish botanists did

not recognize it as a distinct taxon, or treated it as a variety of the commoner species. Druce (1932), for example, simply notes under *A. plantago-aquatica* that 'narrow leaved (*lanceolatum*) and broad leaved (*latifolium*) plants occur'. It is difficult to assess trends in the distribution of *A. lanceolatum* because of the absence of reliable historical records in many areas and it may still be overlooked for the commoner species. In eastern England, Simpson (1982) and Tarpey & Heath (1990) suggest that the species must have declined greatly if the older records can be trusted. *A. lanceolatum* is occasionally found as an introduction, and it is established in a burn in Skye, north of its native range (Murray, 1980).

A. lanceolatum is widespread in Europe, N. Africa and W. Asia. It is the commonest species in the genus in the Mediterranean area and Asia Minor, and has a more southerly limit in Europe than *A. plantago-aquatica*, only reaching S. Sweden. The species becomes distinctly calcicolous towards the northern edge of its range.

Björkqvist (1967, 1968) found that the cytotype of *A. lanceolatum* with 2n=26 was widespread in W. & C. Europe, but that with 2n=28 was much rarer, being found in Sweden, Denmark and Poland. He did not examine any British or Irish material, but 2n=28 has been reported from the Cambridge area by Priestley (1953).

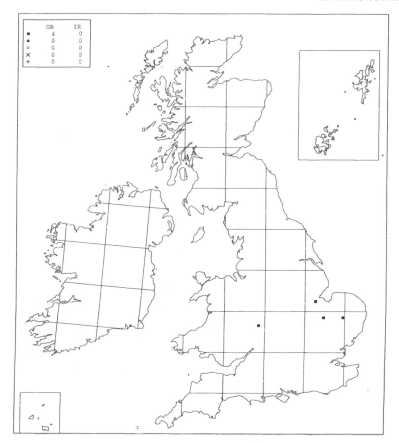

	GB	IR
■	4	0
●	0	0
○	0	0
×	0	0
+	0	0

Alisma gramineum Lej.
Ribbon-leaved Water-plantain

A. gramineum is recorded from shallow, eutrophic water at the edge of lowland lakes and rivers, or in fenland drains. At Westwood Great Pool it usually grows amongst stands of emergents in water less than 10cm deep, although it sometimes grows in open water or on recently exposed mud. Lousley (1957) found it with abundant *Eleocharis palustris* but in recent years most plants have grown amongst *Typha latifolia*. In Lincolnshire it formerly grew in water 0.7–1.3m deep over a soft, muddy substrate in the R. Glen and its associated drains. It has been seen recently in a single ditch in this area, where it appeared in 1991 and 1992 after clearance. It did not persist after *Phragmites australis* recolonized the site, but it presumably survives as buried seed. The other British records are from a fenland ditch in Cambridgeshire (Libbey & Swann, 1973) and Langmere in Breckland (Swann, 1975).

In the shallow water in which it usually grows *A. gramineum* forms rosettes of numerous submerged leaves. Plants with submerged leaves may flower, but mature plants may also develop floating and emergent leaves. The observation in Clapham *et al.* (1952) that the flowers open between 6 and 7.15 a.m. is fictitious: in the wild British plants flower throughout the day, and in cultivation Björkqvist (1967) found that the flowers of this species remained open for 13–15.5 hrs, longer than those of the other species he studied. Cleistogamous flowers may develop and set seed under water. Seed-set is good in the wild, and reproduction by seed occurs although the number of seeds which germinate at Westwood varies greatly from year to year. Many plants in Britain are annuals, but some persist as short-lived perennials.

This species was first discovered in Britain in 1920 at an artificial lake, Westwood Great Pool, but it was initially confused with *A. lanceolatum*. It still persists in this site, where the population size is usually small but can exceed 150 flowering plants in a favourable year. The reason for the fluctuations is unknown. It was discovered in the R. Glen and nearby ditches in 1955, but is now much less frequent in this area. At its other sites it does not appear to have persisted for long, although it may survive as buried seed. It is an inconspicuous species which could be overlooked elsewhere. It has been suggested that the East Anglian populations may be derived from seed carried by wildfowl from Europe (Libbey & Swann, 1973).

A. gramineum is widespread in temperate latitudes in Europe, Asia (except the Far East) and N. America. In Europe it is absent from the Iberian peninsula and the Mediterranean region; it now reaches its northern limit in Denmark, having become extinct in Sweden. In mainland Europe it is uncommon throughout its range, and it is normally found as small populations (Björkqvist, 1967). It is difficult to account for its rarity in western Europe. Its sporadic appearance and apparent dependence on disturbance make it a difficult plant to conserve. By contrast, it has spread along the St Lawrence River in N. America, where it has been described as a 'pesky weed' (Raymond & Kucyniak, 1948; Countryman, 1968).

Damasonium Mill.

This small genus has a remarkably disjunct distribution, occurring in the Mediterranean region in the Old World, and in mediterranean climates in western N. America (*D. californicum* Torr.) and Australia (*D. minus* (R. Br.) Buch). Vuille (1987) recognizes three species in Europe, the tetraploid *D. alisma* and the diploids *D. bougeai* Coss. and *D. polyspermum* Coss. Other authorities (e.g. Tutin *et al.*, 1980) treat all the European plants as a single, variable species, *D. alisma*. *D. alisma* sensu stricto is intermediate in its breeding system between *D. californicum*, which is adapted to cross-pollination, and *D. minus*, which is usually autogamous and can be pollinated when still in bud. In *D. alisma* the stamens are spread at the start of anthesis, exposing the stigma, but self-pollination may occur if cross-pollination fails (Vuille, 1987).

For detailed accounts of the ecology and conservation of *D. alisma* in Britain see Birkinshaw (1994) and Marren & Rich (1993).

Damasonium alisma Mill.

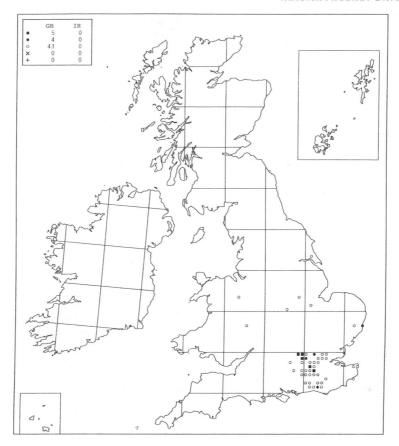

	GB	IR
■	5	0
●	4	0
○	43	0
×	0	0
+	0	0

Damasonium alisma Mill. Starfruit

D. alisma is a plant of shallow water or exposed mud in lowland ponds, where it grows on acidic substrates which are disturbed and sparsely vegetated. Many of its British sites were ponds on grazed commons, where fluctuating water-levels and disturbance by stock ensured that the marginal vegetation remained open. The surviving populations probably rely for their survival on deliberate management by conservationists. It is found amongst sparse emergents such as *Glyceria fluitans* or *Sparganium erectum* at its current sites; other associates include *Apium inundatum*, *Callitriche* spp., *Juncus articulatus*, *J. effusus*, *Potamogeton natans*, *Ranunculus flammula*, *R. peltatus* and *Solanum dulcamara*. Although almost all British records are from ponds, there were isolated occurrences at the edge of lakes and rivers, and in canals and ditches.

D. alisma is an annual or short-lived perennial. A high proportion of fresh seed is dormant, and the germination of additional seeds is stimulated by drying followed by resubmergence. The seeds usually germinate under water in late autumn or early winter and develop into small, submerged plants with linear leaves. Floating leaves may develop in autumn or spring. If water levels fall, terrestrial leaves may be produced by plants with submerged or floating leaves. Plants with floating leaves and those with terrestrial leaves will flower. Although the species will complete its life cycle in water, it eventually becomes outcompeted by aquatic plants such as *Glyceria fluitans* unless the habitat dries out occasionally. Plants vary greatly in size depending on environmental conditions,

and exposed plants will continue to flower only while the substrate remains moist. Aquatic plants may produce hundreds of flowers whereas exposed plants sometimes produce only one. The individual flowers are ephemeral and one or two seeds, one much larger than the other, develop in the conspicuous follicles. The seeds are larger and the seed output less than that of many commoner species that grow on exposed mud (Salisbury, 1970). The smaller seed tends to be released into the water once the fruits are resubmerged, with the larger seed remaining in the follicle. *D. alisma* may fail to appear in years when the climatic conditions are unsuitable. After disturbance, plants may reappear at particular sites after long intervals, suggesting that the seeds may remain dormant for many years.

This southern species was, surprisingly, present in Britain in the last glacial period (Godwin, 1975). In historical times it was never common, but it suffered a catastrophic decline following the enclosure of common land. Some of the ponds in which it grew have been filled in, though most are still present but lack suitable management. They are no longer visited by stock and have become overgrown or surrounded by woodland, or have a stable water level which is artificially regulated for fishing, or are infested by large flocks of wildfowl. Pond restoration has resulted in the reappearance of this species in at least two sites in recent years, and deliberate management of its sites will probably save this species from extinction.

D. alisma has a Mediterranean-Atlantic distribution, occurring in S. Europe, S.W. Asia and N. Africa and extending north along the Atlantic coast to England.

HYDROCHARITACEAE

A family of 19 rather small genera, all of which are aquatics found in freshwater or marine habitats. The members of the family are very variable in both vegetative and floral parts, even in our area. Many species are dioecious, and male and female plants of the same species can have different distributions at both the local and the world scale. It is not, therefore, surprising that for many species vegetative propagation is the main means of reproduction. Detailed accounts of many of the fresh-water genera have been published by C. D. K. Cook and his colleagues in recent years.

Hydrocharis L.

There are three species of *Hydrocharis*: *H. chevalieri* (Blume) Backer (found in tropical Africa), *H. dubia* (De Wild.) Dandy (S. & E. Asia, Australia) and *H. morsus-ranae* (discussed below). For an account of the genus, including a summary of the morphology, ecology and distribution of all three species, see Cook & Lüönd (1982b).

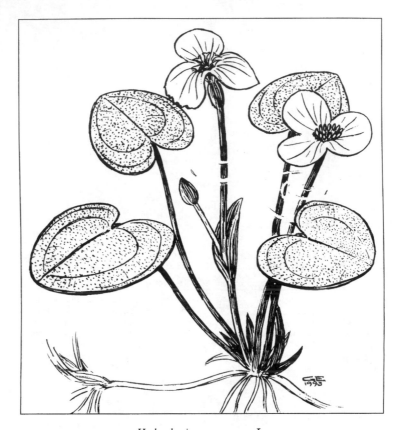

Hydrocharis morsus-ranae L.

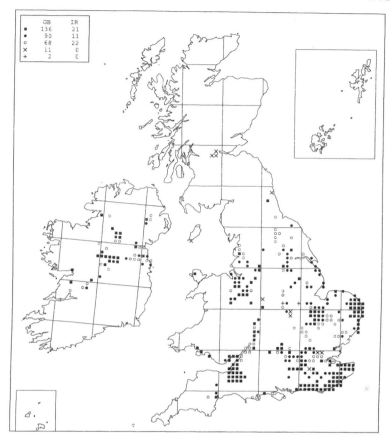

	GB	IR
■	136	21
●	90	11
○	68	22
×	11	0
+	2	0

Hydrocharis morsus-ranae L.   **Frogbit**

The floating rosettes of this species are normally found
in shallow water in the sheltered bays of lowland lakes
(where they often grow amongst the scattered stems of
emergents), ponds, canals and drainage ditches. It
favours calcareous, mesotrophic or meso–eutrophic
water and occurs in water bodies over both peat and
mineral soils. It is a characteristic species of freshwater
ditches in grazing marshes, where it tends to grow in
narrow gravity-drained ditches in water less than 1m
deep. Here emergent species are kept in check by grazing and the water is too shallow to support more vigorous competitors such as *Nuphar lutea* and *Nymphaea
alba*. In such localities *H. morsus-ranae* is often associated with other floating species, including *Lemna gibba*,
L. minor, *Spirodela polyrhiza* and *Wolffia arrhiza*.
Sometimes, as in the Somerset Levels, associated submerged species are few (*Lemna trisulca* and *Ceratophyllum demersum* being the most frequent), but in the
Pevensey Levels and Broadland there is a richer submerged community. The smaller emergents, such as
Alisma plantago-aquatica and *Oenanthe fistulosa*, are frequent associates. *H. morsus-ranae* has successfully colonized some canals. If exposed to terrestrial conditions by
falling water-levels it is able to survive for only short
periods.

The proportion of *H. morsus-ranae* plants in a
population which flowers varies from season to season.
Populations in southern England can flower freely in
warm summers, but fewer flowers are produced in
cooler conditions and some populations in northern and
western sites appear to flower very sparingly. Individual
rosettes bear either male or female flowers. Most plants
are dioecious, producing male or female rosettes, but a
minority are monoecious, producing some male and
some female rosettes (Scribailo & Posluszny, 1984). Ripe
seed is rarely reported from Britain and Ireland, but
fruiting plants are easily overlooked as the peduncle
elongates and bends over after fertilization and the fruit
matures under water. Plants fruited in Britain in the hot
summers of 1947 and 1995 (Gurney, 1949; Preston &
March, 1996). Seedlings have not been reported in
Britain. For a detailed account of sexual reproduction of
the species in Canada, see Scribailo & Posluszny (1984,
1985). Plants multiply by vegetative reproduction, new
rosettes developing on the end of stolons. They overwinter as winter-buds which develop on stolons in the late
summer and autumn, and germinate in spring. Most
winter-buds require two weeks at temperatures approach-
ing 15°C for germination (Richards & Blakemore, 1975).

H. morsus-ranae is still almost ubiquitous in the
Somerset Levels, but has decreased markedly in some of
its former strongholds in recent years. In Cambridgeshire it was described as 'rather common ... throughout
the fens' by Perring *et al.* (1964) but it has since decreased
dramatically and is now restricted to a few sites (Payne,
1989). A similar decline has been documented in
Romney Marsh by Mountford & Sheail (1989). The
main cause has been the conversion of grazing land to

arable, after which those small ditches which have not been completely eliminated become overgrown with emergents. *H. morsus-ranae* survives in such areas only in the small fragments of pasture which remain, or in Cambridgeshire in ponds and ditches in washland with an internal drainage system. Even in areas of Broadland which are still managed as grazing marsh, *H. morsus-ranae* has declined. Some 26% of ditch samples which in 1972–74 had meso-eutrophic species-rich plant communities dominated by *H. morsus-ranae* and *Stratiotes* had other communities in 1988-89 (Doarks, 1990). Many ditches which formerly supported *H. morsus-ranae* are now dominated by *Lemna minor*, *L. trisulca* and filamentous algae, a community characteristic of more eutrophic sites. *H. morsus-ranae* is occasionally found as an introduction and has become well established in the Forth and Clyde Canal, well beyond the northern limit of its native distribution.

H. morsus-ranae is widespread in W. and C. Europe, extending north to S. Sweden and S. Finland and south to N. Italy. It occurs at scattered localities in E. Europe and Asia, extending to the south shore of the Caspian Sea; there are also reports from C. Asia. Its native distribution is mapped by Cook & Lüönd (1982b). The species also occurs as an alien in N. America, where it escaped from a botanic garden at Ottawa into an adjacent canal in 1939 and has since spread over much of S.E. Ontario, S. Quebec and New York State (Dore, 1968; Catling & Dore, 1982). Seed-set is apparently poor throughout the range of the species, perhaps because of the local separation of the sexes.

Stratiotes L.

Stratiotes has a fossil record dating back 48 million years. A number of species have been described from fossil material, including some from Tertiary freshwater deposits in Hampshire and the Isle of Wight. Today only a single species survives, *S. aloides*. For a review of the fossil members of the genus and a detailed account of *S. aloides*, from which much of the information below has been taken, see Cook & Urmi-König (1983).

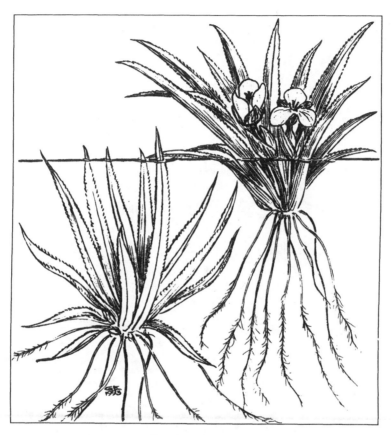

Stratiotes aloides L.

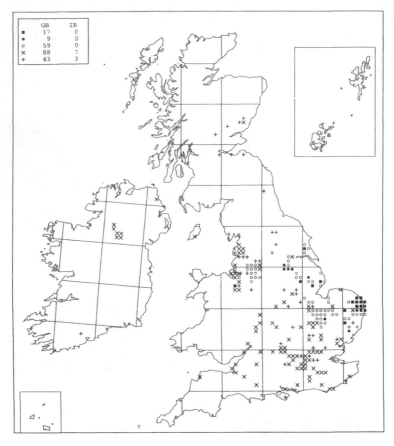

	GB	IR
■	17	0
●	9	0
○	59	0
×	88	7
+	43	3

Stratiotes aloides L.   **Water-soldier**

The unmistakable rosettes of *S. aloides* float on the surface of still or slowly flowing water. In the East Anglian areas where it appears to be native *S. aloides* grows in calcareous, meso–eutrophic water; it was formerly found in lakes and slow-flowing rivers but is now virtually confined to drainage ditches. In Broadland it occurs in the most species–rich grazing marsh ditches, where its associates include *Hydrocharis morsus-ranae*. It is intolerant of brackish conditions and it is also unable to grow on mud, and is therefore absent from sites where there are marked fluctuations in water-level. *S. aloides* is also found as an alien in lakes, ponds, canals, ditches, gravel-, clay- and marl-pits. Most introduced populations are probably derived from material discarded by gardeners. At some sites it has increased rapidly to cover the water surface, as happened in the Basingstoke Canal in the 1950s (Lousley, 1976) and in the Lancaster Canal in the 1960s (Greenwood, 1974). Although some alien populations become well established, others are only transient.

Plants overwinter as rosettes on the bottom of the water. In spring new leaves with gas-filled intercellular spaces develop and the rosettes rise to the surface. In deeper water, or in cool summers in some northern localities, they may remain on the bottom of the water. The flowers are showy and apparently have an unpleasant smell like that of carrion. All British and Irish plants are female (although the female flowers have staminodes which are usually sterile but sometimes produce pollen).

Reproduction here is entirely vegetative: new rosettes (sometimes open, sometimes almost bud-like) are produced on lateral shoots throughout the summer. Parthenocarpic fruits often develop, but contain unripe ovules; the fruits reported from Britain by Geldart (1906) and Crackles (1982) were presumably of this sort as the authors were unable to detect ripe seeds. In autumn, plants become weighed down by dead, waterlogged leaves and sink to the bottom of the water. The species apparently lacks a means of long-distance dispersal over land.

S. aloides was known to 17th century botanists in Cambridgeshire, Lincolnshire, S. Yorkshire and Holderness (Gerarde, 1633; Ray, 1660); it therefore seems likely that it is native in Eastern England and perhaps also in Cheshire. Cook and Urmi-König (1983) argue that the absence from Britain of the dragonfly *Aeshna viridis* Eversmann, which is dependent on *S. aloides*, is a sign that *S. aloides* itself is not native. However, another dragonfly, *A. isosceles* (Müller), is restricted in Britain to Broadland and is there dependent on *S. aloides* (Leyshon and Moore, 1993). The species is less widespread than formerly, and less abundant in its stronghold, Broadland. It is extinct as a native in Cambridgeshire, Huntingdonshire and S.W. Yorkshire, and it has decreased in Lincolnshire and S.E. Yorkshire. It was abundant in Broadland in the late 1940s, probably because nutrient levels had increased in the preceding 50 years and taller and more robust species had replaced the charophytes and smaller vascular plants growing in the hitherto calcareous but infertile waters (George, 1992).

With further increases in nutrient levels in the 1950s and 1960s it virtually disappeared from the broads themselves. It is still widespread in drainage ditches, although the mesotrophic communities in which it occurs are particularly threatened by eutrophication (Doarks, 1990). By contrast, *S. aloides* may have declined since 1967 at Catfield and Irstead Fens because of progressive nutrient depletion at a site which has become more oligotrophic and acidic (Wheeler and Giller, 1982).

S. aloides is widespread in C. & N. Europe, reaching its northern limit at 68° in Finland; it extends east through European Russia to Siberia. It has been grown as an ornamental plant for nearly three centuries, and the native range has been obscured by numerous escapes from cultivation. Cook and Urmi-König (1983), who map its distribution, consider that it is most likely to be native in Sweden, Finland, Denmark, the Low Countries, N. Germany, Poland, the Po plain in Italy and the countries along the Danube. The sexes are not evenly distributed: north of a line drawn from Cherbourg to St Petersburg almost all the plants are female, but males predominate in Belgium and both sexes are present by the Danube. Seed-set is apparently scarce even when both sexes occur together, although fruiting plants may be overlooked as the seeds ripen in October when most botanists have retreated indoors for the winter. *S. aloides* is decreasing in Belgium, the Netherlands and Germany.

Egeria Planch.

There are two species of *Egeria*, both native to S. America. One of them, *E. densa*, has become widely naturalized elsewhere. The genus has been monographed by Cook and Urmi-König (1984).

Egeria densa Planch.

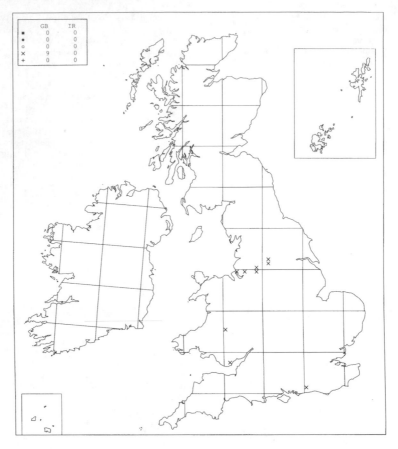

	GB	IR
■	0	0
●	0	0
○	0	0
×	9	0
+	0	0

Egeria densa Planch.   **Large-flowered Waterweed**

Egeria densa was formerly naturalized in a mill stream
and canals in S. Lancashire which received warm water
discharge from cotton mills. Here it grew with other
alien aquatics, including *Lagarosiphon major* and
Vallisneria spiralis. At one locality the plants were
described as 'growing in hot oily water'. It still grows
with *Lagarosiphon major* in a disused canal near St
Helens, in sections which are unheated and in others
which receive hot water from the Pilkington Glass
works. It is also established in the unheated Calder and
Hebble Navigation, where it has persisted for over ten
years, and it has been recorded in the nearby R. Calder.
Plants have also been reported from a pit in Wallasey
and a quarry pool S. of Llanelwedd Rocks, but it is not
known whether these were casual occurrences.

   E. densa overwinters as leafy shoots, which can survive short periods under ice but are killed by prolonged
cold. Plants flower freely at temperatures between 15°C
and 25°C, and in the warm canal water at one
Lancashire site the population flowered continuously
from May to December (Savidge, 1963). The flowers are
borne above the surface of the water. The Lancashire
population is male, and the same is likely to be true of all
the British plants. The species lacks specialized means of
vegetative dispersal. Short stem fragments root readily
from 'bud nodes'. Long-distance dispersal depends on
human agency.

   E. densa was first identified in the Ashton Canal,
Droylesden, in 1953 (Kent, 1955a); it had been known
there for some years but misidentified as *Elodea canadensis*.
The species was discovered in the Calder and Hebble
Navigation and in the nearby R. Calder in 1978, and was
still present in small quantity in the canal in 1992. The
records from Captains Pit, Wallasey, and the quarry pool
at Llanelwedd Rocks were made in 1976 and 1988 respect-
ively. All British populations were probably introduced by
aquarists emptying their tanks into outside waters.

   This species is a native of S. America (Brazil,
Uruguay, Argentina and perhaps Chile), where it is usually found in still water 1–2m deep. Both sexes occur
within the native range but male plants are commoner
than females and seed is very rarely set. This 'is perhaps
the most universally available aquarium plant' (Cook and
Urmi-König, 1984) and is cultivated in tanks and ponds
both for horticultural reasons and as an experimental
subject. Male plants are recorded in natural habitats in
Europe, E. Africa, Japan, Australia, New Zealand and
N. and C. America. Many populations are transient, but
the species is truly naturalized in warm temperate and
cool sub-tropical regions. Unlike some other aquatics it
has not become a serious pest in many areas, but in the
south-eastern states of the USA it can be a troublesome
weed. The native and naturalized populations are
mapped by Cook and Urmi-König (1984).

Elodea Michx.

The genus *Elodea* is native to temperate N.
America and temperate and sub-tropical S.
America; in C. and tropical America it is replaced
by the related genus *Apalanthe*. *Elodea* is taxonomically difficult in its native range: plants are so
variable in vegetative characters that flowers are
necessary for a confident identification. H. St
John recognized 17 species of *Elodea* in a series of
papers published between 1962 and 1965, but his
work proved to be unsatisfactory and in a recent
revision Cook & Urmi–König (1985) have reduced
the number of species to five. Hybrids between *E.
canadensis* and *E. nuttallii* have been created artificially and are believed to occur in the wild.

Three species of *Elodea* have become established in Europe, and all of them have been
recorded from Britain. 'Outside America *E. canadensis* and *E. nuttallii* are remarkably constant, and
it is rare to find plants that cannot be assigned to
one or the other species' (Cook & Urmi–König,
1985). The species of *Elodea* are dioecious and
almost all the plants currently established in
Europe are female, so there is a negligible possibility of *in situ* hybridization. Plants fragment readily
and the fragments produce adventitious roots. The
fragments are presumably distributed by birds,
boats or anglers. The rapidity with which two
species have spread in Britain is remarkable in
view of the fact that they rely on the dispersal of
unspecialized vegetative fragments.

Much of the information cited below is taken
from Cook & Urmi–König's (1985) monograph,
and from the studies of the history, taxonomy and
ecology of *Elodea* in Britain and Ireland published
by Simpson (1984, 1985, 1986, 1988, 1990). For a
review of the biology of *E. canadensis*, see Spicer &
Catling (1988).

Elodea canadensis Michx.; *E. nuttallii* (Planch.) H. St. John

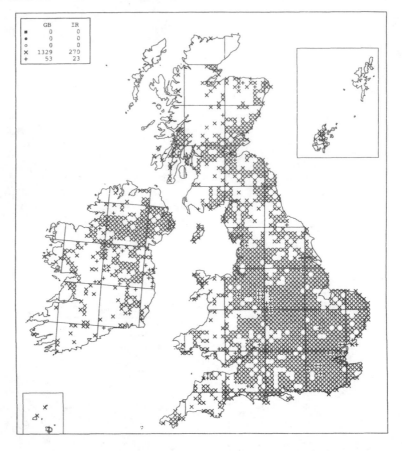

Elodea canadensis Michx.   Canadian Waterweed

A shallow-rooted plant, found in still or slowly flowing, mesotrophic or eutrophic, often calcareous water in lakes, reservoirs, ponds, rivers, streams, canals and ditches. It extends from shallow water to depths of at least 3m, and can form dominant stands in the most deeply colonized zones of lakes (Spence, 1967). It is favoured by high levels of silt deposition, and often replaces *Fontinalis antipyretica* or *Nitella flexilis* in those parts of lakes where most silt is deposited (Spence, 1982). It is one of the few species which can persist in anaerobic substrates, and it became abundant in Rescobie Loch when many of the macrophytes which were formerly recorded there were eliminated by eutrophication (Spence, 1964). It is also able to persist in canals with heavy boat traffic (Murphy & Eaton, 1983). It is primarily a lowland species, but occurs up to 440m at Loch Loch.

Plants of *E. canadensis* die down in the winter, regrowing from underground stems in spring. The female flowers are borne on long hypanthia and float on the water surface. The male flowers are initially attached to long fragile pedicels which usually break so that the flower floats freely. Pollen is released on to the water surface and thus reaches the female stigmas. In the native range of the species fruits are rarely found, probably because the sexes rarely grow together. In Europe all populations currently known are female, although male plants were discovered in a pond on the Braid Hills near Edinburgh in 1879 (Douglas, 1880) and persisted until 1903. Reproduction is entirely by vegetative means. Plants produce rather unspecialized turions in early autumn but reproduction and dispersal is probably also achieved by the spread of undifferentiated fragments.

E. canadensis provides one of the classic examples of the explosive spread of an alien plant. It was recorded in a pond near Waringstown, Co. Down, in 1836 and in a lake (Hen Poo) at Duns Castle, Berwickshire, in 1842. Although there is an unsubstantiated record from the English canal system in 1841, and it has been suggested that it was perhaps present even earlier, the first definite English records were made in 1847 at a reservoir connected to the Grand Union Canal at Foxton, Leics., and at Leigh Park in Hampshire. Thereafter it spread rapidly. From Foxton it spread into the canal system, reaching several midland counties (including Derbyshire, Northamptonshire, Staffordshire and Warwickshire) by 1850. Plants from Foxton were taken into cultivation in Cambridge Botanic Garden in 1847 and placed in a nearby stream in 1848, from whence they spread into the R. Cam and R. Great Ouse and thus colonized Fenland (Marshall, 1852, 1857). In Berwickshire plants were found in the Whiteadder Water in 1848, and by 1850 'had occupied almost every part of the river where the water ran sluggishly almost to choking' (Johnston, 1853). By 1880 *E. canadensis* was widespread in lowland England, S. Scotland and Ireland and expanding into hitherto uncolonized areas. In many newly colonized sites it initially grew in abundance:

drains and sluices were blocked and navigation impeded in some rivers. During the 1880s observers noted that it had declined in abundance in areas where it was once dominant, although it usually persisted in smaller quantity. By 1909 it was certainly less abundant than formerly in areas which had been colonized initially, but still expanding into new localities (Walker, 1912). In recent years it has been replaced in many sites by *E. nuttallii*, although at other localities the two species grow together. For a detailed account of the spread of *E. canadensis* in Britain and Ireland, see Simpson (1984). The origin of the British material is uncertain, but it was probably introduced with American timber (Cook and Urmi-König, 1985).

E. canadensis is a native of temperate N. America. British material was grown in botanical gardens on the European mainland, and from these foci the species spread in the 19th century to become widespread in all regions S. of the Arctic. The species is also naturalized in Australia and New Zealand.

British material was initially described as a new species, *Anacharis alsinastrum*, by Babington (1848) and only later equated with the American *E. canadensis*. It was also given a number of pseudoscientific names, including *Growforevva aquatilis* and *Babingtonia pestifera*, during the period of its initial expansion.

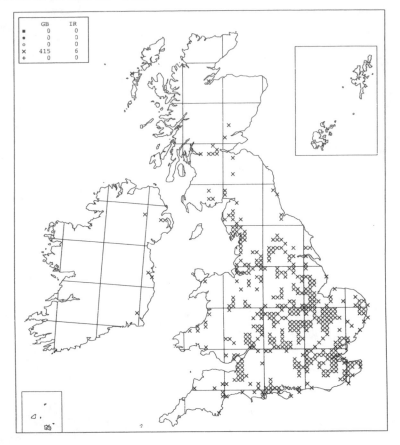

	GB	IR
■	0	0
●	0	0
○	0	0
×	415	6
+	0	0

Elodea nuttallii (Planch.) H. St. John
Nuttall's Waterweed

E. nuttallii favours still or slowly flowing eutrophic water. Like *E. canadensis*, it is found in a wide range of water bodies including lakes, reservoirs, sand- and gravel-pits, ponds, rivers, streams, canals and ditches. It is often found in species-poor macrophyte communities in managed rivers, canals with heavy boat-traffic and newly colonized gravel-pits, where characteristic associates include *Myriophyllum spicatum*, *Potamogeton pectinatus*, *P. pusillus* and filamentous algae. In grazing marshes *E. nuttallii* tends to be found in the more eutrophic main drains whereas *E. canadensis* occurs in the less eutrophic, species-rich field ditches (Mountford & Sheail, 1984). In ponds and shallow streams it can be abundant, sometimes growing in such dense masses that birds such as wagtails can walk from one side of the water to the other. It is resistant to oil pollution (Burk, 1977). Although most frequently found in shallow water, it can occur to depths of 3m or more in Windermere (Simpson, 1990). *E. nuttallii* has only been recorded in lowland habitats.

Plants of *E. nuttallii* overwinter as prostrate shoots at the bottom of the water. New shoots grow upwards in spring, elongating rapidly. If they reach the water surface they branch to form a dense canopy just below the surface. The female flowers, like those of *E. canadensis*, are borne on long hypanthia and float on the water surface. The male flower, however, is released by abscission of the pedicel when still in bud. The bud contains a gas bubble and floats to the surface, where it opens to release the pollen. Only female flowers are known in Britain. Turions are not known in this species, and plants spread by vegetative fragments.

E. nuttallii was first recorded in Britain from a ditch at Beard Mill, Stanton Harcourt, in 1966. It was again found near Oxford in 1969, and from 1970 onwards in an increasing number of sites in England. It apparently spread rapidly, but its appearance caused some taxonomic confusion and the details are therefore less well documented than those of the spread of *E. canadensis*. It was first found in Scotland near Allangrange House, Munlochy, in 1978 and in Ireland at Lough Neagh in 1984 (Simpson, 1985). At some sites it replaced *E. canadensis*, although detailed studies at Amberley Wild Brooks have shown that after an initial period when *E. nuttallii* was dominant, the two species now grow together (Briggs, 1990). *E. nuttallii* is certainly now commoner than *E. canadensis* in S.E. England, and is continuing to spread in Scotland and Wales. Its current distribution is probably under-recorded on the map. British material probably originated from plants grown in aquaria or garden ponds and then discarded into the wild.

E. nuttallii is native to temperate N. America. It has a similar range to that of *E. canadensis*, although with a more southerly bias. It was first recorded in Europe in Belgium in 1939, and was also known in the Netherlands (1941) and Germany (1961) before it was discovered in

Britain. Although most European plants are female, a male colony is known from Germany. In Japan, where the species was first found in the early 1960s and has since spread rapidly, all plants are male.

The belief that *E. nuttallii* is a phenotype of *E. canadensis* which develops in eutrophic water was initially held by some observers, but has been disproved not only by Simpson's (1988) experimental work but by the fact that when both species are found growing together they are easily distinguished. Detatched fragments of *E. nuttallii* root more rapidly than those of *E. canadensis*. Simple growth studies suggest that *E. nuttallii* has a competitive advantage over *E. canadensis* as its stems elongate much more rapidly in the period after planting, and are therefore likely to form a canopy more rapidly (Simpson, 1990). A more detailed study of the competitive relationship of the two species would be worthwhile.

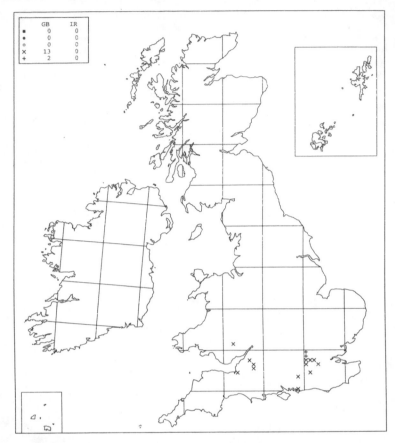

	GB	IR
■	0	0
●	0	0
○	0	0
×	13	0
+	2	0

Elodea callitrichoides (Rich.) Casp.
South American Waterweed

This species has been recorded from still and slowly flowing water in ditches, canals, ponds, ornamental lakes and boating pools in parks and the quieter stretches of lowland streams and rivers. It was present in great abundance in some ditches, canals and rivers when it was first discovered, sometimes to the exclusion of most other aquatics.

The flowering behaviour of *E. callitrichoides* is similar to that of *E. canadensis*, but male flowers remain attached to the parent during anthesis. Only females are recorded in Britain. The species lacks specialized turions, and dispersal is presumably by the spread of vegetative fragments.

Elodea callitrichoides was first collected in Britain in the R. Colne near Harefield in 1948, and in the Longford River at Stanwell in 1950 (Grigg, 1951; Simpson, 1984). Although it originally grew in great profusion at

Stanwell, it was very scarce there by 1964 and has not been seen since (Kent, 1975). It has been recorded since 1950 at a number of scattered localities but most populations have proved to be transient. The longest-established colony is in the Chichester Canal, where it was first found in 1959 and still present in 1995. British populations are believed to be derived from material discarded by aquarists.

Elodea callitrichoides is a native of temperate S. America, where it appears to be uncommon. Unlike *E. canadensis* and *E. nuttallii*, it is not widely naturalized elsewhere. Female plants were first recorded on the European mainland in 1958 near Strasbourg, France, and have since been seen at several sites near the R. Rhine in Germany. The only persistent populations are in thermally polluted water. As in Britain, these populations are believed to derive from plants grown by aquarists: the species is known to have been offered for sale in Germany since 1928.

Hydrilla Rich.

Hydrilla is a monotypic genus. The taxonomy, distribution and ecology of the sole species, *H. verticillata*, have been reviewed by Cook and Lüönd (1982a).

Hydrilla verticillata (L. f.) Royle

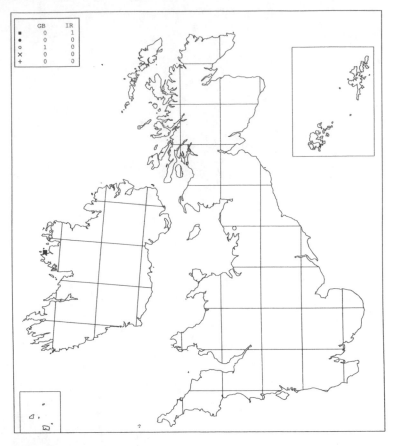

	GB	IR
■	0	1
●	0	0
○	1	0
×	0	0
+	0	0

Hydrilla verticillata (L. f.) Royle
Esthwaite Waterweed

This species is known only from one locality in Britain and from one in Ireland. It was discovered at Esthwaite Water by W. H. Pearsall in July 1914 (Bennett, 1914; Druce, 1916). It grew in slightly peat-stained water some 2–3m deep, with *Callitriche hermaphroditica*, *Najas flexilis* and *Potamogeton berchtoldii* over inorganic, blue-grey clayey mud (Pearsall, 1921). In 1935 it was discovered by Pearsall's son, also W. H. Pearsall, in Lough Rusheenduff (Pearsall, 1936). Here it can grow in water as shallow as 5cm but it is usually found washed up from deeper water, often mixed with *Najas flexilis*.

This species was never seen to flower at Esthwaite, although Pearsall (1921) looked carefully for flowers for six years. The Irish plants have only once been found flowering in the wild, in shallow water during a spell of warm weather, but female flowers are produced abundantly when plants from Lough Rusheenduff are cultivated under glass (Scannell, 1976b; Scannell & Webb, 1976). Reproduction is by turions which develop in the leaf axils, or terminally on stems and branches, and become detached in late summer or autumn. Bulbil-like buds are also formed beneath the substrate at the tip of the stolons.

H. verticillata seems unlikely to have been introduced to Britain or Ireland by human agency. It has been suggested that both populations might arise from turions carried on the feathers or feet of wildfowl (Pearsall,

1921; Webb & Scannell, 1983). The fact that the Irish plants are apparently all female, and could therefore represent a single clone, is consistent with this hypothesis. At Esthwaite Water, Pearsall (1921) considered that it was more abundant in 1920 than when it was discovered in 1914, and it was still present in unchanged quantity twenty years later (Pearsall, 1936). However, it was last seen in 1941. It has not been found subsequently despite frequent searching (Lund, 1979) and is believed to have become extinct. It has been suggested that it disappeared because of progressive eutrophication (Richards & Blakemore, 1975). It still survives in Lough Rusheenduff, although not in the abundance described by the younger Pearsall (Webb & Scannell, 1983).

H. verticillata is native to the Old World, but it is rapidly expanding its range in warmer regions and the original distribution is uncertain (Cook & Lüönd, 1982a). In mainland Europe it is confined as an apparent native to the north-east, where it extends from N.E. Germany through Poland and the Baltic States to White Russia. Here it grows in alkaline, moderately calcareous, mesotrophic or slightly meso-eutrophic waters with associates such as *Ceratophyllum demersum*, *Myriophyllum verticillatum*, *Potamogeton natans* and *Stratiotes aloides* (Pietsch, 1981). It is also reported as an alien from warm springs in Austria. It is widespread in Asia from Iran and Afghanistan southwards and eastwards to Manchuria, Japan and New Guinea. It is probably native in Australasia and Africa, but in both continents it is actively spreading. In N. America it was first recorded in

Florida in 1960 and is now known from several south-eastern states; it has also been recorded recently in California, C. America and Jamaica. As an introduction it tolerates a wide range of water conditions from acidic and oligotrophic to eutrophic or brackish; it 'thrives on many kinds of pollution and nutrient enrichment, and it tolerates a great deal of disturbance' (Cook & Lüönd, 1982a). It has become a pest of shallow waters in the tropics and subtropics, as it grows rapidly and can block fish ponds, irrigation ditches and canals.

H. verticillata is both phenotypically and genetically variable (Verkleij *et al.*, 1983). Material from Britain and Ireland has been included with plants from N.W. Poland in var. *pomeranica* (Rchb.) Druce, and is said to differ from plants growing further east in Europe, but Cook & Lüönd (1982a) do not recognize any intraspecific taxa. For detailed accounts of the Irish plant, see Scannell & Webb (1976) and Czapik (1978).

Lagarosiphon Harv.

There are nine species of *Lagarosiphon*, all natives of Africa. One species, *L. major*, has become natu-ralized in Europe. A monograph of the genus is available (Symoens & Triest, 1983).

Lagarosiphon major (Ridl.) Moss

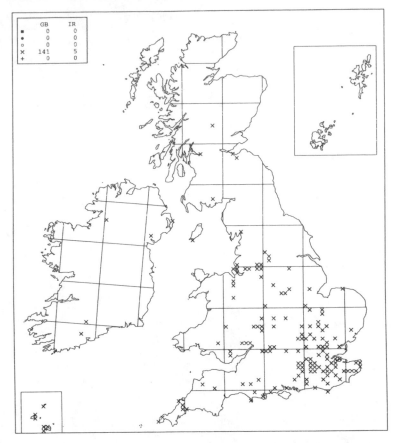

	GB	IR
■	0	0
●	0	0
○	0	0
×	141	5
+	0	0

Lagarosiphon major (Ridl.) Moss
Curly Waterweed

L. major is found in standing water in chalk-, clay- and gravel-pits, disused quarries and in lakes and ponds. In some small ponds it can be present in dense masses. It is also established in canals. It is confined to lowland sites.

The female flowers of *L. major*, like those of *Elodea*, are borne singly in the axils of the leaves. The hypanthia are long and the flowers float on the surface of the water. The male inflorescence produces numerous flowers which are released as buds and float to the surface of the water, where they open. Each flower has a sail composed of three staminodes, and the flowers are blown by the wind across the water surface. Pollination occurs after a male flower encounters a female. Few observers have noted the sex of British populations, but the available evidence suggests that all are female. Plants persist and spread within a water body by vegetative propagation, as small fragments can root and become established. At some sites short lateral branches become detached and washed up in quantity around the edge of the water. There is, however, no evidence that plant fragments or lateral shoots can be carried from site to site by natural agents of dispersal.

L. major was first recorded in Britain from a chalk-pit near Arlesey in 1944, where it has persisted for over 40 years, (Kent, 1955b; Belcher and Swale, 1990). It has subsequently been found in numerous localities, and at many sites it has become well established. It is clearly able to persist through the coldest winters. *L. major* is widely sold as a plant for aquaria and garden ponds (often as 'Elodea crispa' or even 'Canadian pondweed'), and British populations almost certainly result from the deliberate or accidental release of such material.

L. major is a native of southern Africa, where it is 'often an obnoxious water pest' (Obermeyer, 1964). It has become naturalized locally in W. Europe and much more extensively in New Zealand, where it was first recorded in 1950 and is now a troublesome weed (Mason, 1960).

Vallisneria L.

A small genus of at least three closely-related species. As with many dioecious plants, the taxonomy of *Vallisneria* is complicated by the fact that taxa have often been described on the basis of only one sex. The genus has been reviewed by Lowden (1982) who recognizes two species, *V. americana* Michx. and *V. spiralis*, each with two varieties. In addition to these, *V. caulescens* F. M. Bailey & F. Muell. should be recognized as an Australian endemic (McConchie & Kadereit, 1987). Further species may occur in Australia, but are poorly known. *V. spiralis* var. *spiralis* is native to Europe and occurs in Britain as a naturalized alien.

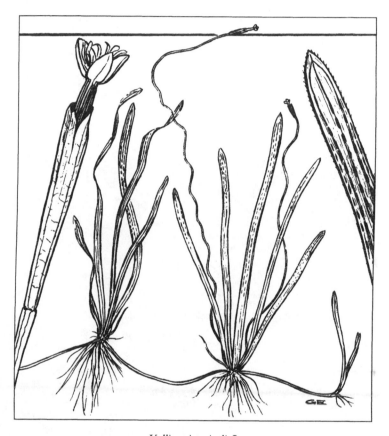

Vallisneria spiralis L.

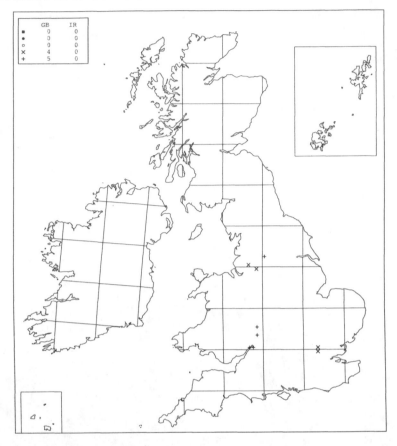

	GB	IR
■	0	0
●	0	0
○	0	0
✕	4	0
+	5	0

Vallisneria spiralis L.   Tapegrass

V. spiralis is recorded from several canals and a brick-pit.
It is well naturalized in the R. Lea Navigation canal,
where it was first found in 1961 (Harris & Lording,
1973). Here it grows in water of negligible flow over a
substrate of mud and gravel. In shallow water at the edge
of the canal it forms dense clumps and can be the dominant macrophyte, but it is absent from water deeper than
1m in the central channel and the vicinity of locks. This
canal does not receive heated effluent. In the Reddish
Canal (now drained), *V. spiralis* formerly grew in water
heated by mill effluent to 29°C, but it was absent from
the hottest stretches where temperatures could reach
38°C (Savidge, 1963).

The solitary female flowers are borne at the end of
long pedicels, and float on the surface of the water. The
male inflorescence bears hundreds of flowers which are
released as buds and float to the surface, where they
open. Pollination occurs after the male flowers encounter the female (Faegri & Pijl, 1971). After pollination the
pedicel spirals and the flower is drawn back under the
water. British populations flower freely; Sledge (1942)
noted that in a canal near Halifax plants flowered twice a
year, in May and September. Plants are stoloniferous
and reproduce vegetatively. In the R. Lea Navigation
canal dispersal may be aided by dredging, which releases
large rafts of plants that drift downstream (Harris &
Lording, 1973).

V. spiralis was first recorded in Britain in 1868, when
it was collected in stagnant water at Knapp's Brickyard,
Northwick (Amphlett & Rea, 1909). This population
was apparently only transient but in 1906 *V. spiralis* was
discovered in the Reddish Canal, where it persisted for
at least forty years. It also became established in other
thermally polluted canals in industrial N. England, but
the only records since 1970 have been from the R. Lea
Navigation canal. The species is also recorded from the
Berkeley Canal and the R. Avon at Tewkesbury but
there is no evidence that it persisted at either site. The
British populations presumably result from plants discarded from cultivation; early this century it was
'frequently grown in aquariums and bowls by the botan-
ically inquisitive' (Amphlett & Rea, 1909) and it is still a
popular aquarium plant.

Vallisneria spiralis var. *spiralis* occurs as a native in S.
Europe, north to N. France; it is reported as an alien
from scattered localities further north. It is also native in
S.W. Asia, and occurs as an introduction in the West
Indies. It is replaced by var. *denseserrulata* Makino in
Africa, Asia and Australia. A similar species, *V. americ-
ana* Michx., is frequent in eastern N. America; a detailed
account of its ecology is provided by Korschgen &
Green (1988).

The pollination mechanism of *Vallisneria* was
included by Paley (1802) in his 'evidences of the existence and attributes of the deity, collected from the
appearances of nature'.

APONOGETONACEAE

This family contains a single genus, *Aponogeton*.

Aponogeton L. f.

This genus contains 43 species, all of which are aquatic plants. The genus is native to the tropical and subtropical regions of Africa and Asia, with its centre of diversity in Madagascar. The species are found in a range of habitats from shallow, seasonal pools and brackish lakes to fast-flowing torrents. They are perennials which survive unfavourable seasons as rhizomes or starchy tubers: the roasted tubers are sometimes eaten by man. The plants have submerged and/or floating, and occasionally emergent, leaves. Their inflorescences are held above the water-level on peduncles which are inflated and floating or rigid and erect. The plants are monoecious or dioecious, and have flowers which appear to be adapted to pollination by insects. Some species are thought to be apomictic: these may have perfect flowers or may lack stamens, depending on the species. The inflorescence sinks after anthesis and the fruits develop under water. *A. undulatus* Roxb. is viviparous. The basic source of information on the genus is a monograph by Van Bruggen (1985).

Several species are popular aquarium plants which are normally grown from imported tubers. *A. distachyos* is widely grown outdoors in temperate climates and has become established in our area.

Aponogeton distachyos L. f.

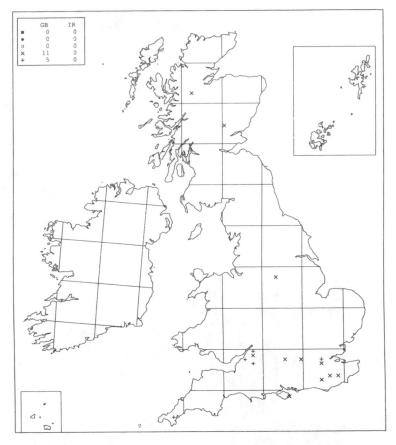

Aponogeton distachyos L. f.   **Cape-pondweed**

This species persists as a relic of cultivation in water up to 2m deep in lowland lakes and ponds.

A. distachyos is a tuberous perennial, with floating leaves and an inflorescence of sweet-smelling flowers. The seeds float until the fleshy testa decays (Ridley, 1930). There are few observations on its reproductive biology in Britain, but seedlings have been found at a site near Bristol. *A. distachyos* spreads within a pond, but although some of its sites are remote from houses they may all originate from planted or discarded stock.

It is difficult to specify the point at which a population of this species ceases to become planted and counts as naturalized, and correspondingly difficult to identify the first record of naturalized plants. *A. distachyos* was recorded from a fountain at Avery Hill in 1906 by J. F. Bevis (Grinling *et al.*, 1909), where it was presumably planted. It was collected at Keston, Kent, in 1909, where it still persists (Burton, 1983). There were also early records from Tremough, Cornwall, in 1910, and the Bristol area, where White (1912) recorded it from two localities. The first Scottish record appears to be from a lochan at Hilton Lodge, where it was first seen in 1953 (McCallum Webster, 1978) and still present in 1975.

A. distachyos is native to the Cape Province of S. Africa, where it grows (sometimes in abundance) in shallow pools and slow-flowing streams. It has been naturalized at a site in France for over 150 years, and is also established as an alien in S. America, Australia and New Zealand.

In S. Africa the inflorescences are stewed with mutton and eaten; they are canned commercially and sold as 'waterblommetjies'. We have eaten the canned product and find it of more interest as a curiosity than a delicacy. For a recipe, see Van Bruggen (1985).

POTAMOGETONACEAE

This exclusively aquatic family contains two genera. *Potamogeton* is one of the largest genera of aquatic plants whereas *Groenlandia* is monotypic. The closely related genus *Ruppia* is often included in the Potamogetonaceae, but Stace (1991), following Cronquist (1981), treats it as a separate family, Ruppiaceae.

Potamogeton L.

There are 80–90 species of *Potamogeton*. The genus has a virtually cosmopolitan distribution, with its centre of diversity in eastern N. America and more species in temperate regions than in the tropics. In Europe there are 22 species, of which 21 occur in the British Isles. This is the largest aquatic genus in our area, and it is ecologically important as some species can occur in abundance in a wide range of water bodies.

An overview of the taxonomy of the genus is provided by Wiegleb (1988), but there is no recent world monograph, and there are many taxonomic problems which need to be tackled before one can be prepared. However, the European species are relatively well understood, in part due to the work of J. E. Dandy and G. Taylor, who devoted many years to the study of the genus. Their work is summarized by Dandy's account of the European species (Tutin *et al.*, 1980). Although the *Potamogeton* species have the reputation of being difficult to identify, they are not critical like those in genera such as *Callitriche* and *Ranunculus*. The superbly illustrated monograph of the British and Irish taxa by Fryer & Bennett (1915) is now outdated, and a new account has been prepared (Preston, 1995).

The British and Irish plants can be divided into two distinct subgenera. Subgenus *Potamogeton*

contains the majority of the species. The first nine species in the following account (*P. natans*–*P. perfoliatus*) are included in section *Potamogeton*. They are rhizomatous plants with broad leaves which may be submerged or floating. The next nine species (*P. epihydrus*–*P. acutifolius*) are in section *Graminifolii*. *P. epihydrus* is a rhizomatous plant with narrow submerged and broad floating leaves, but the other members of this section lack rhizomes and have only submerged, linear leaves. They are shallow-rooted plants which are restricted to still or very slowly flowing water. The rhizomatous *P. crispus* is the only member of section *Batrachoseris*. There are two plants in subgenus *Coleogeton*, *P. filiformis* and *P. pectinatus*. They are rhizomatous plants with submerged, linear leaves.

All *Potamogeton* species have bisexual flowers which are individually inconspicuous, but are aggregated into spikes. The members of subgenus *Potamogeton* have emergent inflorescences which are borne on rigid peduncles. The flowers are protogynous and they are traditionally regarded as wind-pollinated. Although this is often true, particularly for the larger species with more robust peduncles, they may also be pollinated underwater or by pollen floating on the surface of the water. If plants are forced to flower under water, perhaps after a sudden rise in water-levels, an air bubble forms when the anthers of the submerged flowers dehisce, and the pollen collects on its surface. As the bubble expands the pollen is carried to stigmas on the same flower. Pollen shed from aerial inflorescences may collect on the water surface and be deposited on stigmas as inflorescences are drawn below the water by strong currents, or released again to resurface.

The peduncles of the species in subgenus *Coleogeton* are flexible and the inflorescences float on the surface of the water. Pollen is transferred

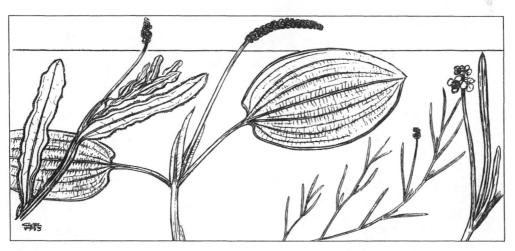

Potamogeton crispus L.; *P. coloratus* Hornem; *P. pusillus* L.

on the surface of the water. These species may also flower under water, and pollen can then be transferred to the stigmas on air bubbles.

There are few detailed studies of the reproductive biology of *Potamogeton* species. Some are known to be self-compatible, but there is evidence that others may be self–incompatible. Many species set fruit freely, and others sparingly. The seeds are hard-coated and will not germinate unless this coat is ruptured. *Potamogeton* seeds are often eaten by birds, and those which are not totally digested are shed in the droppings and germinate more freely than intact seeds. It is, however, very difficult to find seedlings in the wild. It seems likely that new populations may be established by seed, but that plants in existing colonies usually reproduce vegetatively.

The members of *Potamogeton* section *Potamogeton* may spread by rhizomatous growth, or by establishment from fragments of stem. Some also reproduce by buds which develop on stolons in the axils of the leaves. Plants persist through the winter as leafy stems, buds formed on the rhizomes or swollen rhizomes. The linear-leaved species in section *Graminifolii* have specialized turions which develop in the leaf axils. The mature plants decay in the autumn and the species overwinter as turions. These plants are therefore functionally (but not genetically) annual. In subgenus *Coleogeton* plants overwinter as tubers which develop on the rhizomes, and sometimes as leafy shoots.

The genus *Potamogeton* is notable for the occurrence of sterile hybrids, many of which are known to be able to persist vegetatively for many years. The eight most frequent hybrids in our area are treated here. A further 18 hybrids are recorded in Britain and Ireland, and many of these can be found in the absence of one and sometimes of both parents. For details of these hybrids, see Dandy (1975) and Preston (1995).

The Quaternary history of *Potamogeton* is well documented as the fruits are frequently found in deposits laid down in lakes and other aquatic habitats. Many species were apparently frequent in the last glacial and the early stages of this interglacial. As most hybrids do not produce fruits their history is not documented, but some of those now found in the absence of one or both parents may be relics dating from a period when the parents grew together.

As the taxonomy of *Potamogeton* was not clarified until the 1930s, records made before 1940 have been accepted only if supported by herbarium material. Most of the older records on the maps come from a card index of 20,000 herbarium specimens determined by J. E. Dandy & G. Taylor, or by Dandy alone, and now held by the Natural History Museum. For the period after 1940 we have accepted localized records from reliable observers, but we have not included records from 10-kilometre squares or tetrads if we do not know the locality or the recorder. The excluded records include those collected by 'square-bashing' for the *Atlas of the British Flora* and those plotted on maps in 'tetrad' floras. A few relatively common and easily identified species (*P. crispus, P. natans, P. pectinatus, P. perfoliatus, P. polygonifolius*) may therefore be under-recorded in some areas where botanists have not made localized records. The comments on the ecology and distribution of the taxa are based solely on records accepted for the maps.

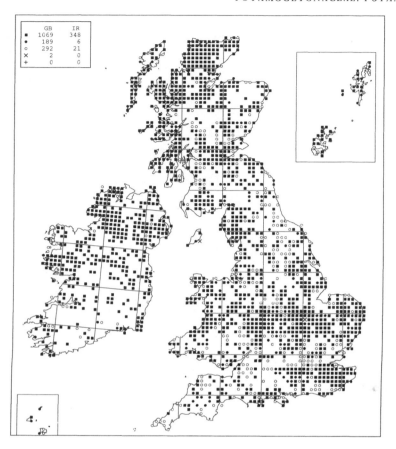

	GB	IR
■	1069	348
●	189	6
○	292	21
×	2	0
+	0	0

Potamogeton natans L.   Broad-leaved Pondweed

P. natans is often present in abundance in the still waters of sheltered lakes, ponds or flooded quarries, or in slowly flowing canals, streams and ditches. It can also grow in the quiet backwaters of more swiftly flowing rivers, and sometimes persists in the more rapidly flowing reaches. Its floating leaves may cover the entire surface of the water, or the species may grow with other floating-leaved species (e.g. *Nymphaea alba*) or in open stands of emergents such as *Carex rostrata*, *Equisetum fluviatile* or *Schoenoplectus lacustris*. It tolerates a very wide range of water chemistry, growing in acidic and oligotrophic lakes and ponds, and in calcareous waters which may be nutrient-poor, mesotrophic or eutrophic. It is found over a correspondingly wide range of substrates, including peat, sand and soft mud. It is most frequently found in water of moderate depth (1–2m), but it can occur as dwarf plants in very shallow water, or on moist ground where lakes or ponds have dried out. It is also found in deeper water, down to depths of 5.4m in particularly clear water in a rain-fed quarry at Castlederg, and it was reported from depths of 6m by West (1910). It occurs from sea-level to 760m (below Stob Ban).

This species overwinters as plants with rather moribund stems, or as rhizomes. These plants initially produce submerged phyllodes in early spring, then floating leaves develop at the surface of the water. In strongly flowing streams and rivers the phyllodes may persist throughout the season. The inflorescence is borne on a robust peduncle and is wind-pollinated. In still or slowly flowing water plants usually flower and fruit freely. However, recent isozyme studies of a population in ditches at Wareham which fruits freely have failed to detect any genetic variation, suggesting that it does not reproduce sexually (Hollingsworth, Preston & Gornall, 1995). Plants spread by growth of the rhizome, and perhaps by stem fragments (Sauvageau, 1893–94). They usually lack specialized means of vegetative reproduction, but fascicles of phyllodes are occasionally produced in the leaf axils and act as propagules (Fryer, 1886). The species is an effective colonist of isolated water bodies (Grime *et al.*, 1988), presumably reaching them as seeds transported by birds.

Fossil evidence shows that *P. natans* is a long-established native which persisted through the last glacial period and has been widespread since the late glacial. It is still a frequent species in suitable habitats throughout Britain and Ireland. As the most frequent species of *Potamogeton* in lowland ponds and ditches, *P. natans* has doubtless decreased in many areas as many of these water bodies have been lost. Local studies of a Norfolk grazing marsh which was converted to arable showed that *P. natans* persisted in the drainage ditches, but in reduced quantity (Driscoll, 1985a).

P. natans is widespread in the boreal and temperate regions of the northern hemisphere. In Europe it extends south from Iceland and northern Scandinavia, but is scarce in southern Europe where it tends to be replaced by *P. nodosus*.

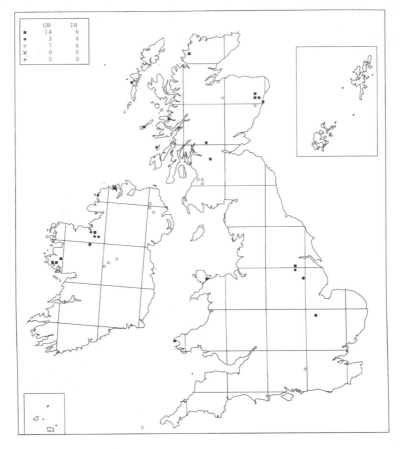

	GB	IR
■	14	6
●	3	4
○	7	6
×	0	0
+	0	0

Potamogeton × sparganiifolius Laest. ex Fr. (*P. gramineus* L. × *P. natans* L.)

Ribbon-leaved Pondweed

P. × sparganiifolius is found in non-calcareous, oligo-trophic or mesotrophic water, in some sites which are calcareous but nutrient-poor, and in at least one eutrophic ditch where *P. pectinatus* is abundant. It can be abundant in shallow and rather rapidly flowing water of rivers and streams, where it may be the dominant macrophyte. It also grows, although not so abundantly, in deeper and slower rivers. Other habitats include shel-tered lakes and fenland drains. The Anglesey population grows subterrestrially in moss carpets and in shallow fen pools, as do some plants on Tiree. There is a single record from the Basingstoke Canal. In still waters it is usually found in the vicinity of its parents, but in run-ning waters it may occur in the absence of one or both putative parents. It is confined to the lowlands.

This is a variable hybrid. Plants in running water can have long, ribbon-like submerged leaves resembling those of *Sparganium emersum*, with which it sometimes grows. In still water the submerged leaves are short, narrow and phyllode-like, and floating leaves predomin-ate. Some populations flower very sparingly (Heslop Harrison, 1949); others flower more frequently but the hybrid is sterile and fruits do not develop. It is a peren-nial which spreads by rhizomatous growth, and detached fragments can perhaps become established as new plants. Like both of the putative parents, *P. × sparganii-folius* lacks specialized means of vegetative dispersal. It can be very persistent in some sites, still surviving, for instance, in the Bealanabrack River at Maam where it was first collected in 1853.

P. × sparganiifolius was first collected in our area at Maam by T. Kirk, and identified by Babington (1856). At this period the existence of *Potamogeton* hybrids had not been recognized, and *P. × sparganiifolius* was regarded as a species. Fryer & Bennett (1915) suggested that it was probably a hybrid, and it is now accepted as the hybrid between *P. gramineus* and *P. natans*. For many years it was regarded as a very rare plant, and even Pearsall (1931), writing over 70 years after its discovery, knew it only from Maam. However, Dandy & Taylor (1939b) reported some additional localities, and subse-quent fieldwork has revealed further sites. The plant has been found in several new localities in recent years, and others doubtless remain to be discovered. R. C. L. Howitt considered that the Nottinghamshire population was threatened by the use of herbicides (J. O. Mountford, pers. comm.).

This is a frequent hybrid in Scandinavia, and is also recorded from the Netherlands, European Russia and Siberia (Dandy, 1975). As in Britain and Ireland, its dis-tribution lies within the range of its rarer parent, *P. gramineus*.

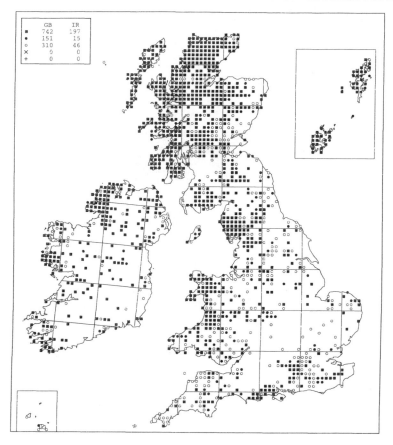

	GB	IR
■	742	197
●	151	15
○	310	46
×	0	0
+	0	0

Potamogeton polygonifolius Pourr.

Bog Pondweed

P. polygonifolius is a calcifuge, which grows as an aquatic in oligotrophic and mesotrophic water and in moist subterrestrial habitats. As an aquatic it is typically found in water less than 1m deep at the edge of lakes, and in the backwaters of rivers, moorland streams and ditches, flooded peat-cuttings and roadside ditches. Unlike other European *Potamogeton* species *P. polygonifolius* frequently grows subterrestrially in flushes, sphagnum lawns and 'brown moss' communities dominated by *Scorpidium scorpioides*. Its many associates include *Callitriche hamulata*, the aquatic form of *Juncus bulbosus*, *P. natans* and *Myriophyllum alterniflorum* in deeper waters, and *Carex nigra*, *C. rostrata*, *Eleogiton fluitans*, *Menyanthes trifoliata*, *Potentilla palustris*, *Ranunculus flammula* and *Utricularia intermedia* sensu lato in shallow water or on damp ground. *P. polygonifolius* is most frequent over peat but also grows on mineral substrates. Although normally strictly calcifuge, it has been found in highly calcareous flushes and fenland pools in N. England and W. Ireland. It occurs from sea-level to 780m at Lochan an Tairbh-uisge.

In deep or rapidly flowing water *P. polygonifolius* may be found as vegetative plants with submerged leaves. In other habitats it usually possesses floating or aerial leaves, and flowers and fruits freely. Plants spread by rhizomatous growth, but lack specialized vegetative propagules.

Fossil evidence suggests that this species has been present in Britain and Ireland since at least the middle of the last glacial period (Godwin, 1975). In the acidic peatlands of northern Britain and western Ireland this is the most frequent *Potamogeton* species, which in many areas can be found in almost any roadside ditch and flooded peat-cutting. The map probably underestimates its frequency in some of these areas. In the south and east, *P. polygonifolius* has decreased as its scattered acidic habitats have been drained and taken into agricultural use. It is extinct in some counties, including Middlesex and Cambridgeshire, and it is rare and vulnerable in others, including Kent, Essex, Suffolk and Lincolnshire.

P. polygonifolius is an amphi-atlantic species, known from Europe, N. Africa and N. America. It is most frequent in W. Europe, from C. Norway southwards to Spain, Portugal and the Azores, but it extends eastwards to Finland and Bulgaria. In N. America it is confined to Newfoundland and some nearby islands.

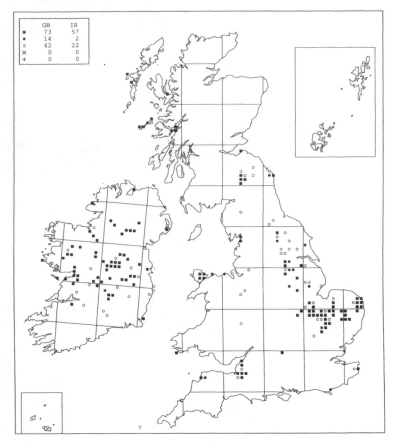

	GB	IR
■	73	57
●	14	2
○	42	22
×	0	0
+	0	0

Potamogeton coloratus Hornem.   **Fen Pondweed**

Although this species is closely related to *P. polygoni-folius* it is a strict calcicole, and is therefore quite distinct ecologically. It is usually found in shallow, calcium-rich but nutrient-poor water at the sheltered edges of lakes on limestone, and in slow-flowing streams, fenland pools and ditches and pools on calcareous sand dunes. It has also colonized flooded clay- and marl-pits. It is found over a range of substrates including peat, fine calcareous marl, limestone gravel, sand and clay. It is often associated with charophytes, including *Chara hispida*, *C. pedunculata*, *C. rudis* and *C. vulgaris*. Although it normally grows in different sites to *P. polygonifolius*, the two species can be found together towards the edge of the range of *P. coloratus*, as in swamps on Lismore. *P. coloratus* is confined to the lowlands.

P. coloratus is a rhizomatous perennial which overwinters as leafy shoots. Plants flower freely in shallow water, and are self-compatible. The species usually produces numerous, small fruits. In periods of drought *P. coloratus* may persist as terrestrial plants, or as buds on the rhizome (Fryer, 1887b; Fryer & Bennett, 1915).

Although there is a single fossil record of this species from the last glacial period, it has not been recorded from late glacial deposits. As in previous interglacials, it

only appears as fossil fruits once the climate becomes more temperate (Godwin, 1975). In the historic period the British range of *P. coloratus* has contracted as some of its Fenland sites have been drained, or following the conversion of grazing land with a managed ditch system to arable. The species has also been adversely affected by eutrophication. Many of its remaining semi-natural habitats are protected as nature reserves or SSSIs, and are therefore safeguarded from habitat destruction, if not from drying out or eutrophication. *P. coloratus* has also colonized disused clay-pits, but these sites are vulnerable to natural successional changes or to use as rubbish tips. The species is still locally frequent in C. Ireland.

P. coloratus is confined to W., C. and S. Europe, N. Africa and S.W. Asia. It reaches its northern limit in Scotland and S. Sweden. It is local or rare in many parts of its range and in the Netherlands it is listed as a 'most vulnerable' species (Weeda *et al.*, 1990).

British *P. coloratus* appears to be rather uniform both morphologically and genetically. Hollingsworth, Gornall & Preston (1995) found no evidence for isozyme variation within ten of the twelve populations they studied, nine of which appeared to share the same genotype. Two genotypes were detected at Wicken Fen, and seven in ditches at Gordano.

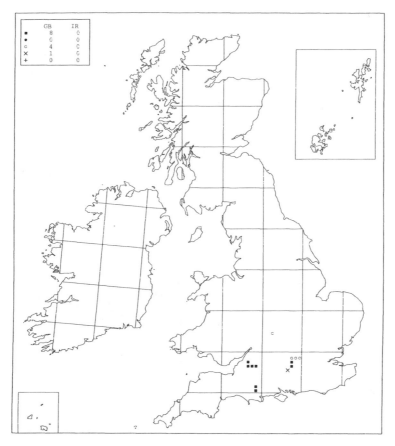

Potamogeton nodosus Poir.   Loddon Pondweed

This species is confined to a few calcareous and moderately but not excessively eutrophic rivers in lowland England. It is found in both shallow and relatively deep water, growing in stretches of moderately rapid flow with species such as *Elodea nuttallii, Lemna minor, Myriophyllum spicatum, Nuphar lutea, P. crispus, P. pectinatus, Sagittaria sagittifolia, Schoenoplectus lacustris* and *Sparganium emersum*. In the R. Loddon it is often particularly abundant in well-aerated stretches of the river below weirs and sluices (Archer, 1987). In Dorset it grows in fairly shallow water over a gravelly substrate, and cannot be found where there are soft clay sediments.

P. nodosus is a rhizomatous perennial which appears to vary in abundance from year to year. It flowers freely in summer when growing in shallow water. Plants do not appear to set fruit in the wild but will do so if cultivated in small containers, suggesting that the apparent sterility is caused by environmental factors. Buds develop on short stolons in the leaf axils in late summer and act as vegetative propagules. Established populations overwinter as buds on the rhizome, which develop in response to short days (Spencer & Anderson, 1987). In S. Europe this species grows in a much wider range of

habitats than it does in England, and fruits freely.

This species was first discovered in England in the R. Loddon by G. C. Druce in 1893 and named *P. drucei* by Fryer in 1898 (Preston, 1988). It was subsequently discovered in the Bristol Avon, R. Stour (Dorset) and R. Thames. However its correct identity continued to puzzle British botanists until Dandy & Taylor (1939a) established that it was the widespread continental species *P. nodosus*. The species has apparently become extinct in the Thames, where it was once locally frequent (Lousley, 1944a) but perhaps disappeared as a result of eutrophication and increasing boat traffic in the 1950s. We have been unable to confirm recent reports of its rediscovery. It cannot now be found in the Warwickshire Stour at Alderminster, where it is known from a single 19th century specimen. However, it survives as vigorous stands in the R. Avon, R. Loddon and the Dorset Stour, and has recently been planted in additional sites in the Loddon and in the nearby R. Whitewater and Blackwater River.

This is the most frequent broad-leaved *Potamogeton* in the Mediterranean region, and extends north in Europe to England, the Netherlands and Estonia. It also occurs in Africa south to Angola and Madagascar, temperate and tropical Asia, and N. & S. America.

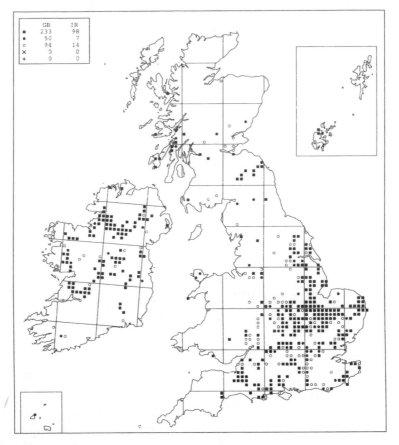

	GB	IR
■	233	98
●	50	7
○	94	14
×	0	0
+	0	0

Potamogeton lucens L.   **Shining Pondweed**

P. lucens is a calcicole which grows in relatively deep water in lakes, slow-flowing rivers, canals, flooded chalk-, gravel- or borrow-pits and major fenland drains. It is found in the crystal-clear water of unpolluted limestone lakes as well as more eutrophic water in rivers and fenland lodes. Although sometimes found in shallow water at the edge of lakes it has little resistance to desiccation, and plants exposed above the water dry up without producing a modified terrestrial form. It is most frequent in the lowlands, ascending to 380m at Malham Tarn.

The leafy shoots of *P. lucens* die in autumn, and the plant persists through the winter as the rhizome, which develops short, thick internodes (Sauvageau, 1893–94). Plants in still or sluggish water may flower freely, with the robust inflorescences supported by a raft of stems floating near the surface of the water (Fryer, 1887a). The flowers are usually wind-pollinated but if drawn below the water surface they may also collect pollen floating on the surface, at least under experimental conditions (Guo & Cook, 1989). Mature fruits develop, and may be the way in which the species colonises new habitats. *P. lucens* lacks specialized vegetative propagules, but may spread by rhizomes uprooted during floods and perhaps by detached stem fragments.

In England *P. lucens* has decreased in some areas, including Gloucestershire (Holland, 1986) and Surrey (Lousley, 1976). In some areas of arable land it has been replaced by narrow-leaved *Potamogeton* species such as *P. pectinatus* and *P. pusillus*. Nevertheless, it is still locally frequent in strongholds such as the Somerset Levels, East Anglian Fenland, Broadland, and in the R. Tweed and its tributaries. There has been a more marked decline in eastern Scotland, where it once grew in lowland lochs such as Balgavies Loch, Rescobie Loch and Loch Gelly which are now very eutrophic, or, in the case of Kinghorn Loch, have been contaminated by alkaline run-off from industrial waste (Young & Stewart, 1986). There is little evidence of a serious decline in Ireland, where it remains plentiful in calcareous waters. However, it has apparently been lost from its most south-westerly site, Lough Leane, probably because of eutrophication.

P. lucens is widespread in temperate regions of Europe, where it extends north to northern Scandinavia, and western Asia; its distribution further east in Asia is uncertain. It is also known from N. Africa and an outlying site in Uganda. In N. America it is replaced by the related *P. illinoensis* Morong.

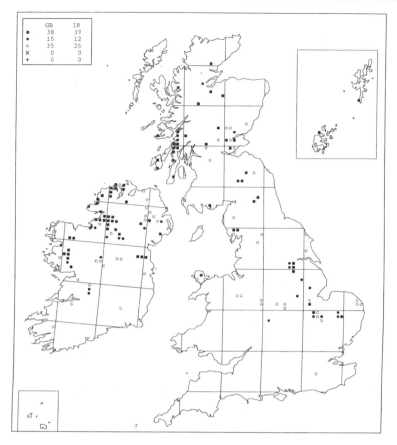

	GB	IR
■	38	39
●	15	12
○	35	25
×	0	0
+	0	0

Potamogeton × *zizii* W. D. J. Koch ex Roth (*P. gramineus* L. × *P. lucens* L.)

Long-leaved Pondweed

This hybrid grows in a range of water bodies, including lakes, Breckland meres, rivers, streams, fenland lodes and ditches. In lakes it is usually found in water over 0.5m deep, with dense stands tending to grow in rather deeper water. In northern and western areas of Britain and Ireland it is found in mesotrophic waters which have some base-enrichment, but are apparently not sufficiently base-rich to support *P. lucens*. In these sites it may grow with *P. gramineus*, or in the absence of both parents. In the south and east it is much less frequent, usually growing in the vicinity of both parents in calcareous but not excessively eutrophic water. In these areas *P.* × *salicifolius* is the commoner hybrid. It is recorded from sea-level to 395m at Loch na Craige.

P. × *zizii* dies down in the winter, regrowing from the rhizome in the spring. Most populations are of plants which have only submerged leaves, but some populations produce floating leaves in shallow water; whether or not these are genetically distinct is unclear. Although some populations of *P.* × *zizii* are consistently sterile, others may be at least partially fertile and some plants set well-formed fruit (Fryer, 1892). Their fertility has not been critically studied and we do not know whether the fruits are capable of germination, but

there is no reason to suppose that they are inviable. The hybrid lacks specialized means of vegetative dispersal, but spreads by growth of the rhizome and almost certainly by detached plant fragments. Vegetative propagation is almost certainly more frequent than any reproduction by seed that may occur. In times of drought, *P.* × *zizii* can persist as a terrestrial form with aerial leaves (Fryer, 1887b).

This is a persistent hybrid. It is still present at Cauldshiels Loch, where it was first recognized in our area (Baker, 1879), and it has also been known for over a century at other sites, including Coniston Water and Sessiagh Lough. It can no doubt persist indefinitely in suitable habitats. It is still frequent in parts of Scotland and Ireland, where it may be under-recorded. Like *P. gramineus* it has declined in southern and eastern England and southern Scotland, where the main reasons for its decline have been eutrophication and the loss of species-rich ditches in areas where pasture has been converted to arable.

P. × *zizii* is widespread in both N. & C. Europe.

P. × *zizii* is a hybrid which behaves almost as a species, having a capacity to produce well-formed fruits and a distribution which in some areas is apparently independent of that of its parents. Fryer (1890, 1892) considered that in the Cambridgeshire Fenland it was capable of backcrossing with *P. gramineus*, and this possibility deserves further investigation.

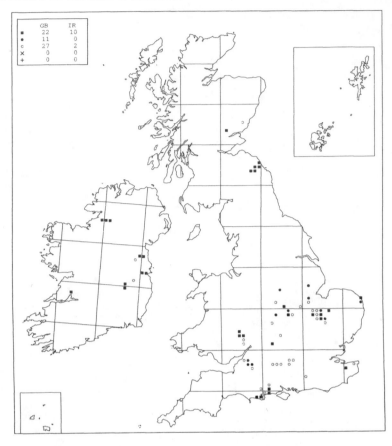

	GB	IR
■	22	10
●	11	0
○	27	2
×	0	0
+	0	0

Potamogeton × salicifolius Wolfg. (*P. lucens* L. × *P. perfoliatus* L.)

Willow-leaved Pondweed

Like both parents, this hybrid grows in relatively deep water which is not subject to marked fluctuations in depth. It is found in moderately or strongly calcareous, meso-eutrophic or eutrophic water. It is recorded from lakes, but most records are from slow-flowing rivers, canals and large fenland drains, perhaps because these are the usual deep-water habitats in the areas where it is most frequent. It may be present as large stands, often growing in the vicinity of both parents. It is confined to the lowlands.

P. × *salicifolius* dies down to the rhizome in winter, regrowing in spring. It has submerged but not floating leaves. It is a sterile hybrid which may flower but never sets well-formed fruit. Although it lacks specialized means of vegetative dispersal, it spreads by growth of the rhizome or by the spread of plant fragments. Detached stem fragments send out roots when floating in the water. Like other *Potamogeton* taxa which lack floating leaves, it is intolerant of desiccation and is unable to survive periods of drought as a terrestrial form.

This is a local hybrid which is well established in a number of river systems, including the R. Frome, R. Wye, R. Teviot and the lower reaches of the R. Tweed. It may be very persistent in these rivers: it was, for example, first collected in the R. Tweed at Birgham in 1830. It is tolerant of eutrophic water and still survives in many of the water bodies where it has been recorded. However, it has not been seen recently in the Kennet & Avon Canal, where it was first recognized in our area by Baker & Trimen (1867), and it has been lost from the excessively eutrophic Balgavies and Rescobie Lochs. It can be similar to *P. lucens* in appearance, and might therefore be under-recorded.

This is one of the more frequent *Potamogeton* hybrids in both N. & C. Europe, and it is also recorded from S.W. Asia.

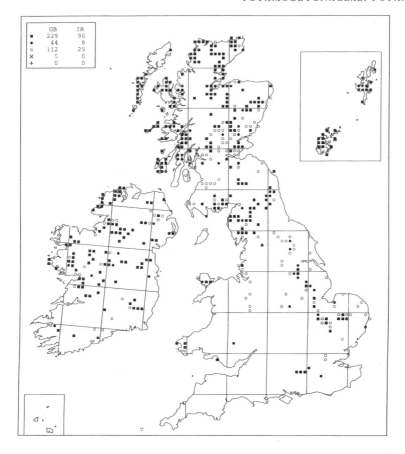

	GB	IR
■	229	90
●	44	8
○	112	25
×	0	0
+	0	0

Potamogeton gramineus L.
Various-leaved Pondweed

This very variable species occurs in a wide range of habitats, including lakes, reservoirs, rivers, streams, canals, ditches and flooded sand- and gravel-pits. It is primarily a plant of shallow water, usually found at depths of less than 1.5m although occasionally descending to 3m or more. Although absent from the most acidic and oligotrophic sites it is more tolerant of base-poor and nutrient-poor water than most other broad-leaved pondweeds, often growing in such sites with *Apium inundatum*, the aquatic form of *Juncus bulbosus* and *Myriophyllum alterniflorum*. It extends into meso-eutrophic and moderately base-rich sites, and it also grows in highly calcareous but nutrient-poor waters over chalk and limestone. Spence (1967) classifies it as ubiquitous with regard to its chemical tolerance in Scotland, but in areas where there is a greater range of eutrophic waters it is absent from the more nutrient-rich sites. It is found over a range of fine-grained and coarse-grained substrates, and at altitudes from sea-level to 915m (Meall nan Tarmachan).

P. gramineus is a rhizomatous perennial, which dies down in winter to buds on the rhizomes (Spencer & Ksander, 1990). It initially has submerged leaves, later producing floating leaves unless it grows in deep or fairly rapidly flowing water. Plants usually flower and fruit freely. The species lacks specialized means of vegetative dispersal. Terrestrial forms develop in sites where plants are exposed above the water-level.

Godwin (1975) concludes that *P. gramineus* has been present in our area since at least the middle of the Pleistocene, and its fruits are well represented in deposits from the last glacial period and through the current interglacial. It is still relatively frequent in the northern and western parts of our area. However, it has decreased during the last century in the south and east, where eutrophication and the unsympathetic management of ditches in areas which were once grassland and are now predominantly arable has led to its decline.

This species has a circumpolar distribution, being found in boreal and temperate regions throughout the northern hemisphere. In Europe it is found from Iceland and northern Scandinavia south to C. Europe, with scattered localities further south; it is absent from much of the Mediterranean area.

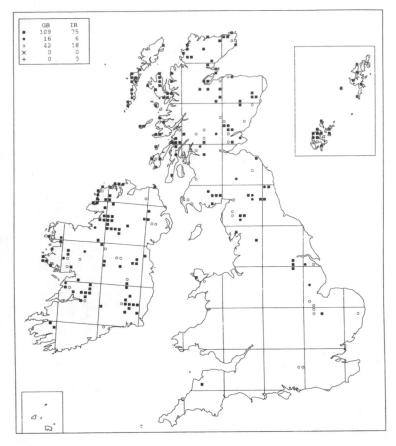

	GB	IR
▪	109	75
●	16	6
○	42	18
×	0	0
+	0	0

Potamogeton × nitens Weber (*P. gramineus* L. × *P. perfoliatus* L.)   **Bright-leaved Pondweed**

This hybrid is found in more or less mesotrophic water in lakes, reservoirs, rivers, streams, canals, fenland lodes and ditches, and occasionally in flooded sand- and gravel-pits. In rivers such as the Aberdeenshire Don it can be present in large beds with *Myriophyllum alterniflorum, P. alpinus, P. crispus* and *P. perfoliatus*. It may also be abundant in standing waters, where it is most luxuriant at depths of 0.5–1.5m, and sometimes grows around the margins of reservoirs subject to moderate variations in water-level. It may grow with one or both putative parents or in the absence of both; in S.W. England its isolated locality in the R. Torridge lies outside the range of *P. gramineus*. *P. × nitens* is recorded from sea-level to 640m in Loch Oss.

P. × nitens is a rhizomatous perennial which dies down in autumn. It is phenotypically plastic: it may possess only submerged or both submerged and floating leaves, and vigorous terrestrial forms can be found on exposed mud at the edge of lakes and reservoirs. Plants may flower freely but they are sterile and ripe fruit does not develop. Short stolons sometimes develop in the leaf axils, often just below the inflorescence. They are easily detached and must act as propagules, but the factors which trigger their formation are unclear.

This is the most frequent *Potamogeton* hybrid in our area. Once its appearance is known it can often be identified without difficulty, but it is usually overlooked by inexperienced observers and it may be more frequent in Scotland and Ireland than the map suggests. In S.E. England it was always an uncommon plant, and it is now very rare. It has not been seen in drainage ditches in the East Anglian Fenland since 1949, and it is extinct in the Basingstoke Canal. However it was recently discovered in a gravel-pit in Cambridgeshire which had been colonized by both parents.

P. × nitens is widespread in both N. & C. Europe. It has also been reported from temperate Asia and N. America (Dandy, 1975).

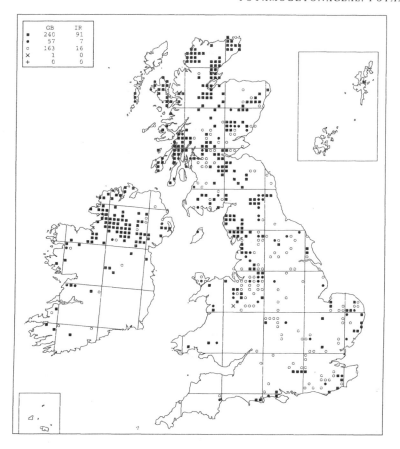

	GB	IR
■	240	91
●	57	7
○	163	16
×	1	0
+	0	0

Potamogeton alpinus Balb.   Red Pondweed

P. alpinus is a plant of still or slowly flowing water in lakes, rivers, canals, ditches and flooded clay-, gravel-, sand- and marl-pits. It usually grows over a relatively deep silty or peaty substrate, often in sites where silt accumulates such as the point where an inflow stream enters a lake, or the sheltered backwater of a river. The most vigorous stands are found in water approximately 1m deep. *P. alpinus* thus favours deeper water than *P. gramineus* but shallower places than *P. perfoliatus* and *P. praelongus*, although there is considerable overlap in the depth range of these broad-leaved pondweeds and all can be found growing with *P. alpinus*. *P. alpinus* is a plant of oligo-mesotrophic or mesotrophic, often neutral or mildly acidic water. It is found from sea-level to 945m (Meall nan Tarmachan).

P. alpinus dies down in the winter to buds produced on the rhizome. The stems are unbranched and in addition to the submerged leaves they usually have a few, small floating leaves if they reach the surface. Populations often flower and fruit abundantly. Plants will regenerate from detached leafy shoots (Brux *et al.*, 1987), and short stolons terminated by buds sometimes develop in the lower leaf axils in the autumn, and provide an additional means of perennation and vegetative

propagation. Brux *et al.* (1988) found that there was no regeneration from seed in the German rivers they studied. The distribution of the species in these rivers was rather unstable; stands often disappeared from sites if the perennating buds were washed away, then recolonized in subsequent years. In ponds germinating seeds were found only once. In aquaria, freshly-collected seed germinated, but the seedlings were much smaller than plants grown from buds and remained small for many months.

There is clear evidence that this species is a long-established native which was present throughout the last glacial period in both England and Ireland (Godwin, 1975). It is fairly frequent in the northern parts of Britain and Ireland, where there is little evidence of anything other than local decline, but the map suggests a marked contraction of range in S. England. Many of these records are 19th century records from ponds, clay-pits, ditches and other small water bodies. In more recent years it has been lost from some rivers, including the R. Stort (Burton, 1983), presumably because of eutrophication.

P. alpinus is a circumboreal species, which is found in arctic and boreal regions throughout the northern hemisphere. In Europe it extends south to the Pyrenees and the mountains of C. Europe.

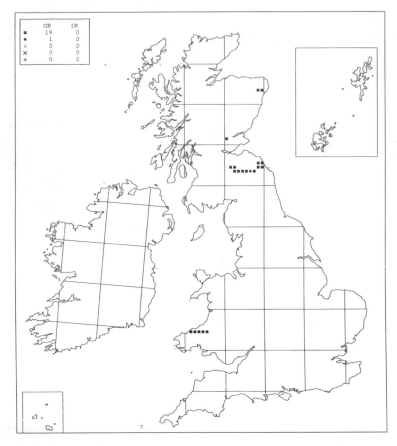

	GB	IR
■	19	0
●	1	0
○	0	0
×	0	0
+	0	0

Potamogeton × olivaceus Baagøe ex G. Fisch. (*P. alpinus* Balb. × *P. crispus* L.)   Graceful Pondweed

This hybrid is established in a number of mesotrophic lowland rivers in Britain, where it usually grows in swiftly but smoothly flowing water over a gravelly substrate. In the R. Earn beds of this hybrid grow with beds of *P. × zizii*. *P. × olivaceus* also grows in a pool on the flood-plain of the Teifi, where it is found in shallow water with *Callitriche hamulata*, *Elodea canadensis*, *P. berchtoldii* and *P. natans*.

This rhizomatous perennial persists through the winter as leafy shoots. Plants flower but are highly sterile and show no sign of developing fruit. Unlike *P. crispus* and its hybrids *P. × cooperi* and *P. × lintonii*, *P. × olivaceus* does not produce specialized turions. However, stems are easily detached and *P. × olivaceus* probably spreads as fragments of stem or rhizome. As the hybrid is winter-green, it is particularly likely to be dispersed by winter floods.

P. × olivaceus was first collected in the Leet Water, a tributary of the R. Tweed, by R. D. Thomson in 1831, although the specimen was misidentified for over a century. It was collected from the R. Ythan in 1900 and found in the R. Earn in 1915, when it was first recognized as a British plant (Dandy & Taylor, 1942). It was not discovered in the Teifi until 1972. It is clearly able to persist in rivers for long periods. Neither *P. alpinus* nor *P. crispus* are known from the Teifi. Both parents have a long history in Britain, and are known to have persisted through the last glacial period. This suggests that the Teifi population, and perhaps the others, may be of ancient origin, dating from a period when the parents grew in close proximity.

The only record of this hybrid from outside Britain is from Denmark, where it does not appear to have been found recently (Pedersen, 1976). The hybrid is much rarer in Europe than one would expect from its British distribution.

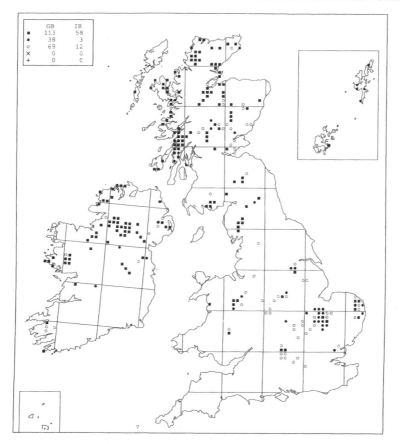

	GB	IR
■	113	58
●	38	3
○	69	12
×	0	0
+	0	0

Potamogeton praelongus Wulfen
Long-stalked Pondweed

This is the species of broad-leaved pondweed which is most characteristically found in deep water. It is rarely found in water less than 1m deep, grows as dominant stands only at depths greater than 1.5m (Spence, 1964) and descends to 6.5m in Loch Borralie (Spence *et al.*, 1984). It is found in clear, mesotrophic water in lakes (where it may often be found in deep, sheltered channels between an island and the shore), rivers, canals and major fenland lodes. Its sites vary from those which are slightly enriched by the presence of outcrops of basalt or calcareous sand in the catchment, to highly calcareous but nutrient–poor lakes over limestone. It is usually rooted in fine silts. It is found from sea-level to 780m at Lochan an Tairbh-uisge.

P. praelongus is a robust, rhizomatous, winter-green perennial which possesses submerged but not floating leaves. It flowers rather earlier than the other submerged pondweeds, and produces relatively few, large fruits. It does not possess any specialized means of vegetative propagation.

The characteristic fossil fruits of this species are frequent in deposits which date from the last glacial period, and are particularly abundant in the late glacial in both Britain and Ireland. The distribution of fossil records,

mapped by Godwin (1975), is remarkably similar to that of the species in historical times. Since 1950, *P. praelongus* has declined markedly in S. England, and the species has disappeared from rivers such as the R. Cam and R. Thames where it had long been recorded. The most likely explanation is that it has been adversely affected by eutrophication, following the increasing use of artificial fertilizers on arable land in recent decades. The species has also declined in the Netherlands, apparently because of pollution and increasing salinity in the waters in which it grows (Weeda, 1976). In northern parts of Scotland and Ireland there is no evidence of a decline. As a species which prefers deep water it is very easily overlooked, and it is almost certainly under-recorded in some areas.

P. praelongus has a circumboreal distribution. In Europe it extends from Iceland and N. Scandinavia southwards to C. Europe, but is absent from S. Europe and the Mediterranean area. It is an endangered species in the Netherlands (Weeda *et al.*, 1990). For further details of its world range, see Haynes (1985).

The rare hybrid between *P. alpinus* and *P praelongus*, *P.* × *griffithii*, is found at a few localities in Ireland, Scotland and Wales in the apparent absence of *P. praelongus* or of both parents; it may be a relic of a period when *P. praelongus* was more widespread (Preston & Stewart, 1992).

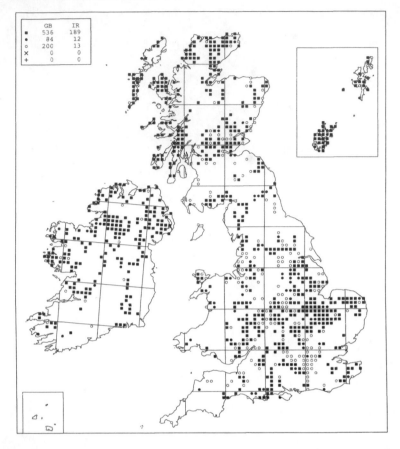

	GB	IR
■	536	189
●	84	12
○	200	13
×	0	0
+	0	0

Potamogeton perfoliatus L.   **Perfoliate Pondweed**

P. perfoliatus is found in larger water bodies such as lakes, reservoirs, rivers, streams, canals, fenland lodes, large ditches and flooded sand- and gravel-pits. Scattered plants of *P. perfoliatus* with prostrate shoots may be found in very shallow water in sites where they are not subject to periodic desiccation, but vigorous stands are found only where the water is at least 1m deep. This species is tolerant of a wide range of water chemistry, being frequent in both mesotrophic and eutrophic waters. Characteristic associates include *Myriophyllum spicatum*, *P. crispus*, *P. pectinatus*, and *Sparganium emersum*. In addition to its occurrence in more nutrient-rich sites, *P. perfoliatus* is also occasionally found in oligotrophic lakes. It is recorded from sealevel to 780m at Lochan an Tairbh-uisge.

P. perfoliatus is a perennial which dies down to the rhizome in the winter. It has submerged but not floating leaves, and often flowers and fruits freely. It lacks specialized vegetative propagules.

Like many members of the genus, *P. perfoliatus* can be shown to be a long-established native which was particularly frequent in the late glacial period (Godwin, 1975). Unlike *P. alpinus*, *P. gramineus* and *P. praelongus*, this species is still frequent in suitable habitats in S. England. This general picture conceals some local fluctuations. Philp (1982) comments that it has increased in abundance in the R. Medway at Maidstone in recent years. However, it has declined in Somerset (Roe, 1981), where it has always been surprisingly scarce, Surrey (Lousley, 1976) and in other areas where grazing marshes have been converted to arable land (Driscoll, 1985a; Mountford, 1994a). Like most other members of the genus it is rare in Wales, S.W. England and S.W. Ireland.

P. perfoliatus is widespread in Europe, from Iceland and Scandinavia southwards to the northern edge of the Mediterranean area. It has an extensive world distribution, being found in much of Asia east to Japan and south to S. India, in eastern N. and C. America, N. & C. Africa and Australia. The very closely related *P. richardsonii* (A. Benn.) Rydb. is more widespread in N. America. For a map of its world distribution, see Haynes (1985).

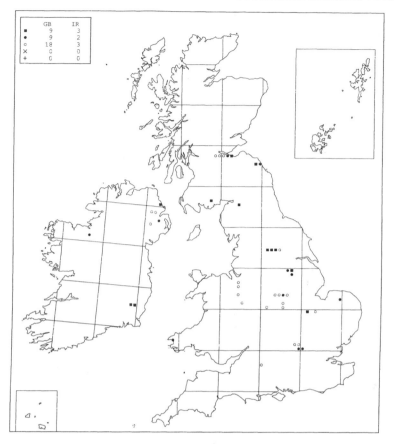

	GB	IR
■	9	3
●	9	2
○	18	3
×	0	0
+	0	0

Potamogeton × cooperi (Fryer) Fryer (P. crispus L. ×
P. perfoliatus L.)   **Cooper's Pondweed**

This widespread but uncommon hybrid occurs in a range of lowland water bodies. Most records are from rivers and canals, but it is also found in lakes, reservoirs, clay-pits and fenland ditches. It is a plant of eutrophic water, characteristically associated with plants such as *Elodea canadensis, E. nuttallii, Lemna minor, L. trisulca, Myriophyllum spicatum, P. crispus, P. pectinatus, P. perfoliatus* and *P. pusillus*. It may be found in shallow water at the edge of lakes and canals, or in deeper water in rivers.

This hybrid is a rhizomatous perennial which retains its leafy shoots through the winter. It does not often flower freely, and when it does the inflorescences are highly sterile. However, turions develop in the leaf axils and are an effective means of vegetative dispersal.

This hybrid is potentially very persistent: it is still found in the R. Wharfe, where it was first discovered in 1881, and the Union Canal, where records date from 1902 (Wolfe–Murphy *et al.*, 1991). In some sites, such as Carlingwark Loch, both parents are present and the hybrid has apparently arisen *in situ*, but it presumably spread vegetatively in the canal system. It has not been recorded recently in some of the rivers and canals where it once grew, and some of the canals are no longer extant. *P. × cooperi* is superficially similar to *P. perfoliatus* and the map may underestimate its former, and perhaps its current, distribution.

This hybrid has been recorded from widely scattered parts of Europe: Dandy (1975) cites records from Czechoslovakia, Denmark, Germany and Romania.

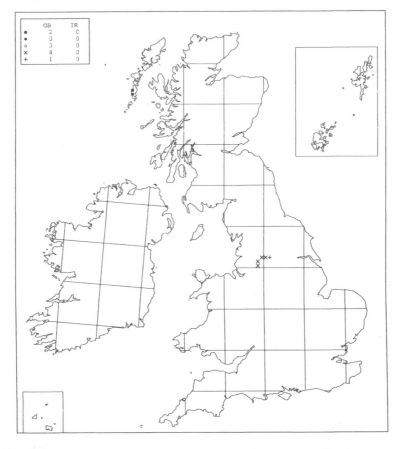

	GB	IR
■	2	0
●	0	0
○	0	0
×	4	0
+	1	0

Potamogeton epihydrus Raf.

American Pondweed

In the Outer Hebrides *P. epihydrus* grows in a few peaty lochans near Loch Ceann a'Bhaigh, where it is found in oligotrophic and base-poor water less than 1m deep. Other species in these lochans include *Eleogiton fluitans*, *Equisetum fluviatile*, the aquatic form of *Juncus bulbosus*, *Littorella uniflora*, *Lobelia dortmanna*, *Nymphaea alba*, *P. natans*, *P. polygonifolius*, *Sparganium angustifolium* and *Utricularia vulgaris* sensu lato. There is also a record from Loch an Duin, a shallow, muddy lochan. Submerged leaves which appear to belong to this species are washed up in late summer at the edge of a large, oligotrophic loch on Skye, but rooted plants have never been found here and the identification remains unconfirmed. However, *P. epihydrus* is locally plentiful in the mesotrophic water of the Rochdale Canal and the Calder & Hebble Navigation, where it grows with *Elodea nuttallii*, *Lemna minor*, *L. trisulca*, *Luronium natans*, *P. crispus*, *Sparganium emersum* and *Nitella mucronata* in water 0.4–1.2m deep. All the known sites are lowland.

P. epihydrus is a rhizomatous perennial. It has both submerged, linear leaves and broader floating leaves, although floating leaves may not develop on plants in relatively deep water. The species is self-compatible and self-pollination takes place once the anthers dehisce if cross-pollination has not taken place (Philbrick, 1983). In the American population studied by Philbrick all the

ovules developed into fruits, and plants with floating leaves flower and fruit freely in both the Outer Hebrides and N. England. In addition, short stolons bearing buds or small fascicles of leaves can be found in the leaf axils in September. They are easily detached and act as vegetative propagules.

This species was originally discovered in Britain in 1907, at Salterhebble Bridge near Halifax (Bennett, 1908). It is still well established in canals on both sides of the Pennines, but it has not been seen recently in the R. Calder, where it was collected in 1942. The English populations are believed to originate from an introduction, although it is difficult to imagine the source of the material. The native populations in S. Uist were discovered by W. A. Clark and J. W. Heslop Harrison in 1943 and 1944 (Heslop Harrison, 1949, 1950).

P. epihydrus is widespread in N. America, where it is found in lakes, pools and streams from S. Alaska and Labrador south to N. California and Tennessee; it is restricted to mountains in the southern part of its range. For a map of its American distribution, see Fernald (1932). The British records are the only known occurrences in Europe.

There are similarities between this species and *Luronium natans*: both have native sites where the water is oligotrophic and also grow in mesotrophic canals. However, *L. natans* spread from its native sites into the canal system, whereas the canal populations of *P. epihydrus* presumably originated independently.

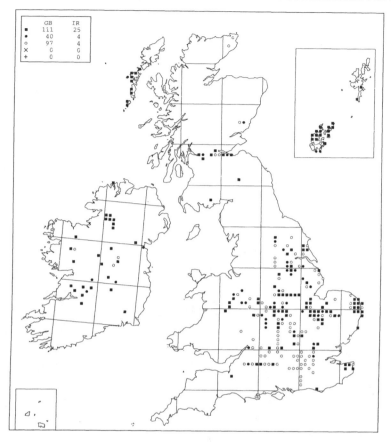

	GB	IR
■	111	25
	40	4
○	97	4
×	0	0
+	0	0

Potamogeton friesii Rupr.   Flat-stalked Pondweed

This is a species of calcareous waters which may be low in other nutrients but are more often eutrophic. It usually grows at depths of 0.5m or more. Most of its British localities are in lakes, sluggish rivers and streams, canals, fenland lodes, ditches and flooded clay- and gravel-pits. It is able to persist in some canals and lodes where there is some boat traffic. In the Outer Hebrides it is restricted to machair lochs, where the water is enriched by bases from calcareous dune sand, and in Shetland it is abundant in a loch influenced by coastal sand. In Ireland, *P. friesii* is found in clear calcareous lakes, canals and in the R. Shannon. It is confined to the lowlands.

Like the other species in section *Graminifolii*, *P. friesii* is an annual species. Its normal mode of reproduction is by robust turions, which may begin to develop in late June but are usually found in July and August. This species appears to flower and fruit rather infrequently.

There are few fossil records of *P. friesii*, although it was probably present in both Britain and Ireland during the late glacial period (Godwin, 1975). It may have been a very local species in Britain until the construction of the canal network, which appears to have allowed it to spread into areas from which it was previously absent. It has subsequently retreated from some of these areas where the canals in which it grew are now dry or have become too heavily used to support it. It fluctuates in response to nutrient levels: it was first recorded in the Thurne valley broads in the 1930s when their nutrient levels were increasing, but died out once they became very eutrophic (George, 1992); it has also been lost from three lochs in Angus for the same reason. It has only recently been discovered at its outlying sites in Roxburghshire and Donegal, and like other linear-leaved pondweeds it may be somewhat under-recorded.

P. friesii is a circumpolar species, recorded from boreal and temperate latitudes throughout the northern hemisphere but rare, or under-recorded, in E. Asia. In Europe it is absent from the Mediterranean area; it extends north to the Arctic Circle although it is rare at the northern edge of its range.

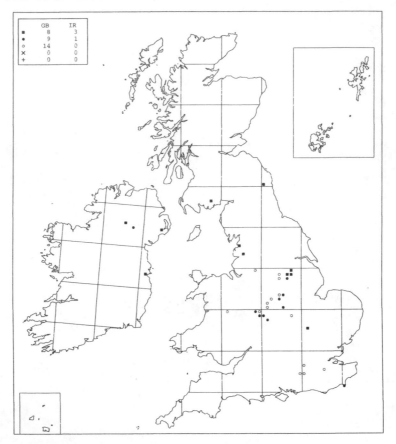

Potamogeton × lintonii Fryer (*P. crispus* L. × *P. friesii* Rupr.)   Linton's Pondweed

This hybrid is found in eutrophic water in both natural sites and man-made habitats. It is found in eutrophic lakes in Ireland and Scotland. In the shallow edges of Carlingwark Loch it grows in water 0.1–0.25m deep, associated with *Callitriche hermaphroditica, Elodea canadensis, Myriophyllum spicatum, P. friesii* and *P. pusillus* (Stewart & Preston, 1990). There are also records from streams and the backwater pools of rivers. *P. × lintonii* is, however, more frequently recorded from canals. In the Chesterfield Canal it grows in shallow water over fine silt at the edge of the canal, together with *Lemna minor, Myriophyllum spicatum* and *P. perfoliatus*. There are also records from a number of other artificial water bodies, including flooded marl-pits and a water storage tank. All sites are in the lowlands.

P. × lintonii is a rhizomatous perennial. It is a highly sterile hybrid which sometimes flowers freely in shallow water, but fails to set seed. Turions are produced in the leaf axils, often in abundance, and presumably provide the main means of dispersal.

P. × lintonii was described, new to science, on the basis of material collected by C. Waterfall in the Chesterfield Canal in 1899. Subsequent research showed that specimens had been collected previously from this and other sites, but misidentified (Dandy & Taylor, 1939c). The isolated population in Carlingwark Loch is almost certainly the result of *in situ* hybridization. There are insufficient historical records to allow us to reconstruct its history elsewhere. We do not know whether it spread into the canals from nearby rivers such as the Trent and the Wey, or whether it arose in the canal system. The extent to which the plants in canals result from the spread of a single clone or from different hybridization events is also unclear. *P. friesii* appears to have declined markedly in recent years as some canals have become disused and dried out and others are now heavily used for recreational purposes. It also appears to be extinct in the Tilling Bourne, where it has not been seen since the stream was thoroughly cleaned out *c*.1950. It might, however, be overlooked in some localities as it can be confused with other linear-leaved pondweeds.

Outside our area the only records of *P. × lintonii* are from Belgium, the Netherlands and the Czech Republic.

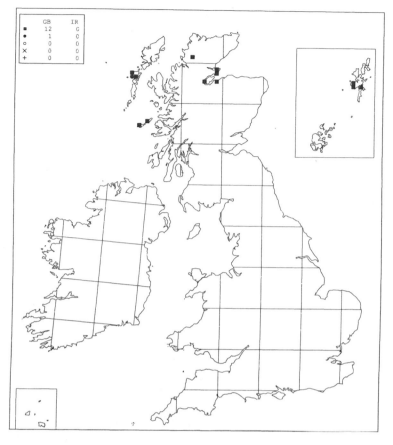

	GB	IR
▪	12	0
●	1	0
○	0	0
×	0	0
+	0	0

Potamogeton rutilus Wolfg.　　　Shetland Pondweed

This species grows in unpolluted, lowland, mesotrophic or eutrophic lochs which receive some base-enrichment; it is also found in adjoining streams. The habitat of *P. rutilus* at its remote British sites has not been studied in detail, partly because it is difficult to find in the deep water in which it often grows, and plants are normally detected after they have become detached and been blown to the side of a loch. In Shetland it grows in water 1–2m deep in mesotrophic lochs, some in limestone and others in sandstone catchments. In the Hebrides the sites are machair lochs at the junction of acidic rock and calcareous dune sand. In a dry summer when water-levels were exceptionally low it was recorded in water only 0.5m deep in Loch Ballyhaugh, growing in a sub-strate of fine silt with *Littorella uniflora, Najas flexilis, Potamogeton gramineus* and abundant *Chara aspera.* In Loch Grogary and Loch Scarie it was found over stony, gravelly, sandy and silty substrates at depths of 0.25–0.75m, also in a dry summer. It grew in species-rich communities which included *Apium inundatum, Baldellia ranunculoides, Littorella uniflora, Myriophyllum alterniflorum, Pilularia globulifera, Potamogeton crispus, P. filiformis, P. gramineus, P. natans, P. × nitens, P. perfolia-tus, P. pusillus, P. × suecicus* and *Chara hispida.* At Loch

Ussie it was first discovered in 1994 in the deep water of this clear, naturally eutrophic loch which supports nine other *Potamogeton* species and one hybrid.

Although flowering material of this species has been collected in Britain, fruiting plants have never been found. Plants overwinter as turions, which develop in the autumn and also provide a means of dispersal within a lake.

This species was completely misunderstood by British botanists until Dandy & Taylor (1938c) revised the existing records and showed that the only correctly determined specimens were from Shetland. *P. rutilus* was discovered in the Outer Hebrides by Clark (1943). All the sites known to Perring & Walters (1962) were in these two archipelagos, but the species has subsequently been found in an increasing number of sites as a result of the more systematic recording of Scottish lochs in recent years. The only site where it is thought to be extinct is Loch Flemington, where it was discovered in 1975 but appears to have been eliminated by pollution from silage effluent and other sources.

P. rutilus is endemic to northern Europe, where it is found from the Arctic Circle south to N. France, Germany and Poland. It is a local plant throughout its range, and it may be the rarest of the European pondweeds at a world scale.

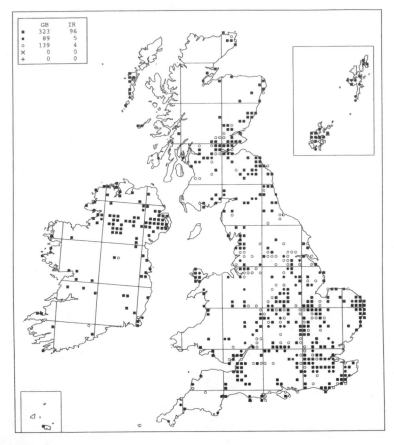

	GB	IR
■	323	96
●	89	5
○	139	4
×	0	0
+	0	0

Potamogeton pusillus L.   **Lesser Pondweed**

P. pusillus grows in standing or slowly flowing water in a wide range of habitats, including sheltered lakes and reservoirs, ponds, rivers, streams, canals, fenland lodes, ditches, and in flooded clay-, sand- and gravel-pits. It usually grows in fairly shallow water over fine, silty substrates. It is found in mesotrophic and mildly acidic water with *P. obtusifolius*, but it is more frequent in eutrophic water. It is also tolerant of slightly brackish water at the seaward side of machair lochs and in coastal lagoons and ditches. Characteristic associates include *Elodea nuttallii*, *Myriophyllum spicatum*, *P. pectinatus*, *Zannichellia palustris* and, in southern England, *Potamogeton trichoides*. It is virtually confined to the lowlands: the highest recorded altitude is 320m at Stoney Middleton.

P. pusillus often flowers and fruits freely in shallow water. As with other *Potamogeton* species there is little direct evidence of reproduction by seed. However, *P. pusillus* is a very frequent colonist of disused gravel-pits to which it is presumably introduced by bird-borne

seeds. The normal mode of reproduction of established populations is by slender turions, which are produced in abundance from July onwards.

The smaller linear-leaved pondweeds were all aggregated as '*P. pusillus*' by Linnaeus. The differences between *P. berchtoldii* and *P. pusillus* sensu lato were set out by Hagström (1916), but the two species continued to be confused in our area until Dandy & Taylor (1938a, 1940a,b) revised the available herbarium material. This taxonomic confusion makes it difficult to draw any conclusions about trends in the distribution of *P. pusillus*, but it may have increased in recent years. It is the most frequent species of *Potamogeton* in disused gravel-pits, and in a detailed local study Driscoll (1985a) demonstrated a dramatic increase in its frequency in ditches following the conversion of the surrounding land from pasture to arable.

P. pusillus is widespread in Eurasia, Africa and N. & C. America. In Europe it is found from the Mediterranean region to Scandinavia. Compared to *P. berchtoldii*, it tends to be commoner in southern Europe and rarer and more coastal in the north.

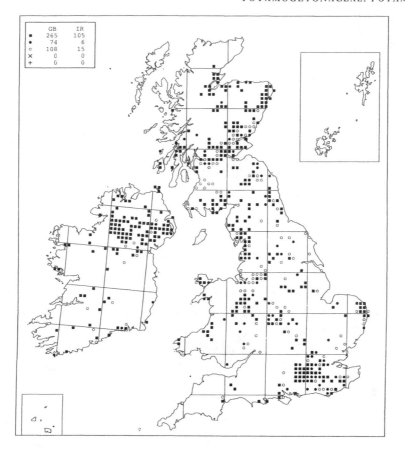

Potamogeton obtusifolius Mert. & W. D. J. Koch
Blunt-leaved Pondweed

P. obtusifolius is normally found in standing waters such as ponds and lakes or in the backwaters of rivers, canals and ditches where the flow is scarcely perceptible. It is also known as a colonist of flooded sand-, gravel-, marl- and clay-pits. It reaches its optimum development in water which is less than 1.5m deep, and it is usually found over fine inorganic or peaty substrates. It is difficult to define its ecological preferences as it tends to be frequent in a range of habitats in some areas but absent from others where there appear to be suitable habitats. It is usually found in mesotrophic or meso-eutrophic, acidic or circum-neutral water with species such as *Elodea canadensis, Lemna minor, L. trisulca, Menyanthes trifoliata, Myriophyllum alterniflorum, Nuphar lutea, P. natans* and *P. pusillus.* The canals in which it grows, for instance, tend to be those where the water is not very calcareous: they include the eastern sections of the Basingstoke Canal, the Lancaster Canal and the Exeter Canal. *P. obtusifolius* is absent from the most acidic and oligotrophic water and from both highly calcareous and highly eutrophic sites (cf. Seddon, 1972). It is a predominantly lowland species which ascends to 395m

north of Bleddfa.

In shallow water *P. obtusifolius* can be abundant, and grows as much-branched plants with masses of leaves in the upper part of the water. It fruits more freely than any other linear-leaved pondweed. However, reproduction is usually effected by turions, which develop from late July onwards.

This is one of the many *Potamogeton* species which can be shown to have persisted in our area since the middle of the last glacial period (Godwin, 1975). In some areas it is one of the commoner members of the genus. It might be more frequent in some areas than the map suggests, and the apparent decline in regions such as Yorkshire could be an artefact resulting from the lack of recent survey work.

P. obtusifolius has a circumpolar distribution, but it is much more frequent in Europe and eastern N. America than in the other parts of its range. It is widespread in Europe, but it is rare in the far north and absent from the Mediterranean area. Reports from Australia, cited by Hultén & Fries (1986), are supported by only a single, unlocalized specimen (Aston, 1973).

P. friesii is a superficially similar linear-leaved pondweed which tends to replace *P. obtusifolius* in calcareous sites.

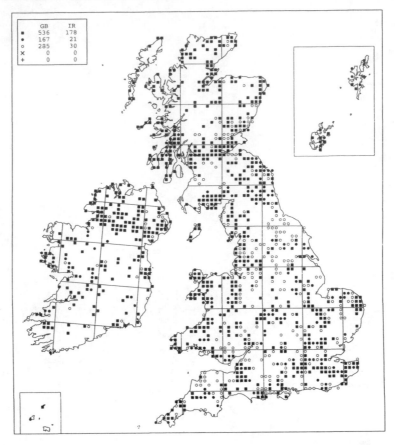

	GB	IR
■	536	178
●	167	21
○	285	30
×	0	0
+	0	0

Potamogeton berchtoldii Fieber   **Small Pondweed**

No other linear-leaved pondweed grows in as wide a range of still or slowly flowing waters as *P. berchtoldii*. It is frequent in a variety of base-rich or base-poor, mesotrophic and eutrophic habitats including lakes, reservoirs, ponds, rivers, streams, canals, ditches and disused mineral workings. It may grow over a range of substrates from calcareous gravel to fine silt, and can also persist in a shade-form in shallow ponds and ditches under trees. It is sometimes found in clear, oligotrophic upland lakes, but it is rare in oligotrophic, peaty water. It is replaced by *P. pusillus* in brackish sites and in some eutrophic localities in S. England. It is found from sea-level to 500m at Llyn Anafon and 505m at Dogber Tarn.

   P. berchtoldii flowers and fruits fairly freely. Philbrick (1983) found that the American population he studied was self-incompatible. Plants with aerial inflorescences and those which flowered under water both set seed, but unlike the self-compatible species the number of seeds which developed was well below the potential maximum. Plants overwinter as turions, which develop from August onwards.

   The confused taxonomic history of this species is outlined under *P. pusillus*. It is an inconspicuous plant which is overlooked by some observers and disregarded by others who are not prepared to attempt to identify linear-leaved pondweeds. There may also be a tendency to label all 'pusilloid' pondweeds *P. berchtoldii* in the belief that this is the commonest species, but the map should not include many of these records. The areas in which there is a preponderance of old records on the map tend to be those from which there is a shortage of recent, critically determined records of all *Potamogeton* species. *P. berchtoldii* is still frequent in suitable habitats in many areas of Britain and Ireland, although it may be declining in some counties because of eutrophication.

   P. berchtoldii is widespread in the boreal and temperate regions of the northern hemisphere. It is found in most of Europe, although it is rare in the Iberian peninsula and absent from the extreme north and some of the Mediterranean islands. Recent American authors (e.g. Haynes, 1974) treat it as a variety of *P. pusillus*, but there is a case for continuing to treat it at specific rank (Preston, 1995).

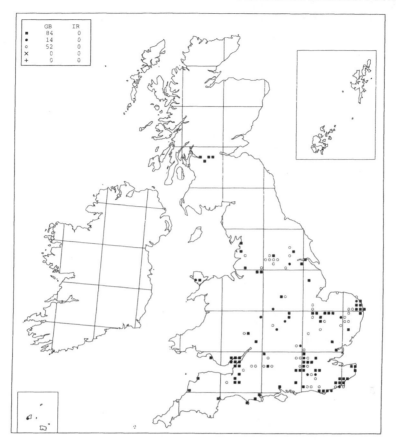

	GB	IR
■	84	0
●	14	0
○	52	0
×	0	0
+	0	0

Potamogeton trichoides Cham. & Schltdl.

Hairlike Pondweed

P. trichoides is found in shallow, still or slowly flowing, meso–eutrophic or eutrophic water over soft inorganic substrates. It grows in a range of habitats, including lakes, reservoirs, ponds, rivers, canals, ditches and flooded clay-, sand- and gravel-pits. It is often found as a colonist of newly available habitats, or in disturbed areas such as recently cleared canals and ditches. It can be found as large plants growing by themselves or inter-mixed with *P. pusillus*, in sparsely vegetated areas, or as scattered stems amongst dense beds of *Elodea nuttallii*. Other frequent associates include *Myriophyllum spicatum* and *Potamogeton pectinatus*. *P. trichoides* is confined to the lowlands.

The turions of *P. trichoides* begin to form from June onwards. They usually require stratification before they will grow, and a higher proportion of stratified turions germinates in light than in darkness. After a cold pre-treatment of ten weeks they will germinate at relatively low temperatures (10°C), ensuring that growth begins in spring (Van Wijk & Trompenaars, 1985). Plants flower and fruit rather sparingly, and seeds are not thought to be important in maintaining established populations although they may permit survival through unfavourable

periods and long-distance dispersal. *P. trichoides* has been rediscovered in sites after a long absence, notably at Malltraeth Marsh where it was collected by H. Davies in *c.*1800 and refound in 1990. This probably reflects the fact that it is easily overlooked, although it may also indicate a capacity to persist as a diaspore bank.

Although this species was first recognized in Britain by Babington (1850), it was subsequently confused with the other 'pusilloid' pondweeds. The older records were revised by Dandy & Taylor (1938b). *P. trichoides* is often overlooked by inexperienced observers and it is some-times misidentified. It is therefore impossible to assess trends in its distribution with any accuracy, but there is some evidence that it has increased in recent years in the Somerset Levels, perhaps because of eutrophication. A similar increase has been noted in eutrophic ditches and streams in the Netherlands in the last 20 years (Bruinsma, 1996). The species is probably more fre-quent in England than the map suggests.

This species has a more southerly distribution in Europe than most *Potamogeton* species, being found from S. Sweden and C. Russia southwards to the Mediterranean area. It is a very rare and threatened species in Sweden (Nilsson & Gustafsson, 1978). It is also found in S.W. Asia, and it is widespread in Africa.

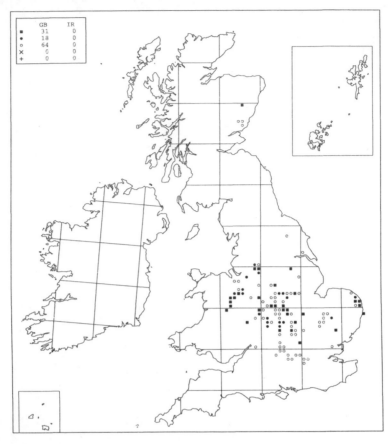

	GB	IR
■	31	0
▪	18	0
○	64	0
×	0	0
+	0	0

Potamogeton compressus L.

Grass-wrack Pondweed

This plant grows in still or slowly flowing, calcareous, mesotrophic water, often in sites which contain a rich assemblage of submerged and floating aquatics. It has been recorded from lakes, sluggish rivers and their associated backwaters, ox-bows and ditches, canals, grazing marsh ditches and a few flooded clay- and gravel-pits. It is locally abundant in the Montgomeryshire arm of the Shropshire Union Canal, growing in clear, moderately deep water, often in aqueducts or other places where the flow is accelerated, and associated with *Callitriche obtusangula, Elodea canadensis, Myriophyllum spicatum, P. berchtoldii, P. friesii, P. natans, P. obtusifolius* and *Ranunculus circinatus* (Sinker *et al.*, 1985). In Broadland *P. compressus* grows in water up to 1m deep in grazing marsh ditches with a wide range of associates, including *Elodea canadensis, Equisetum fluviatile, Lemna trisulca, Oenanthe fluviatilis, Hydrocharis morsus-ranae, Sagittaria sagittifolia* and *Stratiotes aloides. P. compressus* is confined to the lowlands.

P. compressus flowers and fruits rather sparingly; plants in grazing marsh ditches may fruit more frequently than those in other habitats. The normal mode of reproduction is by turions, which begin to develop in late June. This can be established by examination of the lower stems of plants, which usually arise from a characteristically robust turion.

The current scarcity of this species in Britain is the result of a long period of decline. The species was dis-

covered by Ray (1660) in the R. Cam, but it has not been seen in the Cambridge area since 1860 and was last collected in the county in 1912. In the Thames valley it was last seen in Surrey in 1898, Buckinghamshire in 1940, Berkshire in 1944 and Oxfordshire in 1947. In Angus it was formerly known from four lochs: it disappeared from them from the middle of the 19th century onwards and was last collected from Balmadies Loch in 1943. The reasons for the disappearance in Cambridgeshire were not documented, but eutrophication was almost certainly responsible for the other extinctions. The species must have expanded its distribution in the 18th or 19th centuries when it spread into the canal system, but it has subsequently retracted as canals have become disused and dried out or been affected by eutrophication and increasing pleasure-boat traffic. In Broadland it has decreased in recent years in mesotrophic grazing marsh ditches, even in areas where management has been designed to maintain their botanical interest (Doarks, 1990). It has, however, recently been discovered in a new Scottish locality, the Loch of Aboyne.

P. compressus is widespread in the boreal and temperate zones of Eurasia. In Europe it is only absent from the extreme north and from the Mediterranean region. It is replaced in N. America by the closely related *P. zosteriformis* Fernald.

P. acutifolius and *P. compressus* are closely related species. They are the most threatened British pondweeds, and will require positive conservation measures if they are not to be reduced to extreme rarity or extinction.

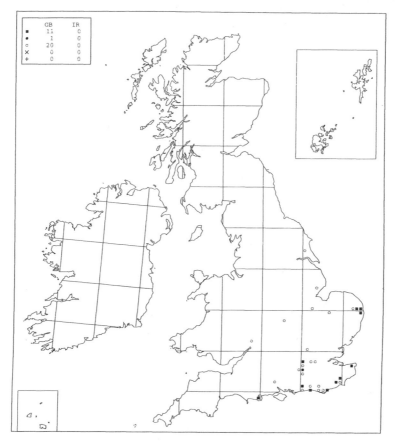

	GB	IR
■	11	0
●	1	0
○	20	0
×	0	0
+	0	0

Potamogeton acutifolius Link
Sharp-leaved Pondweed

P. acutifolius has a narrower habitat range than most aquatics in our area, as it is virtually restricted to shallow, species-rich drainage ditches in lowland grazing marshes. It is found in calcareous, mesotrophic or meso-eutrophic water, where typical associates include *Elodea canadensis, Hottonia palustris, Hydrocharis morsus-ranae, Lemna minor, L. trisulca, Myriophyllum verticillatum, Potamogeton natans, Ranunculus circinatus, Sagittaria sagittifolia* and *Spirodela polyrhiza*. It is also recorded from a pond in Middlesex and there are single records from the Hereford & Gloucester and Oxford Canals. There is, however, no evidence that the species ever became established in the canal system, nor is it known from other newly available habitats such as gravel-pits.

P. acutifolius flowers and fruits more freely than some other linear-leaved pondweeds, including the closely related *P. compressus*. Turions are apparently produced in smaller quantity than in some other species, but they can usually be found from August onwards. The life history and reproduction of the species requires detailed study.

The decline of this species is apparent from the map.

It has not been seen in Yorkshire and Lincolnshire since the late 18th century. It was last seen in Essex in 1700, Hertfordshire in 1846, Warwickshire in *c.*1859, Gloucestershire in 1870, Hampshire in 1898 and Northamptonshire in 1910. In many of these counties the only evidence for its occurrence is a single herbarium specimen. It may be extinct in Surrey, where it was last seen in 1965, and in recent years it has been searched for but not refound at its sole site in Middlesex. Although it still survives in the Wareham area, its distribution there has contracted since the last century and it now occurs in small quantity in one small area; it has also been lost from some of its sites in Norfolk. The reasons for the decline appear to be habitat destruction, the conversion of grazing marshes to arable, and eutrophication. The headquarters of this species is now in Sussex, where it occurs in abundance at Amberley Wild Brooks, Pulborough Brooks and the Pevensey Levels, and has also been recorded recently in other localities. In Norfolk it has been successfully introduced to the Strumpshaw RSPB reserve.

P. acutifolius has a restricted world distribution, being confined to temperate regions of Europe. It is found from S. Sweden southwards to France, Italy and the Balkans, but it is absent from the Mediterranean area.

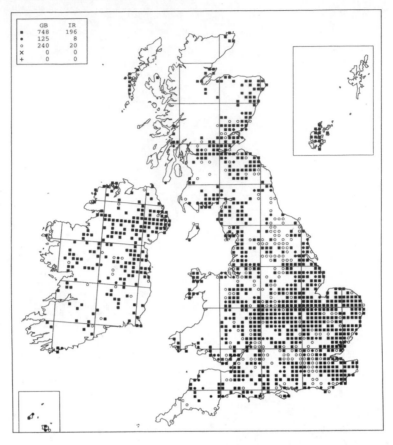

	GB	IR
■	748	196
●	125	8
○	240	20
×	0	0
+	0	0

Potamogeton crispus L.   **Curled Pondweed**

This is a frequent species in mesotrophic or eutrophic, standing or flowing waters. It is sometimes found in water only a few centimetres deep, and usually grows at depths shallower than 1m, but it can be found in water 6m deep (West, 1910). The habitats in which it occurs include lakes, reservoirs, ponds, rivers, streams, canals, fenland lodes, ditches and disused sand- and gravel-pits. It is found over a wide range of substrates from stony gravel or sand in clear rivers with moderate flow, to deep mud in lakes and ponds. It is capable of surviving in very eutrophic water, and can even be found in small quantity under a blanket of *Lemna gibba* and *L. minor*. Driscoll (1985a) found that, like *Potamogeton pusillus* and *Zannichellia palustris*, it increased in frequency in ditches in an area where grazing land was converted to arable. It may also grow in slightly brackish conditions. It is a primarily lowland species which reaches 350m in Drumore Loch.

P. crispus is a rhizomatous perennial which persists through the winter as short, leafy shoots. The longer stems which grow in summer produce turions in late spring and summer in response to a combination of high temperatures and long days (Chambers *et al.*, 1985). These germinate in autumn, and therefore ensure vegetative spread but (unlike the turions of the linear-leaved species) are not a means of overwintering. In shallow water, plants may flower and fruit freely.

P. crispus has very characteristic fruits, which have been found in deposits laid down throughout the last glacial period (Godwin, 1975). It is frequent in suitable habitats in Britain and Ireland. The absence of recent records from Yorkshire is not an indication of decline, but reflects an absence of recent localized records of aquatics from the county.

P. crispus is a native of Eurasia, Africa and Australia. In Europe it is occurs from southern Scandinavia southwards. It is naturalized in N. America, where it was first introduced in the 1840s. It had been recorded in most of the coterminous United States by 1950, and it has become a serious weed in some areas (Stuckey, 1979). It is still spreading in Canada (Catling & Dobson, 1985). It is also recorded as an introduction in New Zealand.

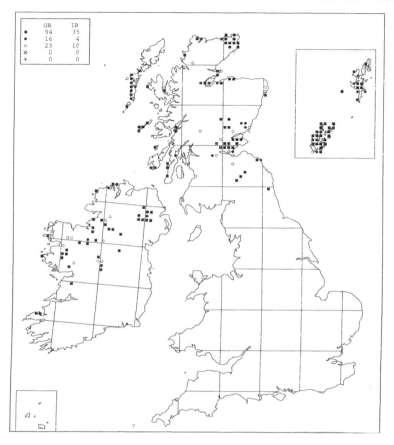

	GB	IR
■	94	35
●	16	4
○	23	10
×	0	0
+	0	0

Potamogeton filiformis Pers.
Slender-leaved Pondweed

P. filiformis is characteristically found in shallow water (less than 1m deep) where the vegetation cover is reduced by a fluctuating water-level or by erosion. It usually grows in water which is base-rich, eutrophic, brackish or possesses a combination of these attributes. It is found over a range of substrates including calcareous gravel and marl in limestone lakes, sand on the seaward side of machair lochs and eutrophic mud at the edge of lakes and reservoirs. Other habitats include streams and ditches, flooded quarries and, rarely, cliff-top rock pools in western Scotland. The open, brackish or base-rich habitats in which *P. filiformis* grows often support a rich charophyte flora, and *Chara aspera, C. hispida, C. vulgaris* and *Tolypella glomerata* can be found growing with it. Another frequent associate is the closely related *P. pectinatus*, but this species reaches its optimum development in deeper water. Although *P. filiformis* is often found in eutrophic water, its biomass in the most eutrophic sites can be reduced by algal growth in the summer months (Jupp & Spence, 1977). *P. filiformis* is primarily a lowland species,

which reaches 350m at Drumore Loch. There is a single herbarium specimen from a loch above Coire Dhubh-chlair, at 735m, but attempts to refind the species here in 1995 were unsuccessful.

P. filiformis is a perennial, and its long pioneer rhizomes can be seen invading open ground in shallow water. Growth begins in March and plants may begin to flower in May (Heslop Harrison, 1949). Plants are self-compatible and mature fruits can be found from June onwards. Plants overwinter as small tubers which develop on the rhizome.

This was a widespread and apparently frequent species in the full and late glacial periods in both Britain and Ireland (Godwin, 1975). Since then its distribution has contracted to the north and west. In Britain it has not been recorded from Anglesey since 1826, but it is frequent in suitable habitats in Scotland. In Ireland it is still present in quantity at its most south-westerly site, Lough Gill. It is probably somewhat under-recorded in both Scotland and Ireland.

This species has a circumboreal distribution, which in Europe extends south to scattered localities in Spain, the Alps, Italy and the Balkans.

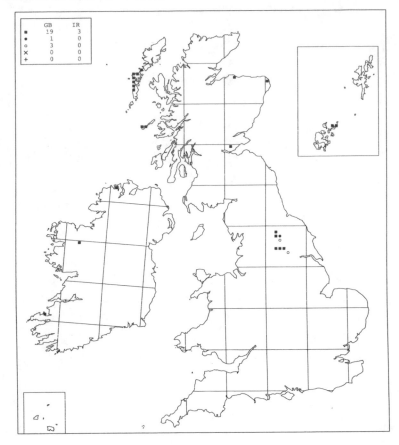

	GB	IR
■	19	3
●	1	0
○	3	0
×	0	0
+	0	0

Potamogeton × suecicus K. Richt. (*P. filiformis* Pers. × *P. pectinatus* L.)   **Swedish Pondweed**

P. × suecicus is a sterile hybrid which is recorded from shallow, standing and flowing waters. The largest and most vigorous stands are found in mesotrophic rivers, where the hybrid may form dense masses in shallow water over sandy, gravelly or stony substrates. Where the flow of water is moderately strong, plants are often rooted under large boulders or the supports of bridges. *P. × suecicus* also grows in shallow water at the edges of lakes or in their outflow streams, growing over sand in some coastal sites and over stony or silty substrates away from the sea. It is also locally abundant in the shallow stream and backwater pools of An Fhaodhail, Tiree, and has also been found in a pool amongst sand-dunes in Donegal. It is restricted to the lowlands.

Populations of this hybrid may occur in the absence of both parents (as in the R. Lossie), or even south of the southern limit of *P. filiformis* (R. Wharfe, R. Ure). The hybrid is a sterile perennial which overwinters as the rhizome, which may be very thick in robust, riverine plants. Vegetative reproduction and dispersal may occur by the production of tubers on the rhizome, or when large masses of material are washed downstream in floods (Dandy & Taylor, 1946). Plants may flower freely but fruits never develop, and old, shrivelled inflorescences and flowers at anthesis may be found on the same plant late in the season. The pollen of *P. × suecicus* initially develops normally, but later the grains split and their contents appear to disintegrate (Bance, 1946). At some localities the population may represent a single clone, but in others it includes more than one genotype (Hollingsworth *et al.*, 1996).

P. × suecicus was first recognized in Britain by Dandy & Taylor (1940c), although some of the specimens they cited had been collected in the 19th century, but misidentified. The populations in the Yorkshire rivers were detected by Dandy & Taylor (1946); plants in the R. Tweed and R. Till referred to this taxon by Dandy (1975) and by Dandy & Taylor (1946) are now thought to be another taxon (Hollingsworth *et al.*, 1996). The hybrid is very persistent: it still occurs at An Fhaodhail, where it was collected in 1897, and in the R. Wharfe, where it was gathered in 1868. The Yorkshire plants are believed to be relicts of a time earlier in this interglacial when *P. filiformis* was more widespread. *P. × suecicus* is easily overlooked as a form of the immensely variable *P. pectinatus*, and tends to be recognized only by experienced observers. In recent years it has been added to the Irish flora (Preston & Stewart, 1994) and found in a number of new Scottish localities. It is almost certainly under-recorded.

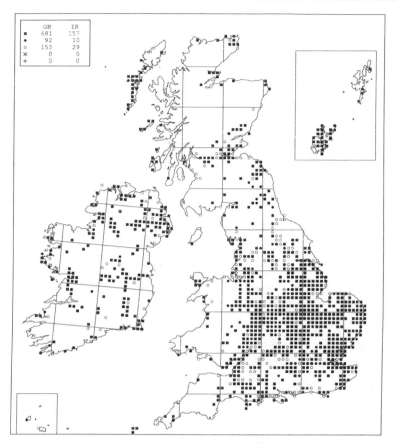

	GB	IR
■	681	157
●	92	10
○	153	29
×	0	0
+	0	0

Potamogeton pectinatus L.   Fennel Pondweed

P. pectinatus is one of the most frequent and character-istic species of eutrophic water. It is found in a wide range of standing and flowing waters, including lakes, rivers, streams, canals, ditches, ponds and flooded sand- and gravel-pits. It grows at a range of depths, but is most abundant at moderate depths of 0.5–1.5m, where it may frequently form dense and dominant stands. It tolerates disturbance in navigable and highly managed waters (Murphy & Eaton, 1983), and is a frequent colonist of gravel-pits and other newly available habitats. It is also tolerant of brackish conditions in coastal lagoons and ditches. It is sometimes found in highly calcareous but nutrient-poor lakes, but rarely occurs in these sites in abundance. Characteristic associates include *Elodea nuttallii*, *Myriophyllum spicatum*, *P. crispus*, *P. pusillus* and *Zannichellia palustris*. *P. pectinatus* is confined to the lowlands.

P. pectinatus is a rhizomatous perennial. Plants are self-compatible and those which reach the surface of the water fruit freely, although less than half the carpels develop into fruits; plants pollinated under water show reduced fruit-set (Guo & Cook, 1989). Tubers develop on the rhizomes and most populations overwinter as tubers, but some are winter-green. In addition to propa-gation by tubers, plants may also regenerate from broken and even apparently moribund stems. Both tubers and fruits are an important source of food for wildfowl. For a detailed account of the reproductive biology of this species, see Van Wijk (1988, 1989).

This species has a long history in our area, and it appears to have persisted through the last glacial period (Godwin, 1975). It is frequent in suitable habitats in Britain and Ireland, and in areas where water bodies are eutrophic and highly managed it is the commonest member of the genus. In northern and western regions it becomes increasingly confined to brackish, coastal sites.

P. pectinatus has a very extensive world distribution: it is found in all continents except Antarctica and it is sometimes a troublesome aquatic weed. It is present throughout most of Europe, although it is absent from the most northerly regions. It is a variable plant, and although much of the variation is environmentally induced some of it is genetically based (Van Wijk *et al.*, 1988).

Groenlandia J. Gay

This monotypic genus is closely related to
Potamogeton. The sole species was formerly
treated as *Potamogeton densus* L., and is still found
under this name in some recent floras.

Groenlandia densa (L.) Fourr.

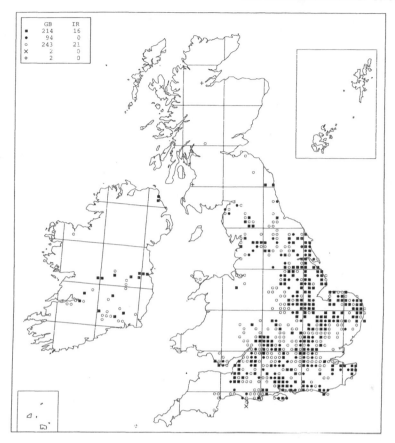

Groenlandia densa (L.) Fourr.
Opposite-leaved Pondweed

G. densa grows in shallow, clear, base-rich water in rivers, streams, canals, ditches and ponds. It is a particularly characteristic species of streams flowing from calcareous springs, where it is often found with *Apium nodiflorum, Berula erecta, Callitriche obtusangula, Zannichellia palustris* and, in larger streams, *Ranunculus penicillatus* subsp. *pseudofluitans*. It is rarely found in lakes and reservoirs, but the isolated site in Co. Antrim is a lake which receives base-enrichment from nearby basalt rocks. Here it is abundant in the shallows, and is present in smaller quantity to a depth of 1.4m. It is rarely found as a colonist of newly available habitats, but it is sometimes introduced to ponds outside its native range by gardeners. It is a predominantly lowland species which reaches 380m in the outflow stream from Malham Tarn.

G. densa is a rhizomatous perennial, although the rhizome is less distinct from the stem in this species than in the rhizomatous *Potamogeton* species. It continues to grow throughout the winter. The flowers are usually self-pollinated, either above the water or below the surface; if a plant flowers under water the pollen is carried from the anthers to the stigmas on the surface of a bubble (Guo & Cook, 1990). There is a possibility that out-crossing may occur if flowers are fertilized by pollen floating on the water surface. Seed-set is normally very high. The species has no specialized means of vegetative reproduction, but it almost certainly regenerates from plant fragments, which are easily detached. The extent to which it reproduces by seed is unknown.

Rather surprisingly, the fossil fruits of this species are often found in full glacial deposits (Godwin, 1975). In the historical period it has declined in both Britain and Ireland. It is now extinct as a native species in Scotland, where it was most frequently recorded near the centre of Edinburgh in habitats which have now disappeared (Preston, 1986). Elsewhere it has suffered from habitat destruction and eutrophication. In Cambridgeshire older records indicate that it was once widespread in the county but it has disappeared from parishes on boulder clay and it is now confined to a few sites on the chalk and in the fens.

G. densa is found in Europe north to Britain and Denmark. It is also recorded from adjacent parts of Asia and N. Africa.

RUPPIACEAE

The Ruppiaceae is closely allied to the Potamogetonaceae. Many authors consider that it is best regarded as a subfamily of that family, but this view has received no support from recent molecular studies (Les *et al.*, 1993). It includes a single genus, *Ruppia*.

Ruppia L.

This genus of aquatic plants is widespread in temperate and subtropical waters. In many areas it is confined to brackish waters near the sea, but it also occurs at saline sites inland and sometimes in fresh water. The species are difficult to identify as the plants (like many aquatics) have a reduced morphology and are phenotypically variable. The wide geographical range of *R. cirrhosa* and *R. maritima* creates additional problems. It is difficult to assess whether or not the differences between plants in different continents have a genetic basis,

especially if only fragmentary herbarium material is available for study.

Although the classification of *Ruppia* on a world scale is not yet resolved, the taxonomy and ecology of the genus in N.W. Europe have received detailed study in recent years (Reese, 1962, 1963; Verhoeven, 1979, 1980a,b; Verhoeven & Van Vierssen, 1978; Van Vierssen *et al.*, 1981, 1984). There is general agreement that there are two species, *R. cirrhosa* and *R. maritima*, which are cytologically, morphologically and ecologically distinct. There are some problems in equating these taxa to those present in S.W. Europe, and critical comparisons with material from other continents have not yet been made.

Before the two species of *Ruppia* were distinguished, all *Ruppia* material was called *R. maritima*. When the species were separated the name *R. maritima* was first applied to the one now called *R. cirrhosa*, and only later to the species that currently bears the name. Historical records therefore require careful interpretation.

Ruppia cirrhosa (Petagna) Grande

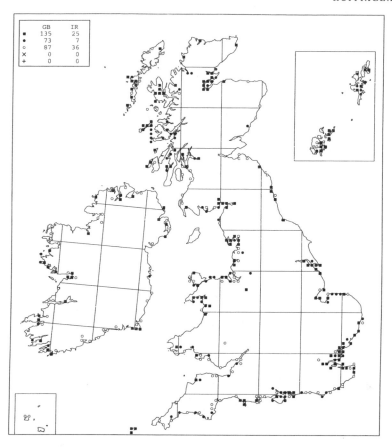

	GB	IR
■	135	25
●	73	7
○	87	36
×	0	0
+	0	0

Ruppia maritima L.   **Beaked Tasselweed**

This species is virtually confined to coastal, brackish waters. It is typically found in lakes, including artificial boating lakes, natural lagoons and sheltered tidal lochs; pools on salt–marshes, raised beaches and shingle beaches; small rock pools which receive salt spray; the tidal stretches of streams and rivers; ditches, creeks and shallow runnels on mud-flats and on the mud-flats themselves. It also colonizes coastal clay- and gravel-pits. It has also been recorded in pools associated with inland salt-marshes in Cheshire (Lee, 1977) and, exceptionally, from a canal some 56km from the sea near Talybont (Barrett, 1885). It is usually found in water less than 0.5m deep, and it can grow on mud which is exposed at low tide. In lakes and ditches it may grow with other salt-tolerant aquatics such as *Myriophyllum spicatum, Potamogeton filiformis, P. pectinatus* and *Zannichellia palustris*, but in the smaller pools and more saline habitats it is often the only vascular plant present.

R. maritima may behave as an annual or as a perennial. In the Netherlands, populations are often annual and reproduce by seed, but plants may overwinter as quiescent vegetative stems. In the milder climate of the Welsh coast, plants may grow and even flower through the winter. This species normally flowers and fruits more freely than *R. cirrhosa*. The flowers are borne on short peduncles and are usually submerged. Plants are self-pollinated by the transfer of pollen within an inflorescence on the surface of air bubbles which develop when the anthers dehisce and which remain attached to the inflorescence for several hours. The seeds lack an impermeable seed coat. Germination is stimulated by cold temperature pretreatments and in the Netherlands plants regenerate from seed in the spring. The growing plant is killed by desiccation but the seeds survive. For further details of the life-cycle of this species, see Verhoeven (1979) and Van Vierssen *et al.* (1984).

The fossil record of *R. maritima* suggests that it has always been restricted to coastal sites in our area. It may be somewhat under-recorded, both because it often has to be searched for deliberately in habitats which are not otherwise very rewarding to the aquatic botanist, and partly because vegetative material of *Ruppia* may not be identifiable to species level. There is little positive evidence of a decline on the stretches of coast where there are no recent records, and further fieldwork is required to establish whether the plant has disappeared from these areas.

In Europe, *R. maritima* is found in coastal habitats and at scattered sites inland from Iceland and northern Scandinavia southwards. It is widespread in Asia, N. Africa, N. America and Australia, occuring in coastal sites and at saline or other alkaline sites inland. The world distribution, which is discussed by Verhoeven (1979), is uncertain because of the rather chaotic state of *Ruppia* taxonomy.

In N.W. Europe, plants of *R. maritima* are diploid (2n=20), and this number has been confirmed for British material by Al-Bermani *et al.* (1993). Both diploids and tetraploids (2n=40) occur in S. Europe (Talavera *et al.*, 1993).

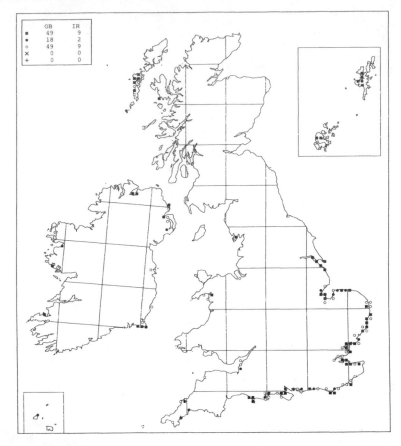

	GB	IR
■	49	9
●	18	2
○	49	9
×	0	0
+	0	0

Ruppia cirrhosa (Petagna) Grande
                                    Spiral Tasselweed

This species grows in brackish water in tidal inlets and in coastal lakes and lagoons which are directly connected to the sea, and in other coastal lakes, ponds, pools on salt-marshes and shingle beaches, and ditches. It may colonize artificial lakes and ponds and flooded gravel-pits. It usually grows over sand or soft mud and, unlike *R. maritima*, it is often found at depths of 0.5m or more. In its less brackish sites associates include *Potamogeton pectinatus* and *Ruppia maritima*. However, *R. cirrhosa* tolerates more saline conditions than *R. maritima*, or any other macrophyte in our area except those in the marine genus *Zostera*. In The Fleet it grows in mixed communities with *Zostera angustifolia* and *Z. noltii* and in brackish lakes elsewhere it may also grow with seaweeds such as *Fucus ceranoides* and *F. vesiculosus*.

R. cirrhosa is a perennial which persists through the winter as quiescent rhizomes or stolons (Verhoeven, 1979). The species flowers less freely than *R. maritima*. The peduncle elongates rapidly until the flowers reach the surface of the water. As it elongates, the anthers are released. They float to the surface, where they burst to release the pollen, which then floats. Cross-pollination is more likely than self-pollination, and the proportion of

carpels which are successfully fertilized is smaller in *R. cirrhosa* than in *R. maritima*. Van Vierssen *et al.* (1984) attempted to germinate the seeds under a variety of experimental conditions but failed to get more than a small proportion to germinate, in marked contrast to the results they obtained for seeds of *R. maritima*. The conditions required for the successful establishment of plants from seed in the wild are unknown.

The map suggests that this species has declined in some parts of its range. One locality in Sussex was destroyed by gravel dredging (Hall, 1980) but there is little other direct evidence of decline. The species is easily overlooked, and a detailed survey of former localities is needed to establish whether it has decreased. It was rediscovered in Yorkshire in 1981 (Crackles, 1983b) and found for the first time in Westmorland in 1991, records which suggest that it might be overlooked elsewhere.

R. cirrhosa is found along all but the arctic coasts of Europe. It is also known from W. Asia, Africa and N. & S. America. Although it is usually found in saline habitats it grows in fresh water in montane lakes in the Andes.

Two cytotypes of *R. cirrhosa* are found in N.W. Europe, a tetraploid (2n=40) and a hexaploid (2n=60). There is no information about the chromosome numbers of plants in our area.

NAJADACEAE

This family contains a single genus, *Najas*.

Najas L.

This is a genus of 40 species, all of which are submerged aquatics. The genus has a cosmopolitan distribution, although most species occur in tropical and subtropical regions. Although there is no recent world monograph, there are detailed treatments of the species in the Old World (Triest, 1988), Africa (Triest, 1987), N. & C. America (Haynes, 1979) and the tropical regions of N. & S. America (Lowden, 1986).

Most species of *Najas* are annuals which reproduce by seed; perennation and vegetative repro-duction appear to be unusual (Triest, 1988). The genus contains monoecious and dioecious species. The flowers are submerged and unisexual, with the male flowers consisting of a single stamen. Pollination occurs under water. The species grow in a range of fresh and brackish water habitats, and no fewer than eleven species are recorded as rice-field weeds in the Old World (Triest, 1988).

Triest (1988) recognizes two subgenera. Two of our species, *N. flexilis* and *N. graminea*, are in sub-genus *Caulinia* (Willd.) A. Br. ex Rendle, which includes species which are usually slender, monoecious plants and lack spines on the stems. The dioecious *N. marina* is the only species in subgenus *Najas*; it is a robust plant which usually has spiny stems.

Najas flexilis (Willd.) Rostk. & W. L. E. Schmidt

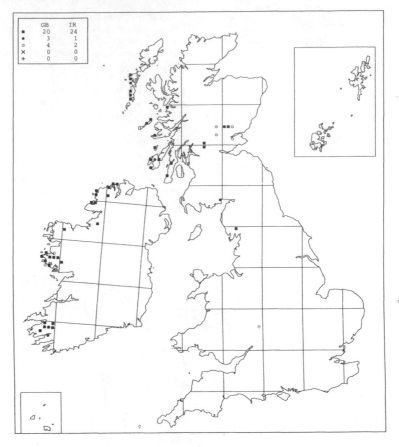

	GB	IR
■	20	24
●	3	1
○	4	2
×	0	0
+	0	0

Najas flexilis (Willd.) Rostk. & W. L. E. Schmidt
Slender Naiad

N. flexilis usually grows in deep, clear water in mesotrophic lowland lakes. It often occurs in sites which have some base-enrichment from nearby outcrops of basalt or limestone, or in machair lochs adjoining calcareous dune sand. Although it can be found in water less than 1m deep, it is usually found at depths of 1.5m or more. It grows over soft, silty substrates, often occurring with *Callitriche hermaphroditica*, *Potamogeton praelongus*, *Nitella flexilis* and occasionally with the tiny *Nitella confervacea*.

N. flexilis is a monoecious annual. The male flowers tend to be at the upper nodes with the female flowers below. Most mature plants appear to set seed, and within a lake the seeds either fall to the bottom or are dispersed when plants are uprooted in autumn gales and blown to the edge. In N. America the stems, leaves and seeds are eaten by wildfowl; Martin *et al.* (1961) describe the genus as 'among the choicest of native eastern duck foods'.

Fossil evidence indicates that *N. flexilis* was present in Britain in the middle of the last glacial period, and was much more widespread than it is today in the late glacial and early Flandrian. Its slender habit and deep-water

habitat make it a very inconspicuous species. It was discovered in Ireland in 1850 (Oliver, 1851) and in Scotland in 1875 (Sturrock, 1876), but not added to the English flora until 1914 (Pearsall, 1915). In recent years more intensive fieldwork, including surveys by divers, have revealed new sites in both Scotland and Ireland. It is, however, vulnerable to eutrophication. It has not been seen recently at a number of sites in E. Scotland, including lochs such as Monk Myre which are now very eutrophic, and it has apparently disappeared from Esthwaite Water, its only English locality, where it was abundant when it was first discovered, was still present in 1982 but could not be refound in 1994. It is threatened by fish farming at one Scottish locality.

N. flexilis has a circumboreal distribution, but it is much more frequent in N. America than in Eurasia. In Europe it has a northerly distribution, extending south to Switzerland. Fossil records, mapped by Hultén & Fries (1986), show that it was once more widespread in Europe, and in many countries it is a rare and declining species. The American distribution is mapped by Haynes (1979). Even in America it is decreasing: in Ohio it was once the most common *Najas* species but it is now becoming rare as many clear lakes have become muddy or turbid (Wentz & Stuckey, 1971).

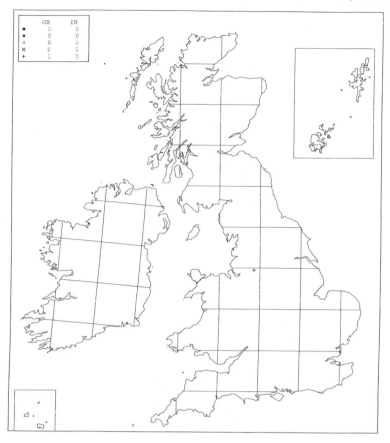

Najas graminea **Delile**

This alien species formerly grew in tepid water over soft, black mud in a section of the Reddish Canal which was warmed by effluent from nearby cotton mills and other industrial works. It grew with *Elodea canadensis, Myriophyllum* sp., *Potamogeton alpinus, P. crispus, P. obtusifolius, P. pusillus* and the alien charophyte *Chara braunii*.

N. *graminea* is a monoecious annual which reproduces by seed. In the British population female flowers were more numerous than male flowers; when present the males were usually found alongside the females in the axil of the same leaf. Mature fruits were produced in abundance (Bailey, 1884).

N. *graminea* was discovered in Britain by J. Lee in 1883. A detailed account of the species was published by Bailey (1884), who believed that it had been introduced to the canal with Egyptian cotton, which was used by one nearby mill. It persisted in the canal until 1947, but disappeared once the mill ceased to discharge hot water into the canal (Savidge, 1963). The canal has now been built over.

N. *graminea* occurs as a native in the subtropical and tropical regions of Africa, Asia and Australia. It is a weedy species which may colonize irrigation canals. In S. Europe it is naturalized in rice-fields and ditches in France (Camargue), N. Italy, Bulgaria and Romania; it also occurs as an alien in N. America (California) and S. America (Brazil).

This species is very variable cytologically, and exists at several different levels of a polyploid series (You *et al.*, 1991).

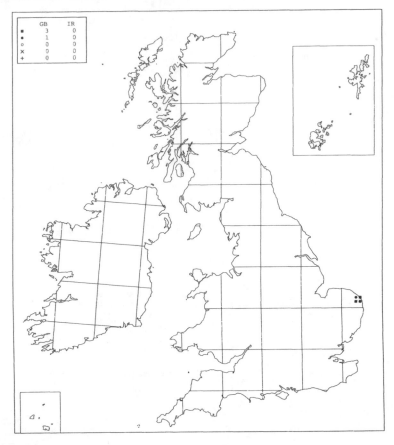

	GB	IR
■	3	0
●	1	0
○	0	0
×	0	0
+	0	0

Najas marina L.   Holly-leaved Naiad

N. marina formerly grew in the clear, calcareous and nutrient-poor water of the unpolluted broads (George, 1992). At Hickling Broad, where there is some saline influence, it formed a low sward with abundant charophytes in water 1–3m deep; Bennett (1883, 1884) listed its associates as *Potamogeton pectinatus, Chara aspera, C. hispida, C. pedunculata* and *Nitellopsis obtusa*. In other sites it colonized bare mud, and could sometimes be found in *Phragmites australis* swamps (Barry & Jermy, 1953). It is now usually found in meso–eutrophic water over deep substrates of silt or peaty mud, growing with species such as *Ceratophyllum demersum, Hippuris vulgaris, Myriophyllum verticillatum, Potamogeton pectinatus* and *Utricularia vulgaris* in water deeper than 0.2m.

N. marina is a dioecous annual. In Britain, female plants set good seed even in the apparent absence of male plants, suggesting that they may be apomictic (A. C. Jermy, pers. comm.). In cultivation plants may complete their life cycle from seed to seed in 100 days (Viinikka, 1976). The germination requirements have not been studied in Britain, but plants from Sweden, the Netherlands and Israel all have similar germination requirements (Agami & Waisel 1984, 1986, 1988). The seeds usually have a hard coat which needs to be broken before they will germinate, although its thickness varies from seed to seed. The seed coat may be ruptured by passage through the digestive system of fish or waterfowl: most seeds eaten by mallard are digested but the remainder show greatly

increased germination compared to intact seeds. Seeds will also germinate after prolonged periods of immersion in cold water. Germination is stimulated by darkness and by anaerobic conditions. Even the most effective pretreatments will not stimulate all seeds to germinate, and *N. marina* has a persistent seed-bank. Seeds stored in dry conditions may remain viable for at least four years. British populations do not have specialized turions, which have been reported from Israel by Agami *et al.* (1986), but they may spread by vegetative fragmentation.

Although *N. marina* was present in Britain in the last glacial period, the fossil evidence indicates that it subsequently increased in abundance. At the 'climatic optimum' it was widespread in England and Ireland, and also recorded from S. Wales and the Isle of Man (Godwin, 1975). It is a most distinctive species, which was discovered as a living plant by Bennett (1883) at Hickling Broad. It was discovered in Martham Broad in 1885, and in several more broads between 1949 and 1952. In the early 1950s its status in Hickling Broad was similar to that described by Bennett (1883), but it declined in the late 1960s in this and other sites as nutrient levels increased (Barry & Jermy, 1953; George, 1992). However, it is frequent in Upton Broad, where phosphorus levels are still low, and it has increased or reappeared in other sites after remedial action was taken to reduce the nutrient levels, probably germinating from dormant seed (George, 1992).

N. marina is a widespread species, found in temperate, subtropical and tropical regions of Europe, Asia,

Africa, Australia and N. & S. America, as well as on oceanic islands such as Hawaii and the Galapagos. In Europe it is widespread north to England and Finland. It is morphologically and cytologically variable. Triest (1988) recognizes twelve subspecies in the Old World. The British plant is subsp. *intermedia*, itself a polymorphic taxon which occurs around the North Sea and the Baltic, in the Alps, and eastwards in Asia to Thailand. The other European subspecies are subsp. *marina*, which is the more frequent plant in western and southern Europe and occurs at scattered localities throughout Asia, and subsp. *armata* (H. Lindb.) Horn of S. Europe, S.W. Asia, Sri Lanka and Australia. Plants in the tropics behave as perennials.

ZANNICHELLIACEAE

This family contains four small and closely related genera of aquatics with submerged, linear leaves. One of these genera, *Zannichellia*, is found in our area and another, *Althenia*, occurs in southern Europe. The remaining genera are found in S. Africa (*Pseudalthenia*) and Australasia (*Lepilaenia*). Five genera of sea-grasses are sometimes included in this family, but are grouped by Cronquist (1981) in the exclusively marine family Cymodoceaceae.

Zannichellia L.

The taxonomy of the genus *Zannichellia* is in a state of confusion: Pearsall's (1933) comment that it is 'by no means easy to reconcile the divergent taxonomic views of authors in this genus' is still true over sixty years after he made it. At one extreme are those authors who recognize a single species which is both morphologically and cyto-logically variable, *Z. palustris*. This was J. E. Dandy's opinion, and it was followed by *Flora Europaea* (Tutin *et al.*, 1980). Most botanists in Britain and Ireland have adopted this view in

recent decades, including Stace (1991) and Kent (1992), although Clapham *et al.* (1952) suggested that our material might be divisible into two sub-species. Continental botanists usually recognize more taxa. Van Vierssen (1982a,b) distinguished four species which differ in morphology, ecology and chromosome number. Three of these occur in western Europe, *Z. palustris* (which is divided into subsp. *palustris* and subsp. *repens* (Boenn.) Uotila), *Z. pedunculata* Rchb. and *Z. major* Boenn.; a fourth, *Z. peltata* Bertol., is found in southern Europe. The European distribution of these species is mapped by Talavera *et al.* (1986), who recognize two additional species, *Z. contorta* (Desf.) Cham. & Schltdl. and *Z. obtusifolia* Talavera, García-Mur. & H. Smit. They cite specimens of *Z. palustris* and *Z. pedunculata* from Britain and *Z. palustris* from Ireland. In eastern Europe totally different characters are used to divide the genus into species, so it is difficult to equate the taxa recognized there with those in the west (Uotila *et al.*, 1983).

According to Van Vierssen (1982b), *Z. palustris* sensu stricto is a plant of permanent, fresh or brackish waters in northern Europe but further south it becomes confined to fresh water. *Z.*

Zannichellia palustris L.

pedunculata grows in coastal habitats, including water bodies that dry out in summer. *Z. major* is found in deep, brackish water in the Baltic.

We have followed Stace (1991) in recognizing a single species, although it is clear that material from our area needs to be reinvestigated in the light of continental work. Most plants from inland habitats in Britain and Ireland are probably referable to the continental *Z. palustris*, whereas some coastal populations correspond to *Z. pedunculata*, which probably includes many of the plants recognized as *Z. palustris* var. *pedicellata* or *Z. pedicellata* in older floras.

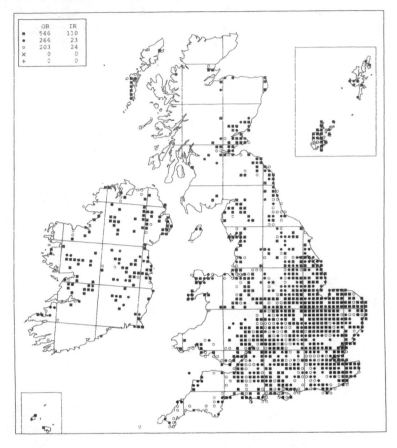

	GB	IR
■	546	110
•	266	23
○	203	24
×	0	0
+	0	0

Zannichellia palustris L.   Horned Pondweed

Z. palustris grows in a wide range of habitats where the water is shallow and calcareous, eutrophic or brackish. It can be found as large patches in clear streams which are highly calcareous but not necessarily nutrient-rich, growing on stony substrates, sand or fine silt in the main channel or near the edge. It is widespread in eutrophic lakes, pools, rivers, streams, canals, ditches and in flooded clay, gravel and other mineral workings. It is often found in water less than 0.5m deep, but it will also grow through stands of more robust aquatics in slightly deeper water. It is one of the few submerged aquatics which will persist in lakes and ponds that are inhabited by large numbers of waterfowl, where it may be found (often in small quantity) in shallow water over deep, eutrophic mud. It is also frequent in brackish lagoons, pools and ditches. It is a predominantly lowland species, which reaches 380m at Llynheilyn.

In some habitats Z. palustris is certainly perennial, and in streams emerging at a constant temperature from calcareous springs it flowers throughout the year. Other populations on lakeshores may regrow from seed after periodic desiccation, and there may be localities in which it is usually annual, although detailed observations in our area are lacking. Plants spread by a creeping rhizome, sometimes forming clonal patches. The species is monoecious and the male flowers release pollen into the

water in small 'clouds' which sink slowly or are carried away by currents (Guo et al., 1990). The anthers may dehisce before or after the funnel-shaped stigmas are receptive, and both self- and cross-pollination occurs. Fruit-set is usually good, and the fruits are 'relished by ducks' (Martin et al., 1961). There is a wide variation in germination behaviour between populations of Z. palustris sensu lato. The seeds of plants attributed to Z. palustris by Van Vierssen (1982a,c) are initially dormant and require stratification before they will germinate, whereas those of Z. pedunculata do not. Seeds which are desiccated for three months retain viability.

Z. palustris has been more or less continuously present in the British Isles since the mid Pleistocene, and persisted through the last glacial period (Godwin, 1975). Unless it is present in abundance it is easily overlooked by inexperienced botanists: it was under-recorded in the Atlas of the British Flora and may still be more frequent than the map suggests, particularly in Ireland. It is frequent in suitable habitats in England, and is increasing in some Scottish counties, including Angus (Ingram & Noltie, 1981) and Perthshire, as a result of eutrophication.

Z. palustris has an almost cosmopolitan distribution. It is widespread in the northern hemisphere, although absent from the more northerly regions, but somewhat less frequent in the tropics and the southern hemisphere. In Europe it extends from Iceland and the coasts of Scandinavia to the southern fringes of the continent.

ARACEAE

This is a large and predominantly tropical family of 106 genera and nearly 3,000 species. It contains several important food plants, and many house-plants and other ornamentals. A recent book, *The Aroids*, provides an excellent illustrated review of the family (Bown, 1988); for a more technical account, see Hotta (1971). Most of the genera are terrestrial, but there are almost 100 aquatic species distributed amongst 20 genera. They include *Pistia stratiotes* L., the water lettuce, which is one of the most troublesome aquatic weeds in the tropics and has been recorded in Britain as a casual (Clement & Foster, 1994), and 50 species of *Cryptocoryne*. None of the aquatic Araceae are native to our area, but two genera are found as naturalized aliens. A third genus, *Lysichiton*, is also naturalized in moist woodland and on damp ground by streams and rivers.

Acorus L.

The genus *Acorus* is an aberrant member of the Araceae, which differs from all the other members of the family in its vegetative morphology, the absence of raphide crystals and the presence in the leaves of ethereal oil cells. Its systematic position has recently been reviewed by Grayum (1987), who lists 13 more characters of *Acorus* that are unique within the Araceae, including 'the much-ballyhooed co-occurrence of the rust *Uromyces spargani* on *Sparganium* and *Acorus*'. Grayum con-cludes that the genus should be moved to its own family, the Acoraceae. The genus contains at least two species, *A. calamus* and *A. gramineus*; some authors recognize additional segregates of both these species. The account below follows the taxo-nomic treatment of Röst (1978, 1979).

Acorus calamus L.

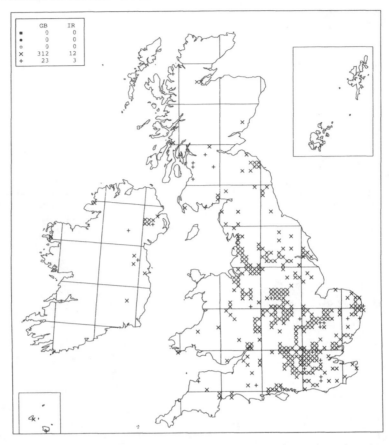

	GB	IR
■	0	0
●	0	0
○	0	0
×	312	12
+	23	3

Acorus calamus L.   Sweet-flag

A. calamus is a tall perennial which grows in water less than 1m deep at the edge of standing or slowly flowing, mesotrophic or eutrophic waters. It often forms narrow stands along the fringes of lakes, rivers, streams, canals, ponds and ditches. It can often be found around lakes in the landscaped grounds of stately homes, where it was originally planted but has become thoroughly naturalized. It may spread into neglected ponds and eventually cover large areas of previously open water. It also spreads naturally along canals and rivers, such as the Lancaster Canal (Greenwood, 1974) and the R. Tweed and its tributaries. Dense stands of *A. calamus* exclude almost all competitors; in more open stands it can be mixed with emergents such as *Solanum dulcamara*, *Sparganium erectum* and *Typha latifolia*, with *Lemna minor* floating on the surface of the water (Rodwell, 1995). It is confined to the lowlands.

A. calamus is a rhizomatous perennial. The plant naturalized in Europe is a sterile triploid cytotype which flowers (although sometimes rather infrequently) but never produces fertile seed (Röst, 1978). Confirmation that this cytotype is present in Britain is provided by Marchant (1973). This plant spreads by direct growth of the stout rhizome, and by the establishment of new colonies from lengths of rhizome washed out in floods or dislodged by human disturbance. Once established, it may be very persistent.

A. calamus has been used in medicine since ancient times, and later used as a fragrant floor covering. It is

believed to have been introduced to Europe in the mid 16th century. Gerarde (1597) knew it as a garden plant, but by 1668 Sir Thomas Browne described it as growing 'very plentifully...by the bankes of Norwich river [R. Yare] ...so that I have known Heigham Church in the suburbes of Norwich strowed all over with it; it hath been transplanted and set on the sides of marish [marshy] pondes in severall places of the countrey, where it thrives and beareth the *Julus* [inflorescence] yearly' (Keynes, 1964). It decreased in abundance in Broadland as a result of grazing by coypu, but increased after the successful attempt to exterminate these rodents (George, 1992). In Ireland it was recorded from Moira Demesne in 1744; it had spread into the Lagan Canal by 1866 and thus reached Lough Neagh (Harron, 1986; Hackney, 1992). It continues to be planted around ornamental lakes and ponds.

The sterile triploid var. *calamus* (2n=36) is widespread in Europe and western Asia, although it is absent from most of the Mediterranean area. It also occurs in the Himalayan region and in eastern N. America. It was introduced to Europe via Turkey; its native range is unknown. The fertile diploid var. *americanus* (Raf.) H. Wulff (2n=24) is widespread in Siberia and N. America; although it can produce seeds it also persists vegetatively in areas where seeds do not mature for climatic reasons. The fertile tetraploid var. *angustatus* Besser (2n=48) is more variable than the other cytotypes. It is found in temperate and subtropical S. & E. Asia. For distribution maps of the cytotypes, see Röst (1978) and Packer & Ringius (1984). These taxa might be more appropriately treated as subspecies.

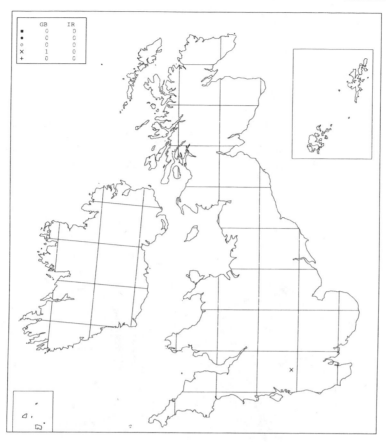

	GB	IR
■	0	0
●	0	0
○	0	0
×	1	0
+	0	0

Acorus gramineus Aiton   **Slender Sweet-flag**

A small clump of *Acorus gramineus* 'Variegatus' grows under trees at the edge of the Basingstoke Canal at Mychett.

Acorus gramineus is a rhizomatous perennial. The species is a fertile diploid, but there is no evidence that the British population reproduces by seed.

This plant was discovered by the Basingstoke Canal by Mrs J. F. Leslie and others in 1986. It is a cultivar which was probably discarded from nearby houses (Mrs J. E. Smith, pers. comm.). It was included in Stace (1991) and Kent (1992) on the basis of this record, but it can scarcely be considered naturalized.

A. gramineus is a variable species which is a native of S. & E. Asia. It is widely cultivated in Europe, but is not listed as a naturalized species by Tutin *et al.* (1980).

Calla L.

This is a monotypic genus which occurs in our
area as an introduction.

Calla palustris L.

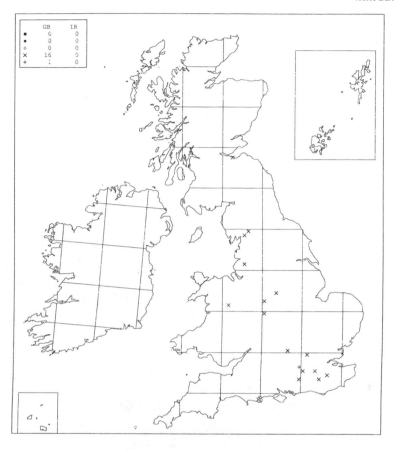

	GB	IR
■	0	0
●	0	0
○	0	0
×	16	0
+	1	0

Calla palustris L.   Bog Arum

C. palustris can grow as a floating plant or rooted in wet terrestrial habitats. It is naturalized at the edges of lakes and ponds, in marshes and in wet alder woodland. In lakes it can be present in considerable quantity. It is confined to the lowlands. In continental Europe it is a plant of rather acidic and nutrient-poor shallow waters, swamps and wet alder woods, and grows with species such as *Carex lasiocarpa, C. rostrata, Equisetum fluviatile, Menyanthes palustris* and *Potentilla palustris*.

This is a perennial which spreads by growth of the long rhizome, and (in floating plants, at least) by fragmentation. The flowers are bisexual or male, and are insect-pollinated. The fruits are red berries, which will float for long periods but are 'greedily eaten by ducks' (Ridley, 1930). In Britain naturalized plants fruit (Caulton, 1961), but we do not know if they reproduce by seed.

C. palustris has a long history in Surrey. It was first recorded at Black Pond, Esher, where it was planted in 1861 and persisted for at least 30 years. It was subsequently found at Bolder Mere, where it 'flourished exceedingly' and 'in 1914 was quite a feature of the aquatic vegetation there, flowering profusely in July' (Britton, 1915). It survived at this site for over 60 years but is now extinct (Lousley, 1976; Leslie, 1987). It was found at Stanmore Common in Middlesex in 1946, where it is also extinct (Kent, 1975; Burton, 1983), and subsequently from a number of scattered sites elsewhere. It appears to have been recorded with increasing frequency in recent years.

This species has a circumboreal distribution. In Europe it is native from Scandinavia and the Low Countries eastwards, but it has distinctly continental tendencies and it is rare towards the western edge of its range.

LEMNACEAE

A family of four genera of small aquatic plants, three of which occur in Britain and Ireland. Five of our six species are free-floating whereas the sixth (*Lemna trisulca*) has submerged vegetative fronds but floating fertile fronds. Vegetative reproduction predominates in the family, with most species flowering only occasionally. Many species are variable cytologically, especially in the tropics and the southern hemisphere, and strains with different chromosome numbers are maintained by vegetative reproduction. The factors which lead to flowering in wild populations are not well understood. In the cool, oceanic climate of the British Isles the duckweeds appear to flower even less readily than they do either in more continental climates at similar latitudes or in the more southerly parts of their range. Martinsson (1984) found that twelve of the 26 *Lemna minor* populations which he studied on Öland (57°N) in the summer of 1982 were flowering. Comparable figures are not available for the British Isles, but it seems most unlikely that they would be as high. Pollination is effected by small invertebrates, which crawl over the surface of the fronds. Dispersal of fronds is by water movement and by animal vectors.

The Lemnaceae of the world have been monographed by E. Landolt and his colleagues (Landolt, 1980, 1986; Lüönd, 1983; Landolt & Kandeler, 1987). Much of the information given below on reproduction, perennation and world distribution is derived from Landolt (1986). There have been few studies of the reproduction and perennation of the British species, and research is needed to relate their behaviour here to that observed in other areas. The literature on the family is covered by bibliographies in Landolt (1980) and Landolt & Kandeler (1987).

Spirodela Schleid.

A small genus of three species, until recently included by British botanists in *Lemna*. The main difference between the genera is the fact that *Lemna* species have a single root on each frond (or, rarely, no roots at all) whereas *Spirodela* species have two or more roots per frond.

Spirodela polyrhiza (L.) Schleid. (Scale divisions represent 1mm)

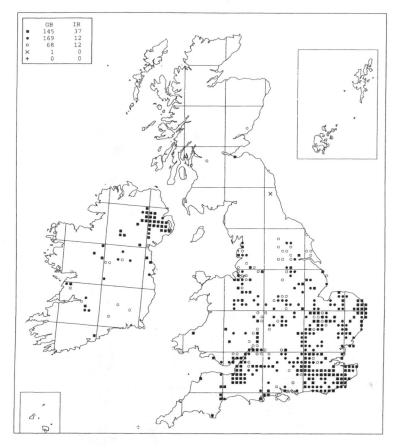

	GB	IR
■	145	37
●	169	12
○	68	12
×	1	0
+	0	0

Spirodela polyrhiza (L.) Schleid.
Greater Duckweed

S. polyrhiza, the largest British duckweed, is most frequent in flat fenland regions which receive drainage from calcareous uplands. It grows in base-rich water in ponds, ditches (especially deeper ditches in grazing marshes), canals and slowly flowing rivers, often in a species-rich assemblage of floating macrophytes which can include *Azolla filiculoides*, *Hydrocharis morsus-ranae*, *Lemna gibba*, *L. minor* and *L. minuta*. It is also recorded from old gravel workings and clay-pits. The presence of several roots on each of its fronds enables *S. polyrhiza* to exploit nutrients in the surface layer of water more effectively than *Lemna* spp. It is tolerant of a degree of eutrophication, and on the surface of nutrient-enriched ditches it may be co-dominant in a carpet of Lemnaceae. It is confined to the lowlands.

Reproduction is almost entirely vegetative. The species flowers very rarely throughout its range, and in Britain has been found flowering only once, at Wedmore, Somerset, in July 1906 (White, 1912). Perennation through the winter is by starch-filled orbicular or reniform turions, which develop in the fronds and sink to the bottom of the water. These are more cold-tolerant than the fronds. Dalgliesh (1926) reported that they were produced as early as 17 April, and when placed in an aquarium began to grow within a week. A detailed account of the life-history of *S. polyrhiza* in Minnesota, USA, is given by Jacobs (1947).

S. polyrhiza has declined in some areas, as ponds have disappeared and grazing marshes have been converted to arable land. Holland (1986) noted that in Gloucestershire it fails to survive pond clearance; this is presumably because the large fronds can be removed from the water relatively easily and no reservoir of propagules remains. If this is generally true, sensitive management of its habitats will be necessary to ensure its survival.

S. polyrhiza is widespread in Europe, north to 66°N. in Scandinavia, but rare in the Mediterranean region. It is recorded elsewhere from all continents except Antarctica, though it is rare in S. America and absent from arid regions. Its world distribution is mapped by Landolt (1986). British and European plants are tetraploid (2n=40), as are almost all the populations in the northern hemisphere for which chromosome numbers are available; triploids (2n=30), tetraploids and pentaploids (2n=50) are found in the tropics (Urbanska-Worytkiewicz, 1980).

Lemna L.

A genus of 13 species with a virtually cosmopolitan distribution.

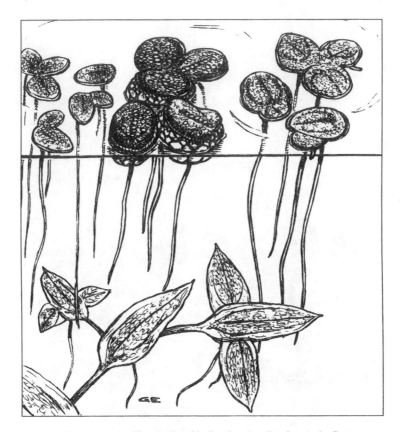

Lemna minuta Kunth; *L. gibba* L.; *L. minor* L.; *L. trisulca* L.

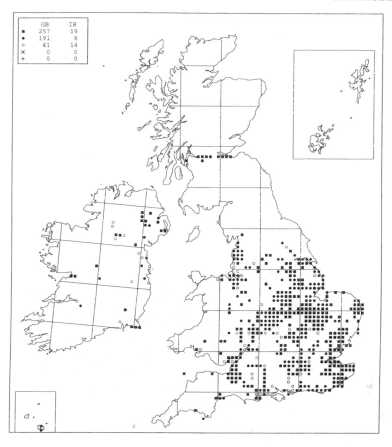

	GB	IR
■	257	19
●	191	8
○	41	14
×	0	0
+	0	0

Lemna gibba L.   Fat Duckweed

A plant of still or very slowly flowing eutrophic water in
lowland ditches, ponds, canals or the quiet backwaters of
streams and rivers. It is usually mixed with *L. minor*, but
very eutrophic ditches (such as those alongside arable
land) can be completely covered by a dense carpet of
pure *L. gibba*, under which no other macrophytes can
grow. *L. gibba* can also be abundant in brackish coastal
ditches. The competitive advantage of this species over
other duckweeds can be attributed to the buoyancy of its
swollen fronds, which resemble small pieces of expanded
polystyrene; these ride over and shade competitors. The
success of *L. gibba* in mixed communities was demonstrated in experimental conditions by Clatworthy &
Harper (1962).

Reproduction is primarily by vegetative budding.
Flowering is initiated by long days. *L. gibba* flowers
more often than most duckweeds, and is probably self-
compatible. In warmer climates reproduction by seed is
important, and the species even grows in a seasonal pool
in Israel where it survives the dry period as dormant
seed (Witztum, 1977). Seed from this population
remained viable after almost five years dry storage
(Witztum, 1986). Flowering in Britain is infrequent,
however. Plants persist through the winter as flat fronds
with a high starch content, which continue to float on

the surface of the water, or as living buds in the pouches
of dead, sunken fronds. In the Netherlands the flat winter fronds become gibbous by the end of May over clay
soils but sometimes persist until August over sand and
peat. In September gibbous plants become less inflated
and eventually produce flat progeny (Lange *et al.*, 1984).

Lemna gibba is locally frequent in lowland England
but rare in Scotland; it reaches its northern limit in the
Forth & Clyde and Union Canals. Field botanists probably mistake the flat winter form for *L. minor*, and *L.
gibba* might therefore be under-recorded in some areas.
Criteria for separating these two species are given by
Lange & Westinga (1979). There is no evidence that *L.
gibba* has declined and its abundance in eutrophic water
suggests that it is well equipped to survive in the modern countryside.

L. gibba is widespread in Europe, north to southern
Scandinavia. Elsewhere it occurs in W. Asia, N., E. and
S. Africa and N. and S. America, being most frequent in
areas with a mediterranean climate. It has been introduced to Japan. For a map of its world distribution, see
Landolt (1986). British and Irish plants are tetraploid
(2n=40) as are most populations in mainland Europe,
although pentaploids (2n=50) are recorded. Heptaploid
(2n=70) and octoploid (2n=80) populations occur in the
southern hemisphere (Urbanska-Worytkiewicz, 1980).

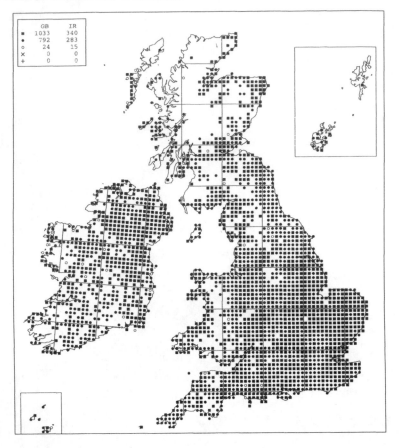

	GB	IR
■	1033	340
●	792	283
○	24	15
×	0	0
+	0	0

Lemna minor L.   Common Duckweed

The most widespread free-floating macrophyte in the British Isles, found in a range of still, mesotrophic to eutrophic, open or shaded waters including ditches, ponds, canals and lakes. In southern England it is often mixed with *Azolla filiculoides*, *Lemna gibba*, *L. minuta* or *Spirodela polyrhiza*; outside the range of these species it is usually found as pure stands. Smaller waters can become covered by a continuous carpet of duckweed, which not only cuts down the light intensity in the water below but reduces the oxygen content, thus retarding aerobic decomposition and leading to an increase in organic matter and hydrogen sulphide. Most submerged macrophytes cannot survive in such conditions, but a few species such as *Ceratophyllum demersum* and *Elodea nuttallii* are sometimes found under *Lemna* swards. In streams and rivers populations of *L. minor* build up where the water flow is impeded, e.g. in quiet back-waters, in stands of emergents or between moored boats. Small populations can be found in terrestrial habitats such as damp mud in swamps or by lakes, or even on the damp rock-faces of cliffs and cuttings. It is usually found in the lowlands, but reaches 500m on Brown Clee Hill.

Reproduction is primarily by vegetative budding. Flowering is initiated by long days. Landolt (1986) described the species as flowering occasionally, on the basis of observations made throughout its range, but in the British Isles flowering is infrequent. Landolt suggests that fruit-set is low, perhaps indicating self-incompatibility. Like *L. gibba*, *L. minor* survives the

winter as resting fronds which float on the surface of the water or as living buds in the pouches of dead, sunken fronds.

L. minor is frequent in the lowlands of Britain and Ireland. Only in areas where oligotrophic waters predominate does it become more local, being restricted to lakes, streams or ponds near the sea. There is no evidence that its distribution is changing.

This species is widespread in Europe, north to 66°N. Elsewhere it is recorded from W. Asia, Himalaya, N. America and N., E. and S. Africa, being most frequent in areas of cooler climate. It is probably introduced to the oceanic islands of Ascension and St Helena and almost certainly introduced to Australasia, where it is local in Australia but frequent in New Zealand. For a map of its world distribution, see Landolt (1986). Plants in Britain, Ireland and mainland Europe are tetraploid (2n=40), although aneuploid populations (2n=42) are also recorded in Europe. The tetraploids and related aneuploids are found throughout the range of the species; diploids (2n=20), triploids (2n=30) and pentaploids (2n=50) are also recorded (Urbanska-Worytkiewicz, 1975, 1980).

Although apparently innocuous, *L. minor* has rather surprisingly become a children's bogey in those parts of north-west England where there are numerous marl-pits. The green carpet of duckweed, known as Jenny Greenteeth, is said to drag venturesome children down into the depths of the water, closing over them to hide all traces of their presence (Vickery, 1983).

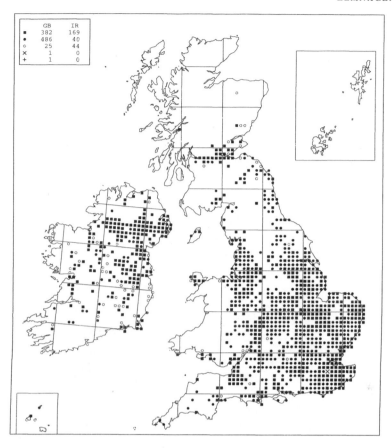

		GB	IR
■		382	169
●		486	40
○		25	44
×		1	0
+		1	0

Lemna trisulca L.   **Ivy-leaved Duckweed**

A species of still or slowly flowing water in lowland ponds, lakes, ditches, canals and in the backwaters of rivers. It is perhaps most abundant in clear, mesotrophic water but extends into eutrophic water and is sometimes found in brackish ditches near the sea. It is found over a range of substrates from peat to clay. Although often found beneath a thin layer of *L. minor* and other congeners, or of *Hydrocharis morsus-ranae*, it cannot survive when the surface of the water supports a dense mass of floating *Lemna*. *L. minor* has a competitive advantage over *L. trisulca* in both eutrophic sites and in shaded ponds and ditches. However, *L. trisulca* dominates where nutrient levels are lower, presumably because its thin, submerged fronds are more efficient at nutrient uptake (McIlraith *et al.*, 1989). *L. trisulca* is also found in lakes which are so large that a continuous cover of floating *Lemna* cannot establish itself because of wind and wave action.

Reproduction of *L. trisulca* is almost entirely by vegetative growth. Plants flower only rarely in the British Isles, and also less frequently than *L. gibba* or *L. minor* elsewhere in the range of the species. The flowering fronds float on the surface of the water; they are shorter than the vegetative fronds, have more air spaces and possess stomata. Plants survive the winter as resting fronds, which sink to the bottom of the water. They are morphologically similar to normal fronds.

L. trisulca is still frequent in lowland England. There is no evidence that its distribution is changing.

L. trisulca is widespread in Europe north to 68°, but rare in the Mediterranean area. It is a circumboreal species, found throughout the temperate northern hemisphere and in E. Africa, S.E. Asia and Australia. Plants from Britain, Ireland and mainland Europe are tetraploid (2n=40). Diploids (2n=20) are recorded from Australia and both hexaploids (2n=60) and octoploids (2n=80) are found in N. America; tetraploids also occur in both these areas (Urbanska-Worytkiewicz, 1980).

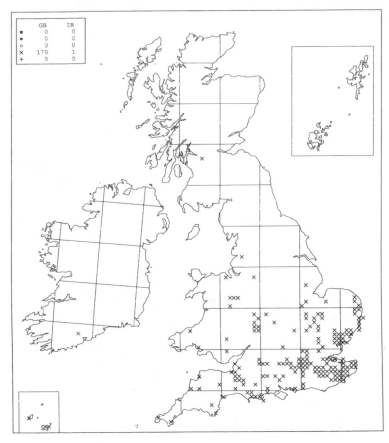

	GB	IR
■	0	0
●	0	0
○	0	0
×	170	1
+	0	0

Lemna minuta Kunth   Least Duckweed

Like other floating Lemnaceae, *L. minuta* is a plant of still or slowly flowing waters. It occurs in lowland ditches, ponds, canals, streams and rivers; more rarely it has been found in flooded peat-diggings and lakes. It is usually associated with *L. gibba*, *L. minor* or both these species, and sometimes with *Azolla filiculoides*, *Spirodela polyrhiza* and *Riccia fluitans*. On occasion *L. minuta* can be much more abundant than any competitor, forming dense stands completely covering the water surface. It appears to be shade-tolerant, and sometimes grows in ponds and ditches which are surrounded by trees. Leslie & Walters (1983) found that *L. minuta* eventually overgrows *L. gibba*, *L. minor* and *S. polyrhiza* when cultivated in pans containing a mixture of the four species.

Reproduction is by vegetative budding. Green fronds tend to persist throughout the winter, although in very much smaller quantity than in the summer months. Flowering has not been observed in Britain, although a small proportion of the fronds in a population in S. France flowers and fruits during the winter months (Jovet & Jovet-Ast, 1967).

L. minuta is an introduced species, first recorded in

Britain in 1977 when it was detected in Cambridge by Professor E. Landolt (Landolt, 1979). By 1981 there were records from 17 10-km squares in ten vice-counties in S.E. England, plus an outlying site in N. Wales (Leslie & Walters, 1983). In subsequent years it was recorded from many more localities, particularly from 1989 onwards when it attained considerable abundance in some areas including Cambridge (Preston, 1991) and Kent (Philp, 1992). Its abundance in canals attracted particular attention (Smith, 1990; Last, 1990; Oliver, 1991; Briggs, 1992). It is almost certainly under-recorded, because of its recent spread and its similarity to *L. minor*. The origin of the British populations is not certain, but they may have been introduced by aquarists releasing the contents of their aquaria into the wild.

A native of temperate N. and S. America. It was first recorded in Europe in France in 1965 (Jovet & Jovet-Ast, 1966), and subsequently found in Germany (1966), Switzerland (1975), Great Britain (1977), Belgium (1983), Ukraine (1983), Hungary (1984), Greece (1988) and the Netherlands (1988). It is also known as an introduction in Japan. The Cambridge population is tetraploid (2n=40), as are all the others for which counts are available (Urbanska-Worytkiewicz, 1980).

Wolffia Horkel ex Schleid.

A genus of ten species, which (unlike *Lemna* and *Spirodela* spp.) lack roots. *W. arrhiza* is the smallest flowering plant known.

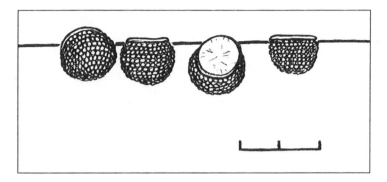

Wolffia arrhiza (L.) Horkel ex Wimm. (Scale divisions represent 1mm)

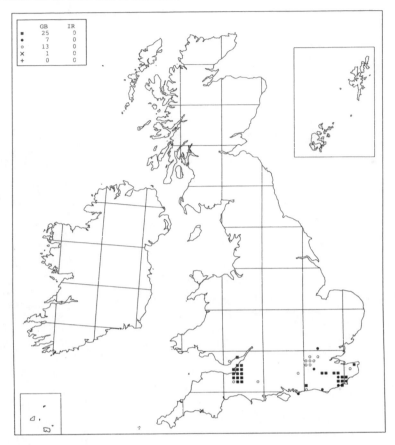

	GB	IR
■	25	0
●	7	0
○	13	0
×	1	0
+	0	0

Wolffia arrhiza (L.) Horkel ex Wimm.
Rootless Duckweed

A plant of shallow, base-rich water in lowland ditches and ponds, usually growing with other Lemnaceae including *Lemna gibba*, *L. minor* and *Spirodela polyrhiza*. It has also been recorded from lakes and gravel-pits. When *W. arrhiza* is abundant it forms distinctive pea-green patches, but it is much less easily seen when it grows as scattered plants amongst a *Lemna* sward. It appears to be favoured by organic pollution, although Landolt (1986) considers that it is at a competitive advantage where the nutrient supply is liable to become temporarily depleted, as it can grow quickly and form turions when nutrient concentrations are low. Both fronds and turions are intolerant of low temperatures and of desiccation. *W. arrhiza* is favoured by warm summers and mild winters.

Reproduction in England and Wales is by vegetative budding; flowering plants have never been observed here although plants occasionally flower in other parts of their range. *W. arrhiza* perennates by resting fronds which sink to the bottom of the water, and by very small, spherical turions with a high starch content.

W. arrhiza was not recognized in Britain until 1866, when it was found in a pond by the R. Thames on Staines Common (Trimen, 1866). It is locally frequent on the Somerset Levels and in the marshes near Winchelsea and Rye, but it has disappeared from its localities in the London area. It was last seen in Middlesex in 1940, in Surrey in 1957 and in Essex in 1968.

W. arrhiza is widespread in Europe but rare almost everywhere, and known to be declining. It is native to the Old World: Asia east to the Caspian Sea, Africa from the Mediterranean to the Cape and Madagascar. It is probably introduced to Brazil (where it has been known from Rio de Janeiro since 1872) and has recently been reported as an established alien in California (Armstrong, 1989). Its world distribution is mapped by Landolt (1986). The species is cytologically variable, and tetraploids (2n=40), hexaploids (2n=60), heptaploids (2n=70) and octoploids (2n=80) occur in Europe (Urbanska-Worytkiewicz, 1980); no cytological data are available for the British populations.

ERIOCAULACEAE

A family of 13 genera and 1200 species, many of which are aquatics.

Eriocaulon L.

A genus of *c.*300 species, most of them plants of marshes, swamps and seasonally inundated habitats in tropical and warm temperate areas. There are *c.*8 species in N. America, and a single species occurs as a native in Europe. A second species, the common pantropical weed *E. cinereum* R. Br., has become established as a rice-field weed in N. Italy and is still spreading (Cook, 1973 & pers. comm.).

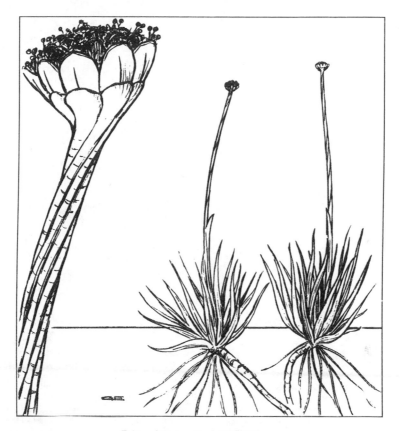

Eriocaulon aquaticum (Hill) Druce

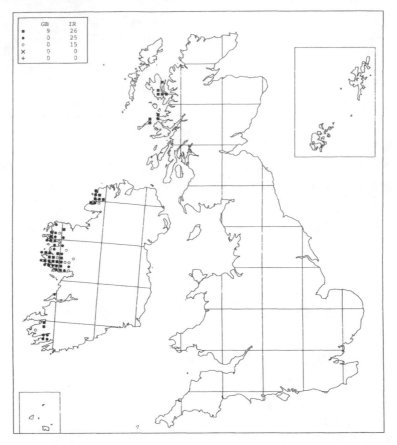

	GB	IR
■	9	26
●	0	25
○	0	15
×	0	0
+	0	0

Eriocaulon aquaticum (Hill) Druce   Pipewort

E. aquaticum grows as small clumps or extensive mats in shallow water, or terrestrially on damp and periodically flooded ground above the normal water-level. It is confined to oligotrophic lakes, and is much more frequent in smaller lakes and pools than in large water bodies. It can sometimes be found in peat-cuttings. *Lobelia dortmanna* is almost invariably found with it; other frequent associates include the aquatic form of *Juncus bulbosus*, *Littorella uniflora* and *Ranunculus flammula*, and scattered stems of emergents such as *Carex lasiocarpa*, *C. rostrata* or *Phragmites australis*. *E. aquaticum* never extends far above the water line (often being replaced at higher levels by *Littorella uniflora*) and descends to depths of at least 2m. It grows over a range of substrates including peat, fine gravel, sand and silt. Although it is usually found in sites where both the substrate and water are acidic, it can occur in slightly alkaline water over acidic substrates (Webb & Scannell, 1983). It is a lowland species, ascending to 300m above Glen Lough (Hart, 1885; Colgan & Scully, 1898).

E. aquaticum flowers from August onwards. Plants can flower in water at least 1m deep, as the peduncles elongate to carry the flowers above the surface. The flowers are structurally complex and are likely to be insect-pollinated or autogamous (Cook, 1988). Little is known about the pollination of plants in Britain or Ireland. Irish plants set viable seed (Löve & Löve, 1958). The species probably spreads vegetatively within lakes in which it

occurs. Large clumps or single plants of *E. aquaticum* are often found washed up at the edge of the water, and it seems likely that the former, at least, could become established as new colonies.

The distinctive rosettes of *E. aquaticum* are unmistakable, even when they are not in flower. Although it has long been known from Ireland, Coll and Skye, the species was only discovered on the Scottish mainland in 1967 (McClintock, 1968; Ferguson & Ferguson, 1971). Further colonies might possibly await discovery in western Scotland. Scottish populations have been surveyed in detail in recent years (e.g. Farrell, 1983). The species grows in remote localities in both Britain and Ireland and its habitats do not appear to be threatened.

E. aquaticum is an amphi-atlantic species, which is found in N. America from Ontario, Minnesota and Indiana to New Jersey and Newfoundland, but in Europe is confined to Britain and Ireland. It is a member of the small but much-discussed North American element in our flora. Its world distribution is mapped by Hultén (1958).

Within the areas in which it occurs *E. aquaticum* is frequent, but it is absent from much apparently suitable habitat both in Ireland (cf. Scully, 1916) and, more particularly, in Scotland. It seems most unlikely that any explanation based solely on current ecological factors could explain this very restricted distribution. *E. aquaticum* must be either a relict species, confined to areas from which it is unable to spread, or a recent colonist. A single grain of *Eriocaulon* pollen from

Hoxnian interglacial deposits in Ireland has been tentat-ively referred to this species (Jessen *et al.*, 1959) and pollen from a postglacial deposit in Connemara has been identified as *E. aquaticum* without qualification. There is therefore no doubt that *E. aquaticum* is a long-established member of the Irish flora (Godwin, 1975; Perring, 1963). Its history in Scotland is less certain. Heslop-Harrison (1953) suggested the possibility that plants might have arrived by long-distance dispersal of seed, and Farmer & Spence (1986) refer to it as 'an historically recent arrival to Scottish freshwaters'. If so, it is curious that a species which could arrive by long-distance transport has failed to colonize new localities in Britain,

for (*pace* Farmer & Spence, 1986) there is no evidence that *E. aquaticum* is spreading. The relationship between the European and the American populations requires further investigation. Löve & Löve (1958) reported a difference in chromosome number between plants from Ireland (2n=32) and those from eastern N. America (2n=64), and also found that they differed in minor morphological characters when cultivated under similar conditions. Fossil and recent pollen from Ireland is larger than recent pollen from America (Perring, 1963). Löve & Löve (1958) suggested that the plants in our area were specifically distinct from the American plants, but this conclusion was not accepted by Tutin *et al.* (1980).

JUNCACEAE

There are over 300 species in the Juncaceae, most of which belong to the two largest genera, *Juncus* and *Luzula*; the rest are in six small genera. The family is well represented in the temperate and cold regions of both northern and southern hemispheres. Two genera contain aquatic species. One of these, *Juncus*, is found in our area. The other is the monotypic genus *Prionium*, an aloe–like pachycaul shrub which grows in mountain streams in S. Africa. It is the only shrub in the family; all the other species are annual or perennial herbs.

Juncus L.

This is a genus of 255 species, of which 24 are found as natives in the British Isles. Many of these grow in permanently damp or seasonally flooded habitats. One species, *J. bulbosus*, may be found as an aquatic. Some authorities recognize two species within the *J. bulbosus* complex, *J. bulbosus* sensu stricto and *J. kochii*. D. E. Allen and P. M. Benoit regard these as distinct species which are 'readily recognisable over a large area of Europe'. They consider that *J. bulbosus* is the commoner species in lowland England, whereas *J. kochii* is more frequent in the mountainous districts of the north and west. Where the two species grow together, they remain distinct. Both species may be found as a modified, aquatic variant. They rarely hybridize in the wild, although hybrids can be synthesized in cultivation (Allen, 1984; Benoit & Allen, 1968; Benoit, 1954). A. Eddy also regards the two species as distinct, but considers that *J. bulbosus* is a rare arctic-alpine, which is replaced over most of our area by *J. kochii* (Department of Botany, British Museum (Natural History), 1970; Jermy & Crabbe, 1978). Other authors, notably Stace (1991), consider that 'plants known as *J. kochii* have been separated in different ways using various characters that are not correlated' and therefore recognize a single species, *J. bulbosus*. Few species in our flora are the subject of such radically different opinions. We have followed Stace (1991) in recognizing a single species, if only because there are insufficient records of the segregates for us to produce adequate accounts of them.

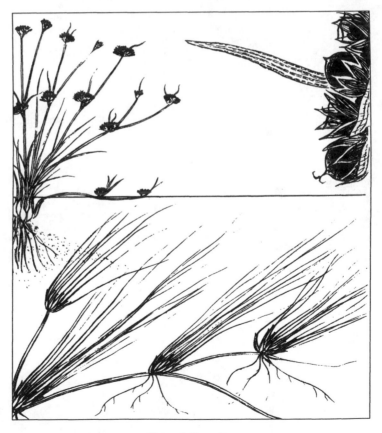

Juncus bulbosus L.

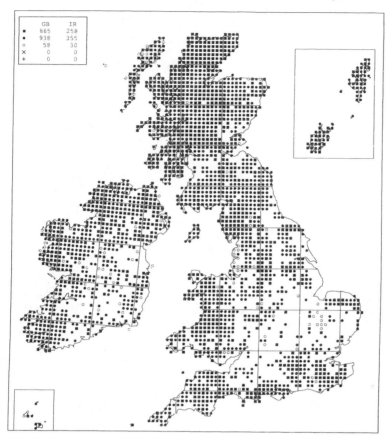

	GB	IR
■	865	258
●	938	355
○	58	30
×	0	0
+	0	0

Juncus bulbosus L.　　　　　　　　Bulbous Rush

This variable species grows as a terrestrial plant or as an aquatic. It is frequent and locally abundant in terrestrial habitats such as base-poor heaths and bogs, often in areas subject to some disturbance (Grime *et al.*, 1988). As an aquatic it grows in base-poor, clear or peaty waters in lakes, reservoirs, rivers and streams, extending from the shallows to depths of at least 2m. It may be found in the most oligotrophic and species-poor lakes, often growing with *Callitriche hamulata*, *Myriophyllum alterniflorum* and *Utricularia vulgaris* sensu lato, or in peaty pools with *Utricularia minor* and *Sphagnum auriculatum*. It may also grow amongst open stands of emergents such as *Equisetum fluviatile* and *Schoenoplectus lacustris*. It is found over a wide range of organic and mineral substrates (Spence, 1964). In some water bodies it can occur in dense masses, sometimes virtually excluding all competitors. It may occasionally be found in highly calcareous but nutrient-poor waters, as in the Burren (Webb & Scannell, 1983). It occurs from sea-level to at least 960m in Caenlochan Glen.

J. bulbosus is a winter-green perennial. In terrestrial habitats it flowers and fruits freely, and it may, like other *Juncus* species, have a persistent seed-bank. Semi-aquatic plants often have proliferous inflorescences (Grime *et al.*, 1988). The aquatic forms are usually vegetative, but sometimes produce buds which fail to develop into flowers. They root freely at the nodes and presumably become established from detached fragments. In a detailed study at Hodson's Tarn, Macan

(1977) found that the aquatic plant varied greatly in abundance from year to year.

In Europe, *J. bulbosus* extends from Iceland and N.W. Norway to Spain, Italy and the Balkans, although it is most frequent in the west and rare in the Mediterranean area and the south-east. It is also found in N. Africa, Newfoundland and (as an introduction) in New Zealand.

J. bulbosus is a common plant in northern and western Britain and Ireland. In S. England it is still frequent in areas where there is much acidic soil, but has declined in regions where such sites are few and far between. It has, for example, become very rare in Bedfordshire (Dony, 1976) and Cambridgeshire (Perring *et al.*, 1964), and in Gloucestershire outwith the Forest of Dean (Holland, 1986), but it remains common in Surrey (Lousley, 1976). The aquatic variants have not usually been recorded separately, although they now tend to be distinguished in lake surveys. The aquatic plant has increased in abundance in the Netherlands and Scandinavia in recent years, apparently a result of the acidification of lakes by atmospheric pollutants (Roelofs, 1983; Svendång, 1990). A similar increase in *J. bulbosus* was noted in a Finnish reservoir which received drainage from sulphur-rich deposits in the catchment, when the pH increased, the quantity of *J. bulbosus* declined (Aulio, 1987). The increase in *Juncus bulbosus* and *Sphagnum cuspidatum* in response to the addition of ammonium sulphate, which creates more acidic and more eutrophic conditions, has been demonstrated experimentally (Schuurkes *et al.*, 1987).

The terrestrial variant of *J. bulbosus* is a small, neat, densely-tufted plant, whereas the most extreme aquatic variant has long, trailing stems and 'tresses of numerous long hair-like leaves' (West, 1905). Intermediate forms grow in habitats in sites where the water is shallow or fluctuating, such as in peat-cuttings, ditches or at the edge of lakes and reservoirs. Plants may change from the terrestrial to the aquatic form or vice versa in response to changes in water-level, and it is therefore inappropriate to give the two variants taxonomic recognition.

CYPERACEAE

This large family of 110 genera and 4300 species is dominated by two genera, the predominantly temperate *Carex* (2000 species) and the predominantly tropical *Cyperus* (600 species). Most of the species are perennial herbs, the remainder being annuals, shrubs or climbers. Cook (1990a) lists 31 genera with aquatic representatives, some of which contain very important emergent species. Most members of the family have small, wind-pollinated flowers. The plants are usually monoecious, although there are some dioecious species.

Mabberley (1987) recognizes three subfamilies in the Cyperaceae, all of which are represented in our area. The Cyperoideae includes *Cyperus* itself, *Eleocharis* and *Scirpus* and allied genera; Rhynchosporoideae includes *Cladium*, *Rhynchospora* and *Schoenus* and Caricoideae includes *Carex* and *Kobresia*.

Generic limits in the Cyperoideae are controversial, and the definition of *Scirpus*, in particular, varies from author to author. Koyama (1958) proposed a very broad definition of *Scirpus* which even included *Blysmus* and *Eriophorum*. Dandy (1958) recognized *Blysmus* and *Eriophorum* as distinct, but nevertheless defined *Scirpus* relatively broadly to include twelve species in our area. The subsequent recognition of several different types of embryo in *Scirpus* sensu lato (Veken, 1965) has encouraged recent students to recognize numerous segregate genera based on both morphological and embryological characters (Wilson, 1981; Goetghebeur & Simpson, 1991). This is reflected by the treatment in Stace (1991) and Kent (1992), which place Dandy's (1958) twelve *Scirpus* species in seven genera: *Bolboschoenus* (1 sp.), *Eleogiton* (1), *Isolepis* (2), *Schoenoplectus* (4), *Scirpoides* (1), *Scirpus* (1) and *Trichophorum* (2). These narrowly defined genera are also supported by a recent cladistic analysis (Bruhl, 1995). However, *Scirpus* has continued to be defined in a broader sense in general works such as Tutin *et al.* (1980) and Mabberley (1987).

The chromosomes of the Cyperaceae, like those of *Luzula* but unlike those of other plants, have a diffuse centromere rather than a single centromere at a particular position on each chromosome. As a consequence, the chromosomes are prone to fragmentation. The individual fragments may behave as whole chromosomes, and are inherited in a normal Mendelian manner, so plants with fragmented chromosome(s) are genetically stable. This phenomenon is known as agmatoploidy and, together with aneuploidy, has probably been an important process in the evolution of the species of *Carex* (Davies, 1956; Faulkner, 1972).

Eleocharis R. Br.

This cosmopolitan genus includes some 150 species. The stems are green and constitute the main photosynthetic organs; the leaves are usually reduced to sheaths. The flowers are bisexual and are clustered in a solitary spikelet at the apex of the stems.

There are seven species in our area. Two are treated as aquatics in this book. The remaining species are also found in fens, marshes, wet heaths, flushes and other damp habitats. *E. multicaulis* is often found in shallow water at the edge of oligotrophic lakes and pools, but usually in seasonally flooded habitats rather than permanent water.

The *Eleocharis palustris* group of species (which in our area includes *E. palustris* and *E. uniglumis*) is taxonomically complex because of the occurrence of polyploidy and hybridization. Much of the information in the following accounts of *E. palustris* and its subspecies is taken from Walters (1949) and Strandhede (1966). The nomenclature of the subspecies used here is that of Strandhede (1960), Clapham *et al.* (1962), Tutin *et al.* (1980), Stace (1991) and Kent (1992). *E. palustris* subsp. *palustris* was formerly known as subsp. *microcarpa* Walters. Some authors (e.g. Clapham *et al.*, 1952) have applied the name *E. palustris* subsp. *palustris* to the plant referred to here as subsp. *vulgaris*.

Eleocharis palustris (L.) Roem. & Schult.

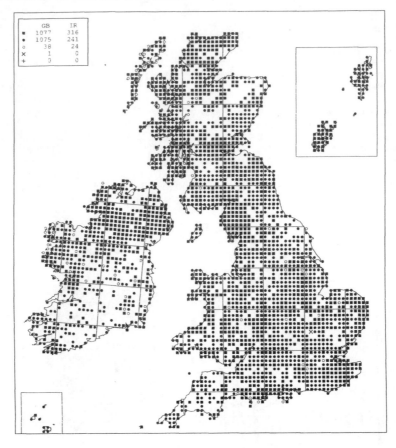

	GB	IR
■	1077	316
●	1075	241
○	38	24
×	1	0
+	0	0

Eleocharis palustris (L.) Roem. & Schult.
Common Spike-rush

Emergent plants of *E. palustris* are found at the edge of lakes, pools, ditches and slowly flowing rivers and streams, where they may be found in dense stands in water up to 0.5m deep. On the sandy or gravelly shores of exposed lakes it may be the only emergent; in these sites it is often accompanied by *Littorella uniflora*. It tolerates fluctuations in water-level and it is often an abundant species in dune-slacks, and in some inland ponds, although it is sometimes absent from ponds subject to great variations in water-level (Grime *et al.*, 1988). *E. palustris* may also grow as a more or less terrestrial plant in fens and moist or seasonally flooded meadows and pastures where taller competitors are eliminated by cutting, grazing or trampling. It is found over a wide range of organic and mineral soils, in both acidic and basic habitats, but it is most frequent in moderately fertile and at least slightly basic habitats, and rarely grows on acidic peat. It also tolerates slightly saline conditions in coastal grazing marshes. It is recorded from sea-level to 550m by the R. Tees near Tyne Head.

E. palustris is a perennial with thick rhizomes. In flooded sites plants often spread rapidly by the growth of the rhizomes, but the rhizomes are less well developed or absent in drier habitats. The flowers are protogynous and wind-pollinated. Seed-set is usually good, although the inflorescence is often infected by the ergot *Claviceps*

nigricans Tul. (Dennis & Ellis, 1956). Freshly sown seed may remain dormant for several months then germinate simultaneously (Walters, 1949). The seeds of *Eleocharis* spp. are eaten, and perhaps dispersed, by birds (Mabbott, 1920). Seedlings are rarely seen in the wild, but the species is an effective colonist of new sites which it presumably reaches by bird-borne seed.

Godwin (1975) suggests that *E. palustris* has been continuously present in our area since the middle of the Pleistocene. It is still frequent in suitable habitats in Britain and Ireland, and even in the London area it has been lost only from the most heavily built-up areas (Kent, 1975; Burton, 1983). However, detailed recording in Essex suggests that it has decreased in frequency in recent years (Tarpey & Heath, 1990) and this may also be true for other areas of high population and intensive agriculture.

Plants referable to the *Eleocharis palustris* complex are very widespread, being found in boreal and temperate regions throughout the northern hemisphere, and (sometimes as introductions) at scattered sites in the southern hemisphere too. The distribution of *E. palustris* sensu stricto is obscured by taxonomic confusion, but its native distribution is centred on Eurasia, where it is widespread, and it is apparently absent from N. America (Strandhede, 1967).

E. palustris shows great phenotypic variation, especially in height. Two cytologically distinct subspecies are dealt with in the following accounts.

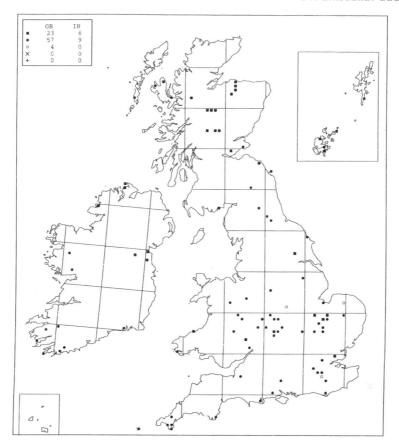

	GB	IR
■	23	6
●	57	9
○	4	0
×	0	0
+	0	0

Eleocharis palustris subsp. *vulgaris* **Walters**

This is much more frequent in our area than subsp. *palustris*, and the description of the ecology of the species is also applicable to subsp. *vulgaris*.

Subsp. *vulgaris* is self-compatible. Normal plants may produce some aneuploid pollen grains, and some individuals are cytologically abnormal (Lewis & John, 1961).

The two subspecies of *E. palustris* were first recognized by Walters (1949). However, few botanists bother to identify them, and both are under-recorded. Subsp. *vulgaris* is known to be the commoner plant, and it probably occurs throughout the British and Irish range of the species.

In Europe, subsp. *vulgaris* has a more restricted distribution than subsp. *palustris*. It is widespread in N., W. & C. Europe, but only extends north to S. Scandinavia (Strandhede, 1961), and is rare or absent from much of S. Europe, including the Mediterranean area.

E. palustris subsp. *vulgaris* is variable cytologically and although 2n=38 is the usual chromosome number, other numbers are not uncommon. Strandhede (1965) has suggested that it evolved from a hybrid between subsp. *palustris* (2n=16) and *E. uniglumis* (2n=46). A hybrid resulting from the fusion of an unreduced gamete of subsp. *palustris* with a reduced gamete of *E. uniglumis* would have a chromosome number of 2n=39. Fertile intermediates between subsp. *vulgaris* and *E. uniglumis*, with 2n=c.42, occur in our area and are particularly frequent in coastal regions of Scotland. They are also known from other parts of N.W. Europe, and similar plants have been synthesized experimentally by Strandhede (1966). The presence of subsp. *vulgaris* and such intermediates means that the distinction between *E. palustris* and *E. uniglumis* is less clear in W. Europe than elsewhere.

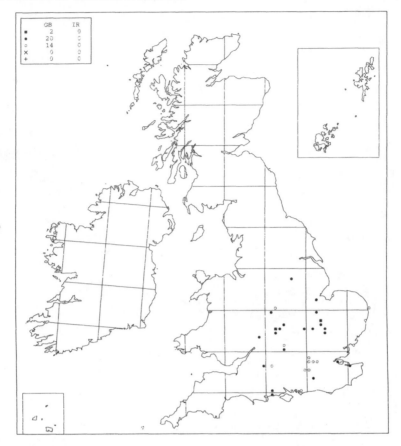

Eleocharis palustris (L.) Roem. & Schult. subsp. palustris

There is apparently no ecological distinction between subsp. *palustris* and subsp. *vulgaris*, and the two taxa may grow together. Subsp. *palustris* is recorded from the edges of rivers, lakes and ponds and from damp meadows, all typical habitats for *E. palustris* in S. England. All records are from lowland sites.

This subspecies is self-incompatible. Normal plants of subsp. *palustris*, like those of subsp. *vulgaris*, may produce some aneuploid pollen grains, and some individuals are cytologically abnormal (Lewis & John, 1961).

Although the subspecies of *E. palustris* are very under-recorded, subsp. *palustris* is known to be much rarer than subsp. *vulgaris*. Most of the records mapped here were collated by S.M. Walters for the map published by Perring & Sell (1968); there have been very few reports of this taxon since then. A survey of its distribution is required to establish whether it is still present at

those sites where it is known only from old herbarium specimens, or from records made in the 1950s.

E. palustris subsp. *palustris* is widespread and frequent in Europe, extending from northern Scandinavia to the Mediterranean region. Its very restricted distribution in the British Isles is surprising in view of its wide continental range. The two subspecies are apparently less easy to distinguish in C. Europe and S.W. Asia than in our area, perhaps because of the presence of other, hitherto unrecognized taxa.

Subsp. *palustris* differs from subsp. *vulgaris* in morphological characters and in its chromosome number. Fertile hybrids between subsp. *palustris* (2n=16) and subsp. *vulgaris* (2n=38) have been found with both parents in Port Meadow. They have a chromosome number of 2n=27, and grow with other plants which may be backcrosses (Lewis & John, 1961). Experimental studies have shown that female flowers of subsp. *palustris* pollinated by subsp. *vulgaris* have almost normal seed-set, and the resulting hybrids are fertile (Strandhede, 1966).

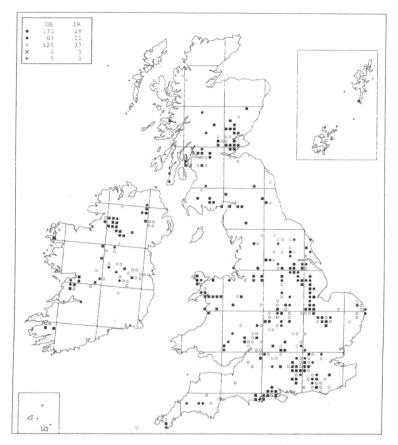

	GB	IR
■	130	28
●	63	11
○	125	33
×	0	0
+	6	0

Eleocharis acicularis (L.) Roem. & Schult.
Needle Spike-rush

E. acicularis usually grows submerged in shallow water or on damp ground exposed by falling water levels. As a submerged plant it is usually found in mesotrophic or eutrophic, mildly acidic to strongly base-rich water less than 0.5m deep. Typical habitats include sheltered lakes and reservoirs, the still backwaters of rivers, slow-flowing streams, canals, fenland lodes and ditches and flooded sand- and gravel-pits. At the edges of such water bodies it grows on damp silt, or on sandy shingle in areas which are flooded in winter but exposed in summer, and it can also occur on mud dredged out of ditches. Both aquatic and terrestrial plants may grow in pure stands, but in the water a wide range of other aquatics are recorded as associates. On land it is often mixed with *Chenopodium rubrum*, *Gnaphalium uliginosum*, *Juncus bufonius*, *Persicaria* spp. and *Rorippa palustris*, most of which are frequently found as dwarf individuals in this habitat. *E. acicularis* is a predominantly lowland species, ascending to 365m at Loch Kinardochy.

E. acicularis is a perennial with very slender stems and shallowly rooted rhizomes. Submerged plants may become quite tall in the clear calcareous waters of canals and fenland lodes. They spread by growth of the rhizome or grow through silt deposited over them, and may persist vegetatively for many years but never flower. In some sites the stems may remain green through the winter. Short lengths of rhizome with attached stems are often washed out of the substrate and doubtless become established in new localities. Plants in terrestrial habitats are much shorter than many aquatic forms, but flower and fruit freely. The seeds have been found on mud on the feet of birds, and may be dispersed in this way (Kerner von Marilaun, 1894). Seeds will germinate readily under water with or without a cold treatment, but germination on moist substrates is stimulated by a cold pretreatment (Rothrock & Wagner, 1975). Germination is rapid and all the seeds which germinate have usually done so in seven days. The species may vary greatly in abundance from year to year, both in aquatic and terrestrial habitats.

This species is often overlooked, especially when growing as an aquatic, and it may be under-recorded. However, the map suggests that it has declined in England and the results of the BSBI Monitoring Scheme also indicate that it has decreased in frequency since 1960 (Rich & Woodruff, 1996). It has been lost from canals which have dried out, and from small ponds which have been filled in, drained or allowed to become overgrown. However, the species has colonized recently flooded sand- and gravel-pits, where it may sometimes be found in abundance.

E. acicularis is widespread in Europe, although absent from the far north and rare in the Mediterranean area. It also occurs in N. Africa, Asia and N. & S. America. Records from Australia are errors for the closely related *E. pusilla* R. Br. (Blake, 1939).

In the Appalachian Mountains of N. America this species is often exceptionally abundant in shallow, very acidic streams polluted by drainage from coal-mines, where it may occur in water with a pH as low as 2.9 (Ehrle, 1960; Rothrock & Wagner, 1975). No other

aquatics and few animals are able to survive in such acidic waters. There do not appear to be any records of *E. acicularis* from very acidic sites in Britain or Ireland.

Bolboschoenus (Asch.) Palla

This segregate of *Scirpus* includes some 16 species, although the genus is in need of a world-wide revision (Goetghebeur & Simpson, 1991). The species are rhizomatous perennials with aerial leaves and relatively large spikelets and fruits. They are often plants of saline habitats in coastal areas or inland deserts. *B. maritimus* is the most widespread species and is the only one which is

found in Europe. In C. Europe some authors recognize two subspecies, subsp. *compactus* (Hoffm.) Heijný and subsp. *maritimus*. Subsp. *compactus* tends to grow in temporarily rather than permanently flooded habitats, and produces more tubers when the two taxa are cultivated under similar conditions (Zákravský & Hroudová, 1994).

Bolboschoenus maritimus (L.) Palla

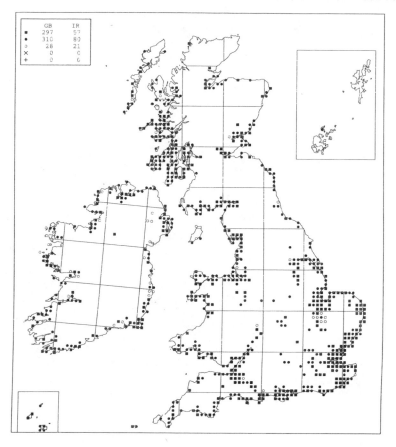

	GB	IR
■	297	57
●	310	80
○	28	21
×	0	0
+	0	0

Bolboschoenus maritimus (L.) Palla

Sea Club-rush

B. maritimus is an emergent which usually grows in dense, almost pure stands on saline ground or in shallow, brackish water. Typical habitats in the coastal zone include the edges of lakes, ponds, borrow-pits, tidal rivers, ditches, creeks and pools, upper salt-marshes and damp pastures. Its sites range from those which are damp and occasionally flooded to those where there is permanent water up to 0.5m deep. It tolerates fluctuating water-levels and occasional total submergence, but is rarely found in constantly flowing water (Rodwell, 1995). It is usually found over mud (which may be black and very anaerobic) but it may grow at the edges of rock-pools and over gravel and shingle. It has a similar tolerance of salinity to that of *Schoenoplectus tabernaemontani* (Ranwell *et al.*, 1964), and the two species are often found by the same water body, with *B. maritimus* growing in a zone nearer the water. In inland sites it has been recorded from pools near saline springs (Lee, 1977) and was recently found in abundance in a lagoon at Smallfield which receives saline drainage from the M23 motorway. However, it also grows inland in a range of freshwater habitats including the edges of lakes, reservoirs, ponds, rivers and grazing marsh ditches, and in flooded clay- and gravel-pits. It is confined to the lowlands.

B. maritimus is a rhizomatous perennial which dies down during the winter. It reproduces by direct growth of the rhizome and by round tubers which are produced on the end of rhizomes. Some new tubers develop into vegetative shoots whereas others remain dormant. The entire rhizome system of a plant can survive for more than one year in a dormant state, and individual tubers are reported to have retained viability for up to eight years (Zákravský & Hroudová, 1994). The tubers may represent 40–60% of the total biomass (Karagatzides & Hutchinson, 1991) and they are important in ensuring survival during prolonged periods of deep flooding or drought. The flowers are bisexual, protogynous and wind-pollinated. They are also visited regularly by syrphid flies which feed on the pollen, but there is no evidence that these are effective pollinators (Leereveld *et al.*, 1981). The roots, tubers and fruits are eaten by wildfowl, especially geese (Cramp, 1977). The fruits float, and may be carried upstream as well as downstream in tidal rivers (Ridley, 1930). Kerner von Marilaun (1894) also found the fruits in the mud on birds' feet.

This species is usually abundant in suitable coastal habitats, and is able to colonize newly available habitats. Its distribution in inland sites is rather sporadic. Some colonies are extensive and long-established, notably those in the Somerset Levels (Roe, 1981) and by the R. Stour in Dorset, where it was known to Pulteney (1813). At other sites, including recently abandoned clay- and gravel-pits, it must be a recent arrival and has presumably been dispersed by birds. However, some inland populations have failed to persist. *B. maritimus* is sometimes planted by lakes and ponds.

B. maritimus is widespread in Europe, where it occurs

from the Mediterranean region north to C. Scandinavia; it has a predominantly coastal distribution in the northern and western part of its range. It extends eastwards to C. Asia and is widespread in Africa. It also grows (as an occasional introduction) in N. & S. America. The species is cytologically variable: it has undergone both polyploidy and aneuploidy and many different chromosome numbers are reported (Stoeva, 1992). Closely related taxa are found in E. Asia, N. & S. America and Australasia.

Schoenoplectus (Rchb.) Palla

This genus of *c.*20 species has a cosmopolitan distribution. Most species are aquatic perennials,
with tall emergent stems, but the genus also
includes some annuals. Aerial leaves are more or
less absent but some species have the capacity to
produce narrow, translucent submerged leaves.
The wind-pollinated inflorescence, composed of
bisexual flowers, is borne at the apex of the stem,
but it appears to be lateral as the lowest bract
continues the line of the stem and overtops the
inflorescence.

There are four native species in our area, the
widespread *S. lacustris* and *S. tabernaemontani*,
and the rare species *S. triqueter* and *S. pungens*.
Typical *S. lacustris* differs from *S. tabernaemontani* in its ability to develop submerged leaves, its
taller, green rather than glaucous stems, and in a
number of floral characters. In some areas (e.g.
Donegal) these two species can be identified without difficulty. However, in other areas plants

which apparently combine the characters of both
species are not uncommon. Intermediate plants
from Cambridgeshire are illustrated by Easy
(1990). These intermediates are not given formal
taxonomic recognition by Stace (1991) or Kent
(1992). Fertile intermediates are also found in the
Netherlands, and Bakker (1954) concluded that
the boundaries between *S. lacustris* and *S. tabernaemontani* had been obliterated in some localities
by introgressive hybridization. He suggested that
S. tabernaemontani should be regarded as a subspecies of *S. lacustris*, a treatment followed by
Tutin *et al.* (1980). *S. triqueter* shares the same
chromosome number as *S. lacustris* and *S. tabernaemontani* (2n=42) and hybridizes with both of
them. *S. pungens* has a different chromosome
number, 2n=78 (Otzen, 1962). Hybrids between
S. lacustris and *S. pungens* have been recorded in
Europe, but only rarely.

Schoenoplectus lacustris (L.) Palla

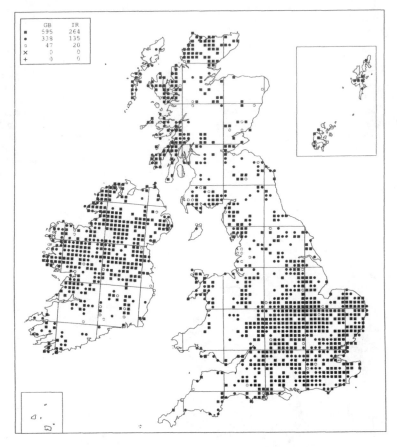

	GB	IR
■	595	264
●	338	135
○	47	20
×	0	0
+	0	0

Schoenoplectus lacustris (L.) Palla
Common Club-rush

S. lacustris is found in deeper water than any other emergent in our area. It is most frequent in lakes and slow-flowing rivers, but it is also found in pits, ponds, canals, large ditches and flooded gravel-pits. It usually grows at depths of 0.3–1.5m, and is often accompanied by aquatics such as *Juncus bulbosus, Littorella uniflora, Nymphaea alba* or *Potamogeton natans*. In lakes *S. lacustris* may be separated from the shore by clear water, and in rivers it may grow across the full width of the channel, or form isolated 'islands' in the centre and dense bands along the edge. It tolerates a wide range of nutrient states, growing in both eutrophic and base-rich sites and in oligotrophic and base-poor lakes, and it can be found over fine silts and clays, peat or coarse gravel. *S. lacustris* is grazed in those sites where it is accessible to stock, and it is also eaten by waterfowl, especially geese (Sinker *et al.*, 1985). It is primarily a lowland species, but reaches 400m at Dock Tarn (Stokoe, 1983).

This is a perennial with a robust rhizome. Submerged leaves develop in clear, standing or flowing water, and in streams and rivers these leaves may predominate. The emergent stems vary in vigour, and may be very slender in clumps which are exposed by falling water-levels, as in the Breckland meres (Coombe *et al.*, 1981). Most colonies spread vegetatively, and they may grow by 1–2m per year in favourable circumstances (Rodwell, 1995). Plants flower and fruit freely, and the seeds are probably dispersed in the mud on birds' feet (Ridley, 1930). Little is known of the circumstances under which the seeds germinate.

S. lacustris has a very long fossil history in our area, where it certainly persisted through the last glacial period (Godwin, 1975). Although there appears to be no evidence for a decline in some areas, in others the species is certainly less abundant than it once was. From 1800 onwards *S. lacustris* spread in Broadland as nutrient levels increased, often expanding at the expense of *Cladium mariscus*. By 1900 *S. lacustris* and *Typha angustifolia* grew prolifically in the broads, but during this century they have decreased and are now rare (George, 1992). *S. lacustris* has also decreased greatly in Shropshire in recent years, perhaps because of the increasing grazing pressure from Canada geese (Sinker *et al.*, 1985) and it is also regarded as a declining species in Surrey (Lousley, 1976) and S.E. Yorkshire (Crackles, 1990).

S. lacustris is widespread in Europe and N. Africa, extending eastwards to C. Asia. In Europe it is found from the Mediterranean to northern Scandinavia, but it is absent from Iceland. It is replaced by the closely related *S. acutus* (Muhl.) Á. Löve & D. Löve in N. America.

This is apparently the true 'bulrush', although now 'bulrush' always means *Typha* to the non-botanist, and the name survives for *Schoenoplectus* only in some botanical books. *S. lacustris* was formerly used to stuff saddles, horse collars and mattresses, and to caulk casks, and is still used to make baskets, 'rush' mats and chair seats. The dialect name 'bumble' or 'dumble' has a long-recorded history in Lincolnshire and Yorkshire, and chairs with rush seats were said to have a 'bumble bottom' (Crackles, 1974, 1983a).

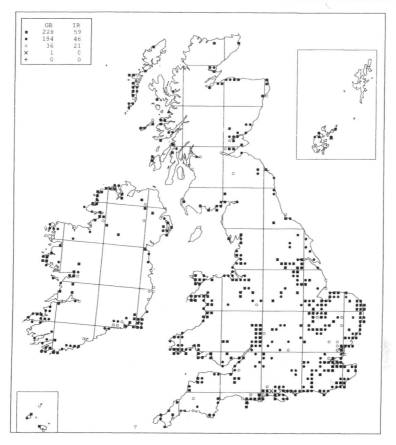

	GB	IR
■	228	59
●	194	46
○	36	21
×	1	0
+	0	0

Schoenoplectus tabernaemontani (C. C. Gmel.) Palla
Grey Club-rush

This species is most frequent in coastal sites, where it grows in shallow, brackish water at the edge of lakes, lagoons, rivers, streams, grazing marsh ditches and borrow-pits. It is also found in small depressions in salt-marshes, where it may grow through an open mat of *Agrostis stolonifera* (Adam, 1981; Rodwell, 1995). It is more tolerant of saline water than *Phragmites australis*, but (unlike *Spartina anglica*) it cannot grow in sea-water and is often absent from the seaward end of salt-marshes (Packham & Liddle, 1970; Ranwell *et al.*, 1964). It also grows in pools and ditches at inland salt-marshes, and in brackish subsidence flashes associated with colliery or other mineral waste (Lavin & Wilmore, 1994; Lee, 1977). However, at most of its inland sites it grows in fresh water at the edge of lakes and slowly flowing rivers and associated oxbows, and in flooded quarries and sand-, gravel-, chalk- and clay-pits. In such localities it is often accompanied by *S. lacustris*. When the water-table is low enough for cattle to reach *S. tabernaemontani*, it may be heavily grazed. The rhizomes of this species, like those of *S. lacustris*, *Scirpus maritimus* and *Typha angustifolia*, are able to survive and produce healthy new shoots even when kept for seven days in totally anaerobic conditions (Barclay & Crawford, 1982). *S. tabernaemontani* is confined to the lowlands.

S. tabernaemontani is a perennial which spreads by growth of the rhizome. It does not produce submerged leaves, and this (coupled with its short stature) is probably one reason why it is unable to grow in deep water or across the entire width of streams and rivers, habitats in which *S. lacustris* is often found. It flowers and fruits freely, but little is known of its reproductive biology.

Although the fossil record of this species is sparser than that of *S. lacustris*, it nevertheless shows that the species was present in our area in the late glacial period (Godwin, 1975). Adam (1981) and Rodwell (1995) have suggested that this species is under-recorded at coastal sites. There is no evidence for any change in its frequency in these habitats. Lousley (1931) regarded it as a very rare plant inland, and was able to cite only nine localities. It is now recorded inland with much greater frequency. The reasons for this apparent change are not entirely clear, but the species may have been overlooked as *S. lacustris* in the past, and it has also spread in recent years in newly available habitats. It has appeared as a natural colonist of a recently flooded gravel-pit in Jersey (Le Sueur, 1984) but it appears to be planted in some Cambridgeshire gravel-pits (Easy, 1990) and in fresh waters in the London area (Burton, 1983).

S. tabernaemontani is widespread in temperate Eurasia, extending from W. Europe and N. Africa eastwards to Japan; it is also recorded from the Sahara and S. Africa. It is replaced by *S. validus* (Vahl) Á. Löve & D. Löve in N. America and Australasia, and by closely related taxa in S. America (cf. Hultén & Fries, 1986).

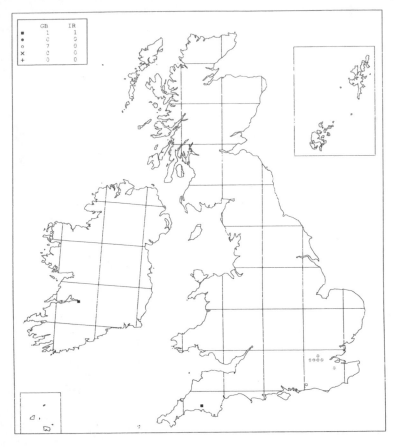

Schoenoplectus triqueter (L.) Palla
                                    Triangular Club-rush

S. triqueter is an emergent which grows on the muddy edge of large, tidal rivers. Clumps may become completely submerged at high tide. They often grow where some fresh water flows into the river: there is a stream flowing into the river at the remaining site by the R. Tamar, and the plant is most frequent by the R. Shannon in sites where fresh water trickles over estuarine mud. These sites are often small bays where there is some shelter from wave action. *S. triqueter* tends to grow on rather bare mud with few associated species, although *Groenlandia densa* grows with it in the flowing fresh water at the Shannon sites. It is usually replaced by dense beds of *Phragmites australis*, *S. tabernaemontani* or *S. maritimus* away from the river's edge.

This is a rhizomatous perennial. The stems start growing later in the year and grow more slowly than those of *S. tabernaemontani*, and flower later, in August and September. Jackson & Domin (1908) noted that they were unable to find ripe fruits of *S. triqueter* by the R. Thames in 1907, and in recent years botanists have been unable to find them in Britain and Ireland.

S. triqueter is extinct at three of its four British sites. It was first discovered in 1650 by the R. Thames near the horse-ferry at Westminster (Kent, 1975) and it formerly grew along a 17km stretch of river from Limehouse to Kew. However, it was gradually eliminated as the muddy edges of the river were replaced by artificial

embankments, and it was last seen in 1946. A major reconstruction of the river wall in 1946–47 destroyed the last mud-bank on which it grew (Burton, 1983; Lousley, 1976). Engineering works probably contributed to the loss of the species from the R. Medway, although some suitable habitat survives there. (Plants reported as *S. triqueter* from the Medway in recent years are now believed to be its hybrid with *S. tabernaemontani*). Apparently suitable habitat also survives by the R. Arun, where the reasons for the disappearance of the species are unknown. *S. triqueter* survives by the R. Tamar, although it has decreased in quantity in recent years and is now reduced to a few clumps. However, it is still locally abundant at its only confirmed Irish localities, by the R. Shannon near Limerick, where it was first recorded by R. D. O'Brien in 1900.

This species is found in C. & S. Europe, N. & S. Africa and W. Asia. It is naturalized in N. America (Lightcap & Schuyler, 1984).

The hybrid between *S. triqueter* and *S. tabernaemontani* (*S.* × *kuekenthalianus*) is more frequent in Britain than that between *S. triqueter* and *S. lacustris* (*S.* × *carinatus*), probably because the salt-tolerant *S. tabernaemontani* is more likely to be found by the estuarine rivers where *S. triqueter* grows. *S.* × *kuekenthalianus* is a vigorous hybrid which is more frequent than *S. triqueter* by the R. Tamar, and survives by the R. Arun and R. Medway. For further details of these hybrids, see Lousley (1931, 1975) and Perring & Sell (1968). In Ireland *S. tabernaemontani* and *S. triqueter* grow in adjacent stands by the R. Shannon but hybrids are apparently absent. Fertile hybrid swarms involving *S. lacustris* and *S. triqueter*, and *S. tabernaemontani* and *S. triqueter*, occur in the Netherlands (Otzen, 1962). In Belgium Vanhecke (1986) notes that the hybrids between *S. lacustris* sensu lato and *S. triqueter* appeared after radical embankment works and have now completely replaced *S. triqueter*.

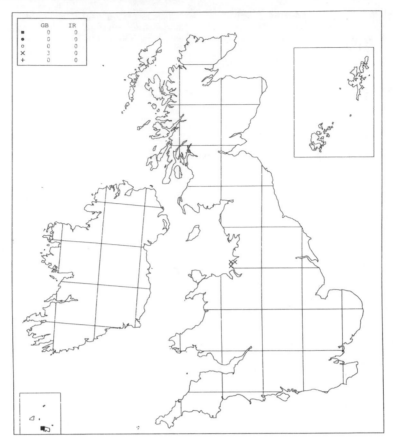

Schoenoplectus pungens (Vahl) Palla

Sharp Club-rush

This species was formerly known from the margins of a coastal lake in Jersey, and from a dune-slack in Lancashire. In Lancashire it grew in the wettest part of the slack, associated with *Eriophorum angustifolium*, *Parnassia palustris*, *Pyrola rotundifolia*, *Salix repens* and *Schoenoplectus tabernaemontani*.

   S. pungens is a rhizomatous perennial. Little is known of the reproductive biology of the British plants. In N. America the seeds of this species are an important source of food for wildfowl, including gadwall and pintail (Mabbott, 1920; Martin *et al.*, 1961).

   This species was first recorded from Jersey by W. Sherard (Ray, 1724). Lester-Garland (1903) described it as abundant at St Ouen's Pond, its only known locality there, and it was still plentiful in the 1940s and 1950s. However, it subsequently decreased in frequency and was last seen in the early 1970s. It appears to have been replaced by *Carex riparia*, which was first seen at this site in 1953 (Le Sueur, 1984). *S. pungens* was collected

near Formby by W. G. Travis in 1909 but not recognized until it was rediscovered by R. E. Baker in 1928 (Druce, 1929; Travis, 1929). In 1928 a large patch grew in a dune-slack. Its origin is uncertain; Druce (1929) concluded that it was unlikely to have been planted as 'one can scarcely credit the most microcephalous sentimentalist endowing this *Scirpus* with sufficient beauty to think he was adding fresh charms to the floral displays of Lancashire'. By 1972 it was noted that the slack was much drier than formerly, and that the *S. pungens* shoots were much shorter, often chlorotic and also subject to rabbit grazing. The species had become extinct at this site by 1978, but the Lancashire stock survived in cultivation and has been planted at other sites in the area (Edmondson *et al.*, 1993).

   S. pungens is found in W. & C. Europe, extending north to the Netherlands and Germany and east to Italy. It is widespread in N. America and Australia, and is naturalized in New Zealand where it is a troublesome weed which 'breaks up footpaths and gutter channels in built-up coastal areas' (Healy & Edgar, 1980).

Eleogiton Link

This genus has a cosmopolitan distribution and includes ten species. Most of the species grow as submerged macrophytes with long, trailing stems and emergent flowers. Each peduncle bears a single spikelet of bisexual flowers. Koyama (1958) considered that the floral characters of the genus suggested that it was allied to *Isolepis*, but the aquatic, predominantly perennial *Eleogiton* has a very different habit from *Isolepis*, which contains many annual, more or less terrestrial species. Only one species of *Eleogiton* is found in Europe.

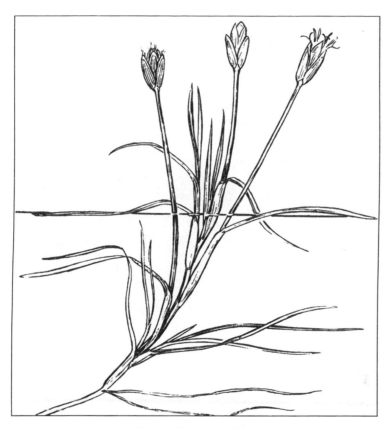

Eleogiton fluitans (L.) Link

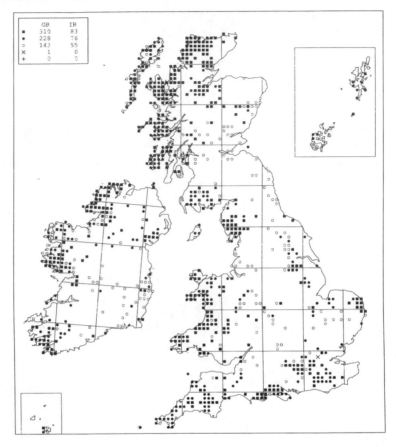

Eleogiton fluitans (L.) Link   Floating Club-rush

This is a plant of shallow, acidic water or moist, terrestrial and often seasonally flooded habitats. It may grow in dense masses as an aquatic in ponds, small reservoirs or at the sheltered edge of lakes, and in moorland streams and ditches. Its other habitats include the backwaters of rivers, flushes flowing into lakes, moist, muddy hollows in grassland and wet heath, and the wet floors of disused quarries and sand- or gravel-pits. It is virtually confined to acidic and oligotrophic habitats, being particularly frequent in peaty areas but also growing over silt and sand. It is often associated with *Apium inundatum*, *Juncus bulbosus* and *Potamogeton polygonifolius*. It is occasionally found in highly calcareous but nutrient-poor sites, such as ditches in Gordano (where it may grow with *Baldellia ranunculoides* and *Potamogeton coloratus*) and fen marl near Lough Fingall in the Burren (Webb & Scannell, 1983). It is recorded from sea-level to 435m at Llyn Gwngu, and 436m at Styhead Tarn (Stokoe, 1983).

This is a perennial species which may remain green throughout the winter, at least when growing as an aquatic. It flowers freely, but little is known about its reproductive biology.

Like many calcifuges, *E. fluitans* was probably always rather local in the eastern part of its British range, where moist, acidic habitats are infrequent, but it has declined as acid heaths and moors have been drained and built over or taken into arable cultivation. The many counties in which it has decreased include Somerset (Roe, 1981), Shropshire (Lockton & Whild, 1995), Essex (Jermyn, 1974) and Suffolk (Simpson, 1982), and it is believed extinct in others, including Durham (Graham, 1988), Northumberland (Swan, 1993) and Berwickshire (Braithwaite & Long, 1990). The map suggests that it has also declined in S.E. Ireland, but there are fewer local studies to substantiate this. It remains a frequent species at low altitudes in peat-covered areas in the north and west, where it may be somewhat under-recorded because of the similarity of vegetative plants to other monocotyledons.

E. fluitans has a disjunct distribution. In mainland Europe it has a markedly oceanic range, extending from the Azores and the western part of the Iberian peninsula through France and the Low Countries to Denmark and S. Sweden; there are a few outlying localities east to Italy and the species is also found in N.W. Africa. It is also known from C. & S. Africa, India, S.E. Asia, Japan, Australia and (as an introduction) from New Zealand; it is a rare species in Asia.

Cladium P. Browne

This is a genus with two species: the widespread and variable *C. mariscus* and the closely related *C. jamaicense* Crantz. The latter is found in southern N. & S. America, and is the main dominant of the Florida Everglades.

Detailed accounts of the autecology of *C. mariscus* have been published by Conway (1936a,b, 1937, 1938a,b) and summarized in a rather brief Biological Flora account (Conway, 1942). This work was based on the population at Wicken Fen, where the results of management experiments also provide an insight into the requirements of the species (Godwin, 1929, 1941; Godwin & Tansley, 1929). For descriptions of the vegetation in which this species grows, see Rodwell (1991, 1995).

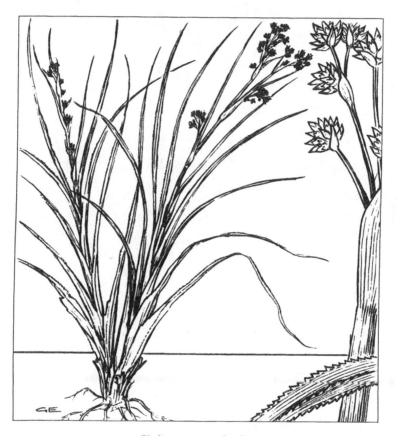

Cladium mariscus (L.) Pohl

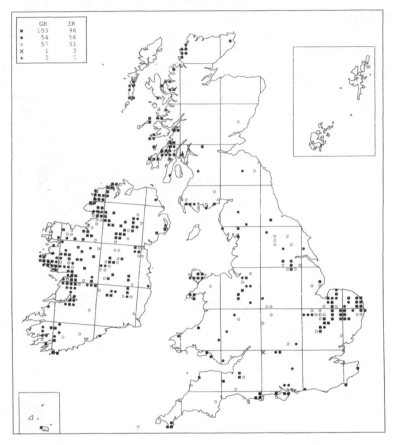

Cladium mariscus (L.) Pohl   Great Fen-sedge

The most vigorous stands of this species grow as emergents in shallow water up to 0.4m deep. Small stands or
extensive, virtually pure swards grow at the edge of lakes
and ponds, and in flooded hollows and disused peat-
cuttings. In peatlands where the soil is moist but only
flooded in winter, if at all, the species grows in more
species-rich communities with associates such as *Angel-
ica sylvestris*, *Eupatorium cannabinum*, *Lysimachia vulgaris*,
Lythrum salicaria, *Molinia caerulea*, *Peucedanum palustre*
and *Phragmites australis*. This vegetation is maintained
only by cutting, which prevents the colonization of
shrubs. Cutting in early summer every 3–5 years favours
Cladium, whereas more frequent cuts reduce its vigour
and it cannot withstand annual mowing regimes. If its
sites are overgrown the species will persist for many
years in fen carr, but eventually dies out. In most parts
of its British and Irish range *C. mariscus* is confined to
calcareous and oligotrophic or mesotrophic habitats,
where it usually grows over peat. In the Breckland it is a
particularly characteristic species of the headwater fens,
which are more nutrient-poor than the valley fens downstream (Haslam, 1965). It also grows in valleys draining
the serpentine plateau in the Lizard peninsula and in
extremely oceanic areas (e.g. Outer Hebrides, Coll,
Connemara) it is found in acidic and oligotrophic lakes
and streams through bogs. *C. mariscus* is an exclusively
lowland species, reaching 260m only at Sunbiggin Tarn.

C. mariscus is an evergreen perennial with a robust rhiz-

ome. The leaf meristems are below ground and the only
aerial growing points are those on the flowering stems.
The leaves grow for two seasons and then remain green
for some months before withering. Their growth rate
increases with temperature, being very low in the winter
months; in summer growth may also be limited by water
stress if the water-table falls below the surface. The dead
leaves remain attached for some time and provide an
important pathway for oxygen to diffuse into the rhizome.
Plants flower throughout our area but often rather sparingly; the number of flowering stems varies from year to
year and is greater in emergent than in terrestrial plants.
Relict plants in fen carr often fail to flower. The spikelets
contain both bisexual and male flowers and are wind-
pollinated. The fruits may float for long periods (Praeger,
1913). Plants are known to fruit in East Anglia but
Conway (1942) never saw seedlings, and was unable to get
the seeds to germinate. The growing points are sensitive
to temperatures below -2°C in winter and below 0°C in
summer, but below-ground temperatures rarely fall to
these levels at the East Anglian sites.

There is no unequivocal evidence that *C. mariscus* survived in our area during the last glacial period, although
fossil remains are often abundant in later deposits
(Godwin, 1975). In the eastern part of its range *C.
mariscus* has declined in historic times because of the
gradual drainage of lakes and fenland sites. It was the
plant that gave Sedgemoor its name and it was still
abundant there in the late 18th century, but it was feared
extinct by 1896, although a few relict patches have since

been discovered (Roe, 1981). In Cambridgeshire, Babington (1860) described it as 'formerly far more abundant than at the present time throughout the Fens' and it has since been lost from some of the sites he knew. It has disappeared from some areas of Broadland because of eutrophication (George, 1992). It is also threatened by scrub encroachment at some sites, but vegetative plants hanging on in shaded sites will respond to scrub clearance (Fitter & Smith, 1979). *C. mariscus* colonizes newly available habitats only in the vicinity of existing stands. Emergent plants in the north and west are easily overlooked, as they are often present as small clumps. The species was under-recorded in W. Scotland and some

parts of W. Ireland by Perring & Walters (1962), and may still await discovery in new sites.

C. mariscus is adapted to warm summers and mild winters. It is widespread in Europe, extending north to S. Sweden and Finland, but it is most frequent in the west and in the Mediterranean region. It also occurs in Africa from the Mediterranean coast to the Cape, S. Asia, Australia and eastern N. America.

This species was once so abundant that it was used as fuel in Fenland, although it was necessary to wear thick gloves when handling it to prevent the serrated leaf edges cutting the skin. It is still harvested for thatching, and is used particularly for the ridge of a roof.

Carex L.

This large genus of some 2000 species has a cosmopolitan distribution. *Carex*-dominated communities occur in wetlands in all climatic zones except the lowland tropics, and include low swards in the tundra, tall 'reedswamps' in temperate wetlands and tussocky vegetation on tropical mountains (Klötzli, 1988). There are 180 species in Europe, and 74 of these grow in the British Isles. Only 12 of these qualify as aquatics by our criteria, but many more are found in habitats such as bogs, fens, moors, wet heaths and flushes.

All *Carex* species are perennials with leafy stems and unisexual, protandrous and wind-pollinated flowers. However, there is considerable morphological diversity in the genus. The three subgenera recognized in Britain and Ireland are described by Jermy *et al.* (1982). The only aquatic representative of subgenus *Vignea* is *C. paniculata*. This has inflorescences composed of spikes which are similar in appearance; most contain both male and female flowers although the male flowers may be missing from some of the lower spikes. The remaining aquatic species belong to subgenus *Carex*. They all have inflorescences with slender spikes made up of male flowers above more robust female spikes. Both these subgenera can usefully be divided into sections of closely related species. The third subgenus, *Primocarex*, contains a heterogeneous assortment of taxonomically anomalous, more or less terrestrial species.

The taxonomy of the wetland *Carex* species has recently been reviewed by Kukkonen & Toivonen (1988). Many species are widespread and variable, and are therefore subdivided into subspecies or varieties. The identification of some populations is made difficult by the fact that some species often fail to flower, or to set well-formed fruit, when growing in shade or when subjected to changes in water-level. Such plants are often mistaken for hybrids. True hybrids do occur, however. At a world scale they tend to be more frequent in northern or montane areas, as the short growing season, open habitats and frequent occurrence of a mosaic of plant communities enhances the possibility of inter-specific pollinations. Vigorous hybrids which are able to spread vegetatively may also be at an advantage in these communities.

Kent (1992) lists 35 *Carex* hybrids from our area. Eight of these are hybrids between the aquatic species and a further ten have one aquatic and one terrestrial parent. The aquatic species are therefore responsible for a disproportionate number of the recorded hybrids. Seven of the 18 hybrids with an aquatic parent involve a member of the critical *C. nigra* group, discussed below. Most of the remainder are rare, and only *C.* × *boenninghausiana* (*C. paniculata* × *C. remota*), *C.* × *involuta* (*C. rostrata* × *C. vesicaria*) and the relict hybrid *C.* × *grahamii* (*C. saxatilis* × *C. vesicaria*) are recorded with any frequency. However, *Carex* hybrids are often overlooked and are probably greatly under-recorded; more hybrid combinations and additional localities for apparently rare hybrids are likely to be discovered in future years.

British botanists formerly thought that the species of *Carex* were difficult to identify, and many were under-recorded in the *Atlas of the British Flora* (Perring & Walters, 1962). Jermy & Tutin's excellent identification handbook, *British Sedges*, was published in 1968 and did much to stimulate interest in the group; it was revised by Jermy *et al.* (1982). Most competent botanists would now identify the majority of the species they encounter without much difficulty. The main exceptions are the species related to *C. flava* and those related to *C. nigra*. The former does not

Carex acuta L.; *C. riparia* Curtis; *C. pseudocyperus* L.

include aquatic species but most of the taxa in the *C. nigra* group are aquatics.

The *Carex nigra* group includes the aquatics *C. acuta*, *C. aquatilis* and *C. elata*, the ecologically tolerant *C. nigra* and the montane *C. bigelowii*. These species have been studied experimentally by Faulkner (1972, 1973). All the species are more or less self–incompatible. *C. aquatilis* and *C. bigelowii* are closely related and as much seed is set when these species are cross-pollinated as when individuals of the same species are crossed. Similarly, *C. acuta* and *C. nigra* are a pair of closely related species with only a slight breeding barrier between them. However, crosses between *C. aquatilis* or *C. bigelowii* and *C. nigra* are less likely to produce good seed and those between these two species and *C. acuta* are almost sterile. The fifth species, *C. elata*, is a link between the two species pairs, as crosses between it and the other species are less productive than those between the members of a related pair, but more fertile than those between species in different pairs.

Faulkner (1973) also showed that the synthesized hybrids in the *C. nigra* group are partially fertile and will backcross with the parents. More seed is set if the hybrid is used as the pollen parent, as the number of seeds set is limited by the number of fertile ovules and the hybrids have fewer than the parents. Although the pollen of the hybrids is also partially fertile, pollen is usually present in excess and there are sufficient fertile grains to achieve good seed-set under experimental conditions. Much of the seed produced when species are cross-pollinated or when hybrids are backcrossed is viable, and most of the resulting plants are as vigorous as the parents.

In the wild, the occurrence of hybrids is limited to some extent by the ecological isolation of the species. However, hybrids between the montane *C. bigelowii* and both *C. aquatilis* and *C. nigra* are recorded in upland habitats. In the lowlands there

are records of hybrids between *C. elata* and both *C. acuta* and *C. nigra*, and between *C. nigra* and *C. acuta*. Both *C. aquatilis* and *C. nigra* have a wide altitudinal and ecological range and their hybrid has been found in both upland and lowland sites. Hybrids in this group of sedges are not always easy to recognize in the field, and Faulkner (1972) found that some apparent hybrids tended to lose their intermediate morphology in cultivation and proved to be species when examined cytologically.

A sixth species is included in the *C. nigra* aggregate by Jermy *et al.* (1982), Stace (1991) and Kent (1992). This is the rare *C. recta*, which is restricted in our area to marshes in four localities in N. Scotland. Faulkner (1973) has shown that the Scottish populations of this species are undoubtedly of hybrid origin, and the parentage is almost certainly *C. aquatilis* × *C. paleacea*. *C. paleacea* Schreb. ex Wahlenb. is a species which has not been recorded in our area. The Scottish *C. recta* sets seed, and at one of its sites there is an extensive hybrid swarm involving *C. aquatilis* and *C. recta*.

The seasonal development of vegetative and flowering shoots of several *Carex* species has been studied, and the relevant papers are cited in the individual species accounts. There is less information on reproduction by seed, perhaps because this is an infrequent event for many of the rhizomatous species. Jermy *et al.* (1982) reported that the seed of most species requires an after–ripening period of 3–12 months, and that this cannot be reduced by low temperature treatment. They also found that abrasion of the testa increases the proportion of seeds that germinate. However, Faulkner (1973) was able to germinate seeds of the *C. nigra* group in the light without any after-ripening period or artificial abrasion of the testa. As Jermy *et al.* (1982) comment, there is much scope for both observational and experimental studies on the germination and establishment of *Carex* species.

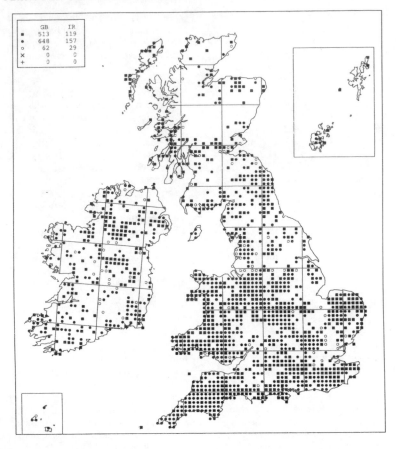

Carex paniculata L.   Greater Tussock-sedge

This species grows in a wide range of shallow waters or seasonally flooded habitats. These include open water at the edge of lakes, ponds, rivers, ditches and canals; open, species–rich stands of emergents such as *Phragmites australis*, *Schoenoplectus lacustris* and *Typha angustifolia*; sedge-dominated fens and swamps with rhizomatous species such as *C. rostrata*; spring-fed marshes; and open fen–carr and swampy woodland dominated by *Alnus glutinosa* or *Salix cinerea*. It is usually found in at least moderately base-rich water, and often in places where there is some water movement. Its sites range from eutrophic to highly calcareous but mesotrophic or even oligotrophic. It is frequent over peat but also grows over other substrates including mud and even boulders (Spence, 1964). It extends from sea-level to 600m on Glychedd.

C. paniculata is a perennial which forms substantial tussocks which often exceed 1.5m in height. It flowers and fruits freely. Seedlings and young plants are sometimes found on the dead and decaying tussocks of the same species (Poore & Walker, 1959).

The tussocks of *C. paniculata* are not easily killed: the population on the Isles of Scilly described by Smith (1907) has survived attempted drainage, toxic chemicals and fire (Lousley, 1971). They may remain alive in dense plantations for several decades. The map does not suggest a significant national decline, but the species has doubtless been lost from many sites because of drainage

or the artificial embankment of rivers and canals. In the Colne valley many populations have been destroyed during gravel extraction, and 'tussock fens' have ceased to develop in Broadland as the emergent communities of *Schoenoplectus lacustris* and *Typha angustifolia* which were formerly colonized by *C. paniculata* have disappeared (George, 1992). The species will colonize old peat-diggings and long-disused sand- and gravel-pits.

C. paniculata is found in Europe, and in adjacent parts of N. Africa and W. Asia. In Europe it is rare in the Mediterranean region but frequent at middle latitudes, especially in the west, and it extends north to S. Scandinavia. The species is divisible into three subspecies: subsp. *paniculata* is the most widespread but is replaced by subsp. *lusitanica* (Schkuhr) Maire in the western part of the Iberian peninsula and N. Africa, and by subsp. *szovitsii* (V. I. Krecz.) O. Nilsson in Crimea and the Caucasus.

Tussocks of *C. paniculata* are colonized by marsh and fen plants, harbour 'a host of hibernating invertebrates' (George, 1992) and support mosses such as *Plagiothecium latebricola* on the shaded bases. Trees also colonize the tussocks, and the presence of this species may accelerate the succession from open water to wet woodland. Once colonized by trees tussocks may press down the soft lake sediments or disturbed peat on which they grow, so that open water is maintained between them (Lambert & Jennings, 1951). The old tussocks were formerly dried out and used as fireside seats in cottages and hassocks in churches (Ellis, 1965).

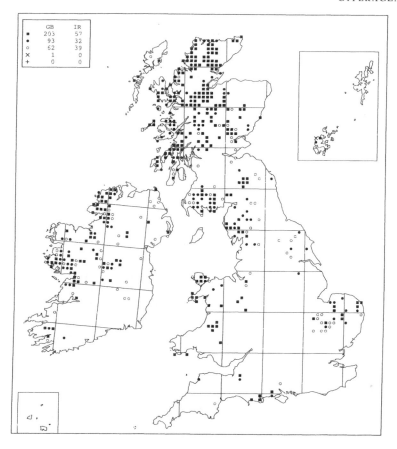

Carex lasiocarpa Ehrh. Slender Sedge

This is a species of shallow, nutrient-poor water. It is most frequent in open stands at the edge of lakes and slow-flowing streams and rivers, in swamps, in flushes and wet hollows in fens and as a colonist of old drainage channels and peat-diggings. It usually grows as an emergent in water up to 0.3m deep, but can also occur as a floating raft over water in sheltered bays and pools, or in fens where the water-table is below the surface of the substrate, at least in summer. It is frequently mixed with other emergents, particularly *Carex rostrata* but also *Cladium mariscus* or *Phragmites australis*, and often accompanied by *Menyanthes trifoliata* and *Potentilla palustris*. In Scotland stands containing *C. lasiocarpa* tend to grow between those of *Carex rostrata* or *Phragmites australis* (which extend into deeper water) and mixed *C. lasiocarpa–Myrica gale* stands and then communities of *Molinia caerulea* and *Myrica gale* on higher ground (Spence, 1964). *C. lasiocarpa* may persist in relatively dry fen on the site of former pools (Proctor, 1974). It is found in both base-poor and base-rich waters, and over a range of substrates including peat and raw, sandy soils at the edge of lakes. It grows from sea-level to 650m near Lochan Achlarich.

This is a rhizomatous perennial which may form extensive, clonal stands. Studies of an American population (Bedford *et al.*, 1988) showed that shoots were produced continually throughout the growing season. Many of the shoots which emerged in summer and autumn overwintered as plants with tightly inrolled leaves, which then expanded in the following spring. The species is known to be a very shy flowerer and 99% of ramets tagged in the American population died without flowering. McClintock & Waterway (1993) found that populations in pioneer stands in open water showed greater genetic diversity and clones were smaller than in closed vegetation, suggesting that reproduction by seed is more frequent in open habitats, but they were unable to find any seedlings in their study sites.

The leaves of *C. lasiocarpa* are very thin (the species is aptly known as 'wiregrass sedge' in N. America). The species is therefore easily overlooked, although, once learnt, its appearance is very distinctive. Systematic surveys of aquatic habitats have revealed that the species is present in many more 10-km squares than were known to Perring & Walters (1962), especially in Scotland. However, the species has certainly declined in eastern England. It is extinct in Suffolk (Simpson, 1982) and Durham (Graham, 1988) and now confined to a single site in Cambridgeshire (Perring *et al.*, 1964). Some sites have been drained and others no longer provide the open, oligotrophic vegetation which this species prefers. In Dorset it is known from three sites, but two are drying out and becoming overgrown by *Molinia caerulea*, and invaded by pine and birch (Pearman, 1994). *C. lasiocarpa* is also extinct in some sites in eastern Ireland, including Lough Neagh (Harron, 1986).

C. lasiocarpa has a circumboreal distribution. In Europe it extends from N. Scandinavia south to the Pyrenees, C. Italy and N. Greece.

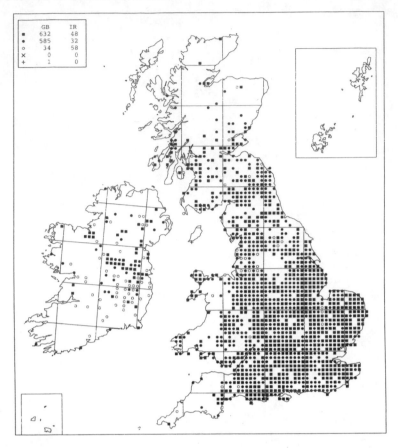

	GB	IR
■	632	48
●	585	32
○	34	58
×	0	0
+	1	0

Carex acutiformis Ehrh.   **Lesser Pond-sedge**

This tall sedge grows in shallow water at the edges of standing and slowly flowing waters, including lakes, ponds, rivers, streams, canals and ditches, and in swamps. In these habitats it often grows with *C. riparia*. It is rarely found in sites which are subject to great fluctuations in water-level, and it is notably intolerant of rivers subject to spate (Haslam, 1978). Like *C. paniculata*, it invaded reed-swamps in Broadland, and as the ground surface rose above water-level it was itself invaded by trees and shrubs. However, it is relatively shade-tolerant, and large stands persist in the resulting swamp woodland and occur on shaded watersides. In addition to these permanently flooded habitats, *C. acutiformis* also grows in sites where the water falls below the soil surface in the summer months. It is often abundant in fen-meadows, where it frequently grows with *C. disticha* and *Juncus subnodulosus*, and also occurs in tall-herb communities in fens. The herbaceous communities in these habitats may be maintained by seasonal mowing or grazing. *C. acutiformis* is a plant of base-rich, mesotrophic or eutrophic waters. It will grow over a wide range of organic or inorganic substrates, including acidic sands if the water is base-rich (Holmes, 1983), and it is present in some slightly saline coastal marshes. It is absent from sites where both water and substrate are base-poor. It extends from sea-level to 370m S. of Garrigill.

C. acutiformis is a rhizomatous perennial which dies down during the winter. Plants spread by growth of the rhizome, and may also regenerate from detached rhiz-

ome fragments or even by large clumps which float downstream (Grime *et al.*, 1988; Ridley, 1930). New shoots emerge in late autumn or early spring; they may survive the winter and have a maximum life span of at least two years (Verhoeven *et al.*, 1988). Plants flower freely in undisturbed, open habitats but may only flower sparingly, if at all, in sites which are mown in summer (Verhoeven *et al.*, 1988) or are shaded (Pearman, 1994). Seed-set is frequently poor (Grime *et al.*, 1988). The fruits float and may be dispersed by water, but reproduction by seed is almost certainly much less significant than vegetative spread. The species rarely colonizes newly available habitats.

This is still a frequent species in suitable habitats over much of its British range. However, it has declined in recent years in N. E. Essex (Tarpey & Heath, 1990), and it may have undergone similar decreases in other areas of dense population and intensive agriculture. According to Allen (1984), it is popularly grown as an ornamental plant in water gardens, and it may occasionally escape from cultivation. *C. acutiformis* has a more restricted distribution in Ireland, where it is 'a characteristic plant of the swamps and rivers of the Central Plain' (Praeger, 1901). Further fieldwork is required to see if it is still present in those Irish localities where it has not been recorded since 1950.

C. acutiformis is widespread in Europe, although it is rare in the Mediterranean region and absent from Iceland and N. Scandinavia. It also occurs in N., C. & S. Africa and eastwards to C. Asia, and as an uncommon introduction in N. America.

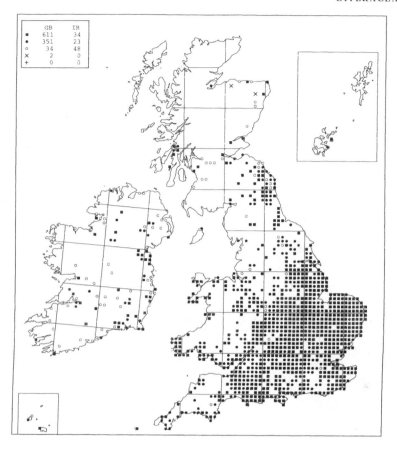

	GB	IR
■	611	34
●	351	23
○	34	48
×	2	0
+	0	0

Carex riparia Curtis   Greater Pond-sedge

This species is often found as pure stands in shallow water at the edge of lakes, ponds, slow-flowing streams and rivers, canals and ditches. It usually grows as a rooted emergent, but it is recorded as a colonist of floating *Glyceria maxima* rafts in Broadland (Lambert, 1946). It sometimes extends into sites where the water-level falls below the substrate in the summer months, such as the banks and flood plains of rivers, tall-herb fens and clearings in willow carr. It is a species of base-rich, mesotrophic and eutrophic sites: in Breckland it is characteristic of the nutrient-rich valley fens rather than the less eutrophic headwater fens. In coastal sites it may grow in brackish ditches and marshes, and on flushed sea-cliffs. It tolerates a wide range of substrates including peat, clay and silt. Unlike *C. acutiformis*, it is not favoured by grazing (Lambert, 1948). It is confined to the lowlands.

C. riparia is a rhizomatous perennial. It dies down during the winter, but the leaves begin to grow before those of *C. acutiformis* (Grime *et al.*, 1988). Plants flower freely in open habitats. The fruits float and may be dispersed by water. As with other rhizomatous sedges, vegetative spread is a more frequent means of reproduction than establishment from seed. Large clumps at the edge of rivers may become detached and float away to start new colonies downstream (Ridley, 1930).

The fossil record of *C. riparia* is sparse, but fruits have been found in deposits from the middle and end of the last glacial period, as well as in later sediments (Godwin,

1975). The species is now frequent throughout much of its British range, even in the London area (Kent, 1975; Burton, 1983). It has doubtless been able to persist in areas where many aquatics have declined because of its ability to grow in eutrophic habitats and to persist for long periods in unmanaged ditches. As a shade-tolerant species, it even survives when pieces of marsh are converted to plantations of poplars or cricket-bat willows (Tarpey & Heath, 1990). However, C. D. Pigott (in Rodwell, 1995) has noted that large stands of this species seem to be declining in the Midlands and East Anglia. It is occasionally planted by lakes and ponds. The distinctly eastern distribution of *C. riparia* in Britain is matched by a corresponding rarity in Ireland.

The world distribution of *C. riparia* is similar to that of *C. acutiformis*. It occurs throughout most of Europe, north to S. Scandinavia, in N. Africa and eastwards to C. Asia.

The ecology of this species is similar to that of the closely related *C. acutiformis*. *C. riparia* will, however, extend into deeper water than *C. acutiformis* (Grime *et al.*, 1988) and is less frequent in fen-meadows and other more or less terrestrial habitats. In the rivers described by Holmes (1983) *C. riparia* was recorded in a wide range of eutrophic rivers whereas *C. acutiformis* proved to be particularly characteristic of calcareous waters. These not only have a distinctive water chemistry but also tend to have a relatively stable flow from chalk or limestone aquifers.

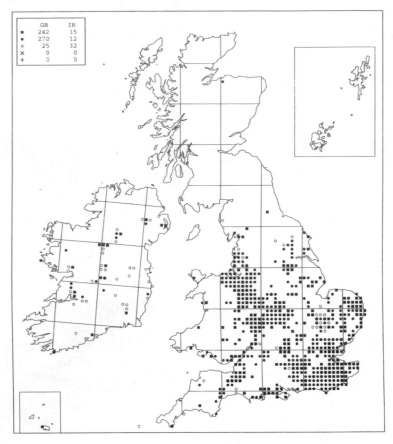

	GB	IR
■	242	15
●	270	12
○	25	32
×	0	0
+	0	0

Carex pseudocyperus L.   Cyperus Sedge

This sedge is a pioneer species of shallow water and wet mud, but is unable to withstand competition from taller, rhizomatous species. It is found in a wide variety of habitats including the edges of lakes, reservoirs, ponds, slow-flowing rivers and nearby ox-bows, canals and ditches. It sometimes grows in open reed-swamps and tall-herb fens, and it may invade floating or semi-floating rafts of *Phragmites australis* or *Typha latifolia* over water or unconsolidated mud (Sinker, 1962; Wheeler, 1980a). It will colonize disused sand-, gravel-, clay- and marl-pits. It is most frequent in sites which are neither very acidic nor very basic, and are mesotrophic or eutrophic, and it grows over peat and over mineral soils. It tolerates fluctuations in water-level and it may be conspicuous when flowering in dried-up farm ponds. Although it is a pioneer species, it is shade-tolerant and will persist in pools and on wet mud between *Carex paniculata* tussocks and in swampy alder woods and willow carr. It is confined to the lowlands.

This is a tussocky perennial with little capacity for vegetative spread. It flowers and fruits freely. Little is known about its reproductive biology, but its ability to colonize open habitats such as recently cleared ditches and newly abandoned gravel-pits suggests that reproduction by seed is not infrequent.

This is a frequent species in some areas, including the Somerset Levels (Roe, 1981) and the meres and marl-pits of Cheshire (Newton, 1971). In much of its British range it is uncommon, but apparently no rarer now than formerly. However, it is declining in certain areas, including the Idle/Misson Levels (Mountford, 1994a) and N.E. Essex (Tarpey & Heath, 1990). It is sold in garden centres and is sometimes planted by ponds and fishing lakes. It was first reported from its northernmost British locality, Sanquhar House, by Graham (1841), but it might be a long-established introduction at this site. In Ireland it is a scarce and apparently declining species.

C. pseudocyperus is widespread in the cooler parts of temperate Europe; it is virtually absent from the Mediterranean region and extends north to S. Scandinavia. It is also recorded in W. & C. Asia, Japan and eastern N. America.

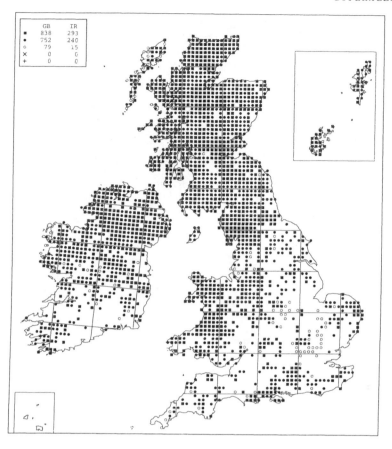

	GB	IR
■	838	293
●	752	240
○	79	15
×	0	0
+	0	0

Carex rostrata Stokes   Bottle Sedge

In the north and west this is one of the most frequent emergents. Its grows in fens and bogs where the water-table usually lies below the soil surface, often over a carpet of *Sphagnum* species or of base-demanding pleurocarpous mosses; in shallow water or as a floating raft, mixed with other rhizomatous species such as *C. lasiocarpa*, *Menyanthes trifoliata* or *Potentilla palustris*; or in water up to 0.6m deep where it may be found as dense, monospecific stands or with *Equisetum fluviatile* in an open sward. It is found by lakes, reservoirs, pools, rivers, streams, springs and ditches, in swamps, marshes, flushes on hillsides and sea-cliffs, in wet meadows and the wettest dune-slacks and in clearings in wet woodland. It tolerates a wide range of water chemistry, growing in sites which range from base-rich to base-poor. It it is most frequent in oligotrophic and mesotrophic waters, and in eutrophic sites it is often restricted to small stands, being replaced as the dominant species by *Glyceria maxima*, *Phragmites australis*, *Typha angustifolia* or *T. latifolia* (Spence, 1964). It is found over a wide variety of organic and inorganic substrates. It is tolerant of a degree of exposure but is absent from the most exposed lake shores. In the south and east it is much less frequent and it is usually an indicator of base-poor conditions, but even here it is sometimes found by nutrient-poor but highly calcareous waters, including those emerging from chalk and limestone springs. At one Dorset locality it grows with

base-demanding sedges on the site of an old watercress bed (Pearman, 1994). It is found from sea-level to at least 930m near Beinn nan Eachan and 1040m in blanket bog near Creag Meagaidh.

C. rostrata is a perennial with a robust rhizome. New shoots emerge throughout the growing season; the leaves die down in autumn but the young leaves are protected by the dead bases of older leaves and the shoot may survive the winter and regrow in spring. Individual shoots have a maximum life of 2–3 years, but few survive as long as this. The species flowers freely and shoots die after flowering. The fruits float and are presumably dispersed by water. Many fail to survive as little as two months' dry storage (Spence, 1964). Reproduction by seed is almost certainly less important in established stands than vegetative spread. For further details of the demography of this species, see Bernard (1976), Gorham & Somers (1973), Hultgren (1988) and Verhoeven *et al.* (1988).

The fossil fruits of *C. rostrata* are recorded more frequently than those of any other *Carex* species, and they demonstrate that the plant was present in our area throughout the last glacial period (Godwin, 1975). It is still an abundant species in suitable habitats in the north and west and, more locally, in areas of acidic soils in the south and east. In these areas it may colonize old peat-diggings and disused sand- and gravel-pits. It has, however, declined in those areas where, in historic times, it has always been an uncommon species. The most marked decline has been in Cambridgeshire, where it was formerly known from a number of widely scattered localities but has not been seen in any of them since 1966 (Perring *et al.*, 1964; Crompton & Whitehouse, 1983). Some sites have been drained and now lack any standing water; at others the species may have been adversely affected by eutrophication.

    C. rostrata is widespread in the arctic, boreal and cooler temperate regions of the northern hemisphere. In Europe it extends south from Iceland and northern Scandinavia to N. Spain, Italy and N. Greece, although it is a rare and predominantly montane species in the southern parts of its range.

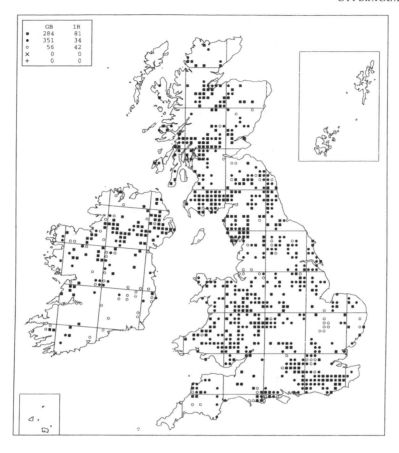

	GB	IR
■	284	81
●	351	34
○	56	42
×	0	0
+	0	0

Carex vesicaria L.   **Bladder-sedge**

This sedge is found in marshes where the water may lie below the soil surface, and as an emergent in water up to 0.4m deep. It grows at the edge of lakes, reservoirs, turloughs, ponds, slowly flowing rivers and streams and their associated ox-bows, canals, and ditches, in wet meadows and damp hollows in pastures, and in wet sallow carr and alder woodland. It is often found in swamps which have developed in silted-up ponds and ditches, or by the inflow streams entering small lakes and reservoirs. It will also colonize disused sand-, gravel- and clay-pits. Stands in very shallow water often grow with a range of associates including *Galium palustre*, *Juncus effusus*, *Mentha aquatica*, *Potentilla palustris* and *Veronica scutellata*; in deeper water the species may grow with *C. rostrata*, *Menyanthes trifoliata* and *P. palustris* or in virtually pure stands (Spence, 1964). The species tolerates a fluctuating water-table. It is often eaten in sites where it is accessible to cattle, and cannot withstand heavy grazing. *C. vesicaria* is a plant of mesotrophic habitats which range from slightly base-rich to highly calcareous. It is found over a range of substrates including mud, clay, silt and peat. It is recorded from sea-level to 455m (Llyn Gorast).

The life-cycle of this rhizomatous perennial has been studied by Soukupová (1988) in Czechoslovakia. The species retains short leafy shoots throughout the winter. In summer vegetative shoots continue to be produced throughout the growing season, and after flowering the vegetative shoots overtop the shoots which bear ripening fruit. The shoots live for up to 18 months, and always die after flowering. The fruits float, and the seeds are dormant when the fruits are shed but germinate in spring. Soukupová (1988) found seedlings at her study site for five successive years, whereas she only once found seedlings of *C. acuta*, which also grew there. At this site *C. vesicaria* is able to persist when mixed with the taller and more productive *C. acuta* because of its winter-green shoots, production of more numerous tillers which are able to invade gaps between the bulky *C. acuta* shoots and more frequent reproduction by seed.

The historic records of this species are sometimes confused with those of *C. rostrata*, which complicates any assessment of changes in its distribution. However, there is little doubt that it has declined in lowland areas of England, including Gloucestershire (Holland, 1986), Essex (Tarpey & Heath, 1990), Cambridgeshire (Perring et al., 1964) and S.E. Yorkshire (Crackles, 1990). Some sites have been drained, but the species may also have disappeared because of more subtle changes in nutrient levels or river management which are not fully understood. In Dorset, Pearman (1994) was unable to detect the species at any of the 25 precisely localized sites where it had been recorded in the 1930s by R. d'O. Good. Most of these were by the R. Frome and R. Stour, where *Phragmites australis* had increased in frequency. This may be a response to eutrophication or to improved drainage of riverside land.

C. vesicaria is a polymorphic species complex which

has a circumboreal distribution. It consists of a number of taxa which are sometimes treated as species, and sometimes as subspecies or varieties (Hultén & Fries, 1986). The typical *C. vesicaria* is found in N. Eurasia and western N. America. In Europe it extends from N. Scandinavia south to the mountains of Spain, Italy and Bulgaria.

This species has similar ecological requirements to those of *C. rostrata*, with which it often grows. However, *C. vesicaria* is restricted to shallower water and it is often frequent on the landward side of stands of dominant *C. rostrata* (Spence, 1964). *C. vesicaria* is a more lowland species and it has a more restricted trophic range: it does not extend into the very oligotrophic habitats in which *C. rostrata* often grows. The hybrid between these two species, *C. × involuta*, is one of the more frequent *Carex* hybrids. It is highly sterile but it may spread vegetatively to form large stands (Jermy *et al.*, 1982; Wallace, 1975). In Shetland *C. × involuta* is present in several sites, although *C. vesicaria* is absent from the archipelago (Scott & Palmer, 1987). *C. vesicaria* also appears to be one parent of the enigmatic sterile hybrid *C. × grahamii*. In our area this plant is restricted to montane habitats in Scotland which are well above the current altitudinal limit of *C. vesicaria*; it often grows near the other putative parent, *C. saxatilis*.

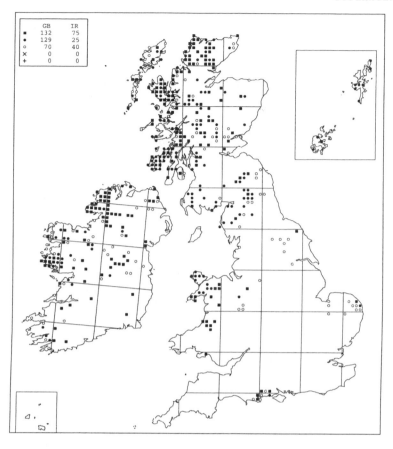

	GB	IR
■	132	75
●	129	25
○	70	40
×	0	0
+	0	0

Carex limosa L.   Bog-sedge

This small sedge is restricted to very wet bogs and fens, and shallow, oligotrophic lakes and pools. It is most frequent in acidic or mildly basic peatlands, where it is found in wet hollows, peat-cuttings and at the edge of bog pools. It often grows in carpets of sphagnum, including the floating masses of sphagnum which extend across the open water of bog pools. It is also found in water up to 0.2m deep at the edge of lakes, where it can be abundant in sheltered bays and sometimes grows in open stands of *Carex rostrata*, *Menyanthes trifoliata* and, less frequently, *Phragmites australis*. Other characteristic associates include *Carex lasiocarpa*, *C. nigra*, *Eriophorum angustifolium*, *Juncus bulbosus*, *Potamogeton polygonifolius*, *Potentilla palustris*, *Utricularia intermedia* and *U. minor*. *C. limosa* can occasionally grow in calcareous basin mires or in highly calcareous fens, especially in vegetation which has colonized old peat-cuttings (Wheeler, 1980a,b). Here it is particularly associated with 'mud bottom' communities in very wet depressions (Wheeler, 1978). It is most frequent in lowland sites, but it is recorded at altitudes up to 830m on Meall nan Tarmachan.

C. limosa is a perennial which spreads by growth of the rhizome. Many stands are made up of numerous vegetative shoots but flower only sparingly, and some upland stands seem to persist without flowering. Those flowers which are produced usually set seed, but little is known of the circumstances under which it germinates.

Sparsely flowering stands of *C. limosa* are easily overlooked, and like many sedges this species was under-recorded in the *Atlas of the British Flora* in Scotland, Ireland and Wales. It may still be under-recorded in some western areas. However, it is a declining species in the southern and eastern parts of its range, where it has succumbed to total habitat destruction or to the drainage of bogs resulting in falling water tables, the drying out of pools and the colonization of woody species. It is extinct in several sites in Northumberland, including all the more easterly localities (Swan, 1993) and in Yorkshire it only survives at Fen Bog (Sykes, 1993). It is restricted to one site in Cheshire (Newton, 1971) and it is extinct in Shropshire (Lockton & Whild, 1995). It was last seen in W. Norfolk in 1924. It survived in E. Norfolk until *c.*1974, but has not been seen since. In Dorset it is still present in several sites but only survives in quantity in one of them (Pearman, 1994). In Ireland *C. limosa* has also been lost from some eastern sites, e.g. in Co. Down (Praeger, 1938a).

C. limosa has a circumboreal distribution. In Europe it extends from Iceland and N. Scandinavia southwards to N. Spain, the Alps and N. Greece. It is an endangered species in the Netherlands (Weeda *et al.*, 1990).

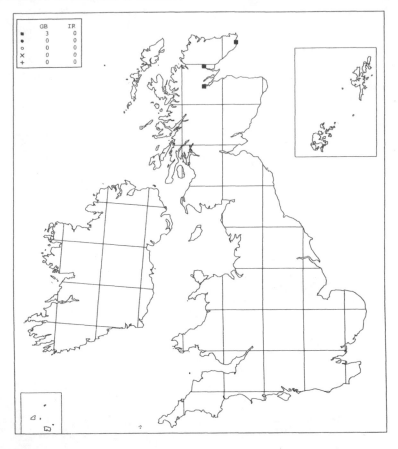

	GB	IR
■	3	0
●	0	0
○	0	0
×	0	0
+	0	0

Carex recta Boott   Estuarine Sedge

This plant grows by estuarine rivers, sometimes extending to sites which are just above the tidal limit. It is found in rough marshy pasture, riverside swamps and on islets. It may grow over large areas, sometimes with *Phalaris arundinacea* and *Phragmites australis*, or with *C. aquatilis* and its hybrid with this species, *C. × grantii*, or as pure stands. It is found over stiff, peaty alluvium or on sandy banks which receive some silt deposition (Jermy *et al.*, 1982).

This is a rhizomatous perennial which spreads vegetatively. The Scottish plants show disturbed chromosome pairing and none of the six plants studied by Faulkner (1972) was a simple euploid. In his experimental studies Faulkner (1973) found that crosses between different plants of *C. recta* resulted in very low seed-set, in contrast to intraspecific crosses of the other species in the *C. nigra* group. When female flowers of *C. recta* were pollinated by other species the resulting seed-set was also low. However, more seed was set when pollen of *C. recta* was used to fertilize the flowers of other species in the *C. nigra* group, the cross between *C. recta* and *C. aquatilis* being the most fertile.

This plant was first recorded from Scotland by J. Grant, who found it on sand-banks by the Wick River (Ridley, 1885). It is now known from three rivers. The largest populations grow by the R. Oykell and the Wick River, and it is by the latter that the species occurs in a hybrid swarm with *C. aquatilis*. There is a smaller population by the River Beauly. Some stands of the species are threatened by wetland reclamation.

The taxonomy of *Carex recta* requires clarification, and it is impossible to give a satisfactory account of its world distribution. The species was originally described from eastern N. America, where it occurs on the arctic and boreal coasts, but the identity of the American and Scottish populations has not been demonstrated, and is perhaps doubtful. Faulkner (1972, 1973) has demonstrated that the Scottish populations are derived from hybrids between *C. aquatilis* and *C. paleacea* Schreb. ex Wahlenb., a plant of brackish marshes and sea-shores which does not grow in Britain but which is fairly frequent around the coasts of Scandinavia. The synthesized hybrid is indistinguishable morphologically from the Scottish *C. recta*. Plants similar in morphology to the Scottish populations are found on the western coasts of Scandinavia, in areas where *C. paleacea* occurs. Other plants referred to *C. recta* in Scandinavia are shorter in stature and are cytologically distinct. They may be hybrids between *C. paleacea* and a smaller species, perhaps *C. nigra*.

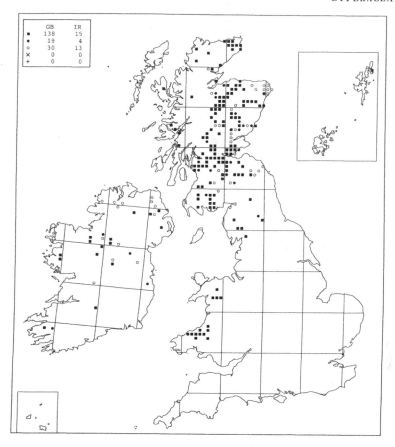

	GB	IR
■	138	15
●	19	4
○	30	13
×	0	0
+	0	0

Carex aquatilis Wahlenb. Water Sedge

C. aquatilis is most frequently found in lowland sites, including the edges of lakes, reservoirs, rivers and streams, in pools, ditches, swamps, marshes and wet meadows on flood plains, in wet woodland and on the muddy margins of east coast estuaries and sea-lochs. The lakes by which it grows are mesotrophic or even eutrophic, and in these habitats it may occur in pure stands or with *C. rostrata* and *C. vesicaria*. In sites with a well-developed zonation it may be found below the level of willow carr or stands of *Phalaris arundinacea*, and it can give way to pure stands of *Carex rostrata* or *Phragmites australis* as the water deepens (Spence, 1964). By rivers it is most frequent along slowly flowing lowland reaches where oligotrophic or oligo-mesotrophic water flows over fine, mesotrophic sediments (Holmes, 1983). *C. aquatilis* also occurs in oligotrophic, base-poor sphagnum bogs over peat, sometimes in the lowlands but more frequently at high altitudes where it may grow with *C. bigelowii*, *C. curta*, *C. nigra* and *C. rariflora*. It reaches 975m in such vegetation at Cairn of Claise and Glen Maol (McVean & Ratcliffe, 1962).

C. aquatilis is a rhizomatous perennial. Both euploids (2n=76) and aneuploids (2n=77) occur; the latter are probably derived by chromosome fission (Faulkner, 1972). Plants in open habitats often flower freely but the species may persist in the vegetative state in shade. Under experimental conditions *C. aquatilis* proves to be as fertile as the other species in the *C. nigra* group

(Faulkner, 1973). However, Dutch populations set very little fruit in the field: an average of one ripe seed per stem and a maximum of five (Corporaal, 1987). The reasons for this are not clear, nor is it known whether any British or Irish populations set so few seeds.

The distribution of this species is rather patchy, as *C. aquatilis* is found along river valleys in some areas but is rare or absent in the surrounding countryside. There are, for example, strong populations along the Afon Teifi, and the species is abundant in Strathspey. It is difficult to assess any changes in the distribution of the species in Britain for two reasons: lowland populations were formerly confused with *C. acuta* and, perhaps partly for this reason, the species was greatly under-recorded until recently. Many new records have been made since the publication of the *Atlas of the British Flora*. The species still appears to be uncommon in Ireland, and has a scattered distribution like that mapped for Britain in the 1962 *Atlas*. This suggests that it may still be under-recorded in Ireland.

C. aquatilis is a circumpolar species which in Europe is frequent at high latitudes and extends south to the British Isles, N. Germany and C. Russia. It is one of the few plants in our area with an arctic (rather than arctic-montane) distribution.

This is a variable species, and even in our area the tall emergent of the lowlands is very different in appearance from the short plants which grow at high altitude. Experimental studies of populations growing on ice-wedge polygons in the Alaskan arctic have shown that genetically distinct ecotypes may occur in the different micro-habitats (ridges, meadows, troughs and ponds) even though the micro-habitats are less than 1m apart (Shaver *et al.*, 1979). However, Faulkner (1972) was unable to find any cytological distinction between lowland plants, dwarf plants from the Eastern Highlands of Scotland (var. *minor* Boott) and dwarf arctic plants sometimes treated as subsp. *stans* (Drejer) Hultén.

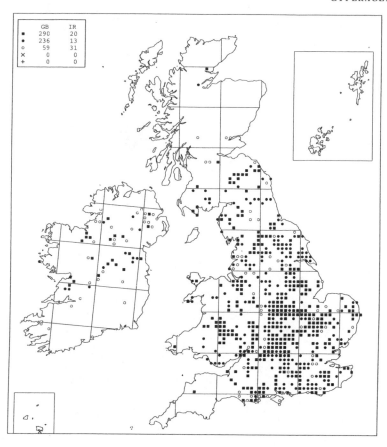

	GB	IR
■	290	20
●	236	13
○	59	31
×	0	0
+	0	0

Carex acuta L.   **Slender Tufted-sedge**

This species is found in shallow water or on wet ground in a wide variety of habitats, including the edges of lakes, reservoirs, ponds, rivers, streams, canals and ditches, and in unimproved flood meadows and washland. The shallow, muddy edges of major rivers provide a particularly characteristic habitat, and the species is tolerant of frequent flooding and is not easily uprooted when rivers are in spate (Sinker *et al.*, 1985). *C. acuta* is often found as isolated tufts, which may grow with *Butomus umbellatus*, *Carex acutiformis*, *C. riparia*, *Glyceria maxima*, *Phragmites australis* and *Sparganium erectum*, but more extensive stands are also recorded by lakes, ponds and rivers. It is usually found by calcareous and mesotrophic or eutrophic waters, and sometimes occurs in brackish marshes. It is shade-tolerant and may grow under riverside willows or in wet woodland. It is confined to the lowlands.

C. acuta is a rhizomatous perennial. Faulkner (1972) found that only 19 out of 39 specimens examined were normal euploids, with 2n=84; the rest had additional chromosome fragments or 2n=83 or 85. The life-cycle of this species has been studied by Soukupová (1988). The shoots die down during the winter, resprouting in spring; they may live for two years, but die after flowering. New shoots begin to develop in autumn but usually emerge above ground in spring. The seeds are dormant when shed and require at least eight months after-ripening. Germination takes place in spring, and a higher proportion of seeds germinate in the light than in the dark under experimental conditions. The establishment of new plants from seed is rare.

This species was under-recorded in both Britain and Ireland in the *Atlas of the British Flora*. It was, for example, not discovered in either the Burren or Connemara until the late 1960s, although both areas are well known to botanists (Webb & Scannell, 1983). Nevertheless, there is evidence that the species is decreasing in some regions. Drainage and the subsequent reclamation of marshy ground has severely reduced the available habitat in the Lough Neagh area (Harron, 1986), and doubtless elsewhere. Populations in Cheshire have been destroyed by canal bank improvements (Newton, 1991). In Dorset all nine of the stands noted by R. d'O. Good in the 1930s have disappeared, and by the R. Stour the species survives only as occasional tufts (Pearman, 1994, 1995). However, it is occasionally recorded as a colonist of disused gravel-pits and borrow-pits.

C. acuta is widespread in the temperate regions of Europe, N. Africa and W. Asia. In Europe it is rare in the Mediterranean region and in northern Scandinavia, and absent from Iceland. It is replaced by the related *C. appendiculata* (Trautv.) Kük. in E. Asia.

Carex acuta and *C. nigra* can be difficult to distinguish in the field, even in the absence of hybrids, as the differences between the species are quantitative and the most luxuriant individuals of *C. nigra* grow in the wet habitats where *C. acuta* occurs (Faulkner, 1973). Intermediate plants are frequent in the wild, and some of these have been confirmed as hybrids by cytological examination. Faulkner (1973) concludes that on balance, the evidence is 'marginally in favour of their retention as independent species, although ... the fertility of the hybrids is certainly high enough to permit either introgression or the forma-tion of hybrid swarms'. In addition to the hybrids with the members of the *C. nigra* group, the variable and par-tially fertile hybrid between *C. acuta* and *C. acutiformis*, *C.* × *subgracilis*, is known from several localities and is proba-bly under-recorded (Jermy, 1967; Crackles, 1984).

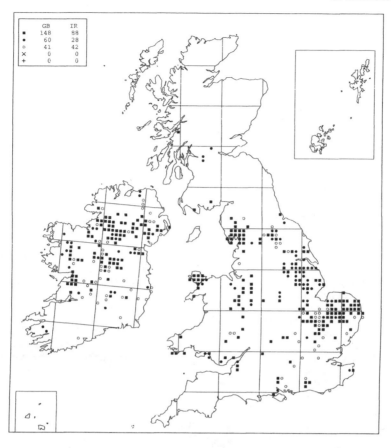

	GB	IR
■	148	88
●	60	28
○	41	42
×	0	0
+	0	0

Carex elata All.   Tufted-sedge

This is a tussock-forming sedge which grows in shallow, calcareous water up to 0.4m deep, or in fens where the water-table usually lies below the soil surface. Typical habitats include the edges of lakes, ponds, canals, fenland lodes and ditches, swamps, old peat-diggings, fen meadows, and wet woodland dominated by willows or alders. *C. elata* often grows with *Cladium mariscus* but it differs from that species in its marked preference for sites with fluctuating water-levels. In Broadland the water-level is more variable at sites where there are mixed stands of these two species than at those where only *C. mariscus* is present, and at Foulden Common small, deep hollows subject to marked changes in water-table are dominated by *C. elata* whereas larger hollows with a more stable water regime are dominated by *C. mariscus* (Haslam, 1965; Wheeler, 1980a). *C. elata* can be one of the pioneer colonists of reed-beds dominated by *Phragmites australis*, and its spread into such vegetation at Esthwaite North Fen was documented by Tansley (1939). Its tussocks, like those of *C. paniculata*, may in turn be colonized by woody species, and *C. elata* can persist in the resulting wet woodland. It is also found in fens and fen-meadows which are maintained by periodic mowing, where it is often associated with *Juncus subnodulosus*. In this vegetation the species may lack its normal tussocky form. It is vulnerable to grazing and at Sunbiggin Tarn it is more abundant in ungrazed than in grazed sites (Holdgate, 1955). It is usually found in oligo-

trophic or mesotrophic sites, but it may grow by eutrophic fenland lodes and ditches, although these sites have probably undergone recent enrichment. It is confined to the lowlands, reaching a maximum altitude of 260m at Sunbiggin Tarn.

C. elata lacks rhizomes and is consequently unable to spread vegetatively. Most plants are euploids with 2n=76, but aneuploids with chromosome numbers of 75 and 77 are recorded. *C. elata* flowers and fruits freely in the open, although flowering is suppressed in the shade. Little is known about the circumstances in which it reproduces by seed.

This species has been confused in the past with tussocky forms of *C. nigra*. Both West (1905) and Spence (1964) reported the species from Inchnacardoch Bay, Loch Ness, but these records are probably erroneous (R. W. David, pers. comm.), and a few mapped records may also be misidentifications. Nevertheless, the available evidence suggests that *C. elata* has decreased in S.E. England. Some sites have been drained, and populations have been reduced or eliminated at others when canals have been dredged and their edges cut away, when ditches and lodes have been widened and deepened, or even when ponds have been dug in areas of base-rich fen. Some of the remaining populations are threatened by similar operations, but many of the *C. elata* sites are species-rich calcareous wetlands which are managed as nature reserves. In some sites all the plants appear to be very old, and an understanding of the reproductive biology of the species is required in order to safeguard these populations. *C. elata* is still locally frequent in some areas of Ireland. The species is sometimes grown in water-gardens, usually as cultivars with discoloured foliage.

C. elata is widespread in Europe, although it is rare in the Mediterranean region; it extends north to C. Scandinavia. It is also recorded from N. Africa and in temperate Asia east to Manchuria. The western and central European plant is subsp. *elata*, which is usually found in base-rich habitats; in eastern Europe it is replaced by the predominantly calcifuge subsp. *omskiana* (Meinsh.) Jalas (Tutin *et al.*, 1980).

The tussocks of *C. elata* are often colonized by fenland plants, as well as woody species. At Sunbiggin Tarn black-headed gulls often nest on them, and even outside the gullery the tussocks support nitrophilous species such as *Sambucus nigra*, *Stellaria media* and *Urtica dioica* (Holdgate, 1955).

POACEAE (GRAMINEAE)

This large and important plant family comprises 785 genera and nearly 8,000 species. The family has a cosmopolitan distribution, and is best known for its terrestrial representatives. These include bamboos and grasses with a wide range of morphology, and numerous ecologically and economically important species. Accounts of the family, with descriptions of all the genera, are provided by Clayton & Renvoize (1986) and Watson & Dallwitz (1992). Cook (1990a) lists 61 genera which contain aquatic species, of which eight are native to our area. Four of these are dealt with in this book. *Agrostis stolonifera* is also listed as an aquatic by Cook. It is frequent in seasonally flooded habitats, and plants which are rooted at the edge of canals and ditches often send out individual shoots which extend over the surface of the water. Three other genera are listed by Cook but do not qualify for inclusion here: *Alopecurus*, *Leersia* and *Spartina*.

Catabrosa P. Beauv.

All the members of this small genus are sometimes included in a single species; other authors consider that they are divisible into as many as four species.

Catabrosa aquatica (L.) P. Beauv.

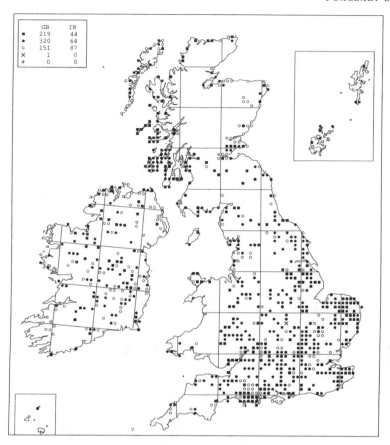

	GB	IR
■	219	44
●	320	64
○	151	87
×	1	0
+	0	0

Catabrosa aquatica (L.) P. Beauv.   **Whorl-grass**

Over much of its British and Irish range this species is usually found at the edge of small streams, springs, ditches and pools. It forms clumps or patches which are rooted in nutrient-rich silt or mud and often extend as floating mats on to shallow, slowly flowing water. It also occurs at the edge of canals and sluggish rivers, at the edge of lakes (particularly at the point where streams enter them) and in flushes and marshy fields. It colonizes sites which are trampled by cattle, but it is a very palatable species and in these localities it may be heavily grazed. It is a rather shade-tolerant grass, and is often noticed under bridges, where it can flower in the absence of grazing (Burton, 1983; Kent, 1975). On the northern and western coasts of Britain and Ireland *C. aquatica* is characteristically found where streams fan out and flow over sandy beaches. It is primarily a lowland species, which ascends to 380m at Malham Tarn. Unusually, it also grows in montane flushes at 710m on Little Fell, with *Epilobium alsinifolium* and *Saxifraga stellaris* (Roberts & Halliday, 1979).

C. aquatica is a perennial which flowers freely in the absence of grazing. Some observers have suggested that it may be rather sporadic in its appearance, disappearing from sites then reappearing after some years (Tarpey & Heath, 1990).

This is rather a local species throughout much of its British and Irish range. It is tolerant of a degree of nutrient enrichment, and has increased at the Loch of Hillwell in recent years following eutrophication (Scott & Palmer, 1987). However, inland habitats are very vulnerable to destruction by the infilling of ponds and ditches, or to unsympathetic management such as dredging which destroys the vegetated edge of the shallow streams and ditches which it favours. It has decreased in many areas, including Kent (Philp, 1982), Durham (Graham, 1988), Angus (Ingram & Noltie, 1981), N. Aberdeenshire (Welch, 1993) and N.E. Ireland (Hackney, 1992). Although the species is distinctive, it is easily overlooked when grazed and in some areas it has been under-recorded until recently (Holland, 1986).

C. aquatica sensu lato is widespread in the northern hemisphere, occurring in Europe, N. Africa, W. & C. Asia and N. America. It is most frequent in temperate regions but extends north to Greenland, Iceland and N. Norway. It also occurs in Argentina and Tierra del Fuego.

Populations of *C. aquatica* on the northern and western coasts of Britain and Ireland are often smaller than those elsewhere and have only one flower per spikelet. These plants are treated as subsp. *minor* by Perring & Sell (1967) and as var. *uniflora* by Stace (1991), and their distribution is mapped by Perring & Sell (1968).

Glyceria R. Br.

This is a fairly small genus of 40 species, all of which grow on wet ground or as aquatics. Two distinct sections of the genus are represented in our area. *G. maxima* in section *Hydropoa* Dumort. is a robust plant with long rhizomes, whereas the other species are all members of section *Glyceria*. These are more slender plants with decumbent stems which often root at the nodes, and sometimes with stolons, but without rhizomes. The three species which are now recognized in our area are superficially similar, and flowering plants are usually needed for a certain identification. They were not distinguished in many floras published before 1940 (Hubbard, 1942), and are still underrecorded in many areas. Consequently there is less information on historical and current trends in their distribution than there is for many more conspicuous species. Two of the species, *G. fluitans* and *G. notata*, are tetraploids; these are ecologically similar and the sterile hybrid between them, *G. × pedicellata*, is fairly frequent. The third species, *G. declinata*, is a diploid and is less frequently found in water than the others.

Mabberley (1987) describes the species as 'luscious pasture grasses for cows'. However, the leaves of *G. maxima* contain cyanogenic compounds which are present at maximum concentration at the start of the growing season. In our area they are usually present at such low levels that they constitute no threat to stock, but in Australia and New Zealand cattle have died after eating this species (Barton *et al.*, 1983; Cooper & Johnson, 1984).

Glyceria maxima (Hartm.) Holmb.

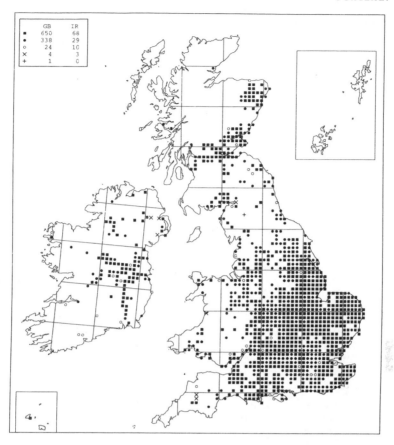

	GB	IR
■	650	68
●	338	29
○	24	10
×	4	3
+	1	0

Glyceria maxima (Hartm.) Holmb.

Reed Sweet-grass

This grass is usually found in dense and often virtually pure stands. It may grow in permanently flooded sites such as pools, ditches, canals and slowly flowing rivers, where it extends into water up to 0.8m deep. It also grows in seasonally flooded habitats, including marsh-land in the flood plain of rivers where the water may descend to 0.6m below the soil surface in summer. It is usually found over substrates which are nutrient-rich and neutral or alkaline; over peat it is found only by flowing waters, which constantly replenish the nutrient supply. It will flourish in ponds which receive treated sewage (Ozimek & Klekot, 1979). Stands sometimes extend as floating masses over the surface of open water, as in the Yare Valley broads or in ponds or disused canals (Lambert, 1946, 1947a). *G. maxima* is confined to open or very lightly shaded habitats. It persists even when regularly mown, and was formerly regarded as a valuable fodder crop. It is one of the most frequent emergents at the edge of heavily used canals, as it is well adapted to survive physical disturbance (Murphy & Eaton, 1983). *G. maxima* favours deeper water than *Phalaris arundinacea* but cannot occur in water as deep as that in which *Phragmites australis* often grows. It is virtu-ally confined to the lowlands, although it grows at 597m at Sprinkling Tarn (Stokoe, 1983).

    G. maxima is a perennial with an extensive but shal-low network of rhizomes. Rooted stands die down in

autumn, regrowing as a dense sward in early spring and continuing to produce new shoots throughout the grow-ing season. Under favourable conditions this enables it to suppress the growth of the later sprouting *Phragmites australis* (Buttery & Lambert, 1965). Floating masses remain winter-green. Plants usually flower well, although the inflorescences may abort in hot summers if plants are droughted (Lambert, 1947b). Only a small proportion (4–18%) of flowers set viable seed. Some seed germinates immediately, but most remains in the seed-bank (Roberts, 1986). Seeds will germinate in spring or summer on damp soil or under water. They grow slowly and seedlings become established only on bare mud. The normal method of reproduction is by vegetative spread. In addition to the direct growth of the rhizome, small clumps of the plant or fragments of rhiz-ome which become detached may float away to become established away from the parent clone (Grime *et al.*, 1988).

    This species is believed to be native in both Britain and Ireland. In Britain it is common in suitable habitats in S. England, even in the densely populated London area (Burton, 1983). In Broadland it expanded into areas for-merly occupied by *Phragmites australis* during a period when *P. australis* was preferentially grazed by coypu, but it retreated as coypu numbers fell (George, 1992). It is sometimes planted as an ornamental species at the edges of lakes and ponds. In N. England and Scotland it has colonized a number of water courses in the last 150 years, often spreading from sites where it was originally introduced. In Aberdeenshire it spread along the R. Don in the mid 19th century, and has subsequently become more widespread (Trail, 1923; Welch, 1993). In the early years of this century it spread along the Forth & Clyde Canal (Lambert, 1947c), and it has subsequently colonized the Tweed (Swan, 1993). In Ireland it has spread along the canal system from its apparently native sites in the south-east (Lambert, 1947c) and in N.E. Ireland it is being recorded as an introduction with increasing frequency (Hackney, 1992). The history of *G. maxima* in our area parallels that of *Butomus umbellatus*.

G. maxima is a member of a complex of subspecies or closely related species which occurs in temperate regions throughout the northern hemisphere (Hultén & Fries, 1986). *G. maxima* sensu stricto is widespread in Europe, although it is rare in the Mediterranean region and is native only as far north as S. Scandinavia, although it is naturalized further north. It is also found in W. & C. Asia. It was deliberately introduced to Australia and New Zealand, where it is now thoroughly established.

These notes are based on Lambert's (1947c) excellent account of the autecology of this species.

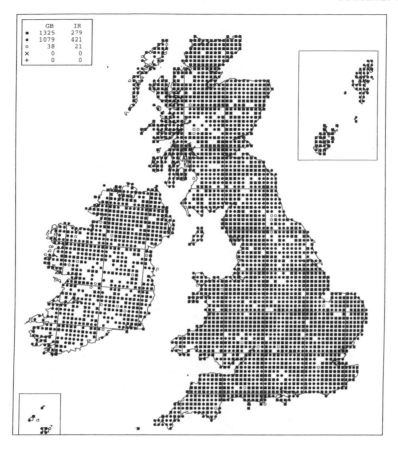

	GB	IR
■	1325	279
●	1079	421
○	38	21
×	0	0
+	0	0

Glyceria fluitans (L.) R. Br.

Floating Sweet-grass

This species frequently grows in dense masses on wet ground, on mud at the edge of water bodies or as rooted plants or floating rafts in shallow water. The most vigorous stands are found in wet depressions in pastures, shallow ponds, ditches and slowly flowing streams, and at the edge of sheltered lakes and canals. *G. fluitans* is tolerant of a fluctuating water-table, and can grow as decumbent or more or less erect stems with aerial leaves above the water, or as submerged stems with floating leaves. It is frequent in winter-flooded fen meadows. It is found in both acidic and calcareous sites, although Grime *et al.* (1988) noted that in N. England it avoids both the most acidic and the most calcareous soils. It requires substrates which are at least moderately fertile, and vigorous stands may be found in very eutrophic sites such as farm ditches. In the oligotrophic lakes of the north and west its presence sometimes indicates places where waterfowl nest or roost. *G. fluitans* is also tolerant of moderate levels of disturbance such as ditch clearance. It is also fairly shade tolerant. It is recorded from sea-level to 720m east of Knock Fell.

G. fluitans is a perennial species. The large anthers are borne on relatively long filaments and hang below the florets. Plants are slightly protandrous, and cross-pollination results in the production of more numerous viable seeds than self-pollination (Borrill, 1958b). The presence of a seed-bank at depths of 25cm in peaty soils has been demonstrated by Chippindale & Milton (1934). In addition to direct vegetative spread by the growth of established patches, detached shoots may be washed away and readily re-root in new localities (Grime *et al.*, 1988).

There is no evidence that *G. fluitans* is decreasing at a national scale. Grime *et al.* (1988) note that it is favoured by disturbance and eutrophication, and suggest that it may be increasing in ponds and ditches in some areas.

This species is found throughout Europe, except in arctic areas, and extends into adjacent regions of W. Asia and N. Africa. It is also recorded as an introduction in N. & S. America, Australia and New Zealand.

G. fluitans is the most variable of the three species in section *Glyceria*. Populations differ genetically in characters such as their habit, which may be compact or spreading, the length and number of florets in their spikelets and the extent to which plants are self-compatible. When plants from these different races are crossed they are usually genetically incompatible (Borrill, 1958a,b).

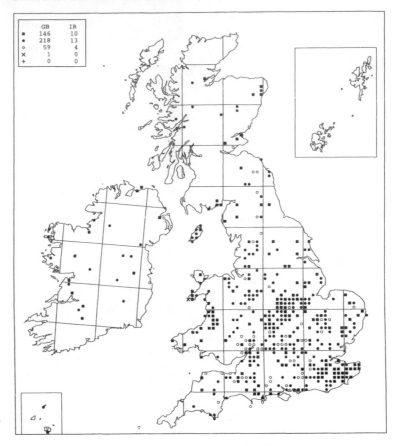

	GB	IR
■	146	10
●	218	13
○	59	4
×	1	0
+	0	0

Glyceria × *pedicellata* F. Towns. (*G. fluitans* (L.) R. Br. × *G. notata* Chevall.)   **Hybrid Sweet-grass**

Small colonies or large and persistent stands of this hybrid can be found at the edges of lakes, ponds, rivers, streams and in ditches or wet meadows. It grows in shallow water or on damp, often cattle-poached mud. It can often be found in swiftly flowing streams, and plants with submerged leaves grow in profusion in Baldwin Reservoir (Allen, 1984). It is normally found in nutrient-rich and often in calcareous sites. It may occur in the absence of one or both parents. It is found from sea-level to 550m at Moor House.

This is a stoloniferous perennial. Its pollen is highly sterile and the anthers fail to dehisce. This sterility can be attributed to the failure of many of the chromosomes to pair at meiosis (Borrill, 1956b). However, *G.* × *pedi-* *cellata* reproduces vegetatively by the spread of established plants and by detached ramets which may be carried in moving water to new sites.

This hybrid was described by Townsend (1850) as a species, on the basis of material from Cambridgeshire and Derbyshire. It is, however, easily overlooked and it is still under-recorded. There is insufficient evidence to assess changes in its distribution or abundance in Britain or Ireland.

G. × *pedicellata* is widespread in N., W. & C. Europe (Tutin *et al.*, 1980).

This is a variable hybrid between two tetraploid species. There are differences between individual clones, some of which overlap in their morphology with *G. fluitans* and others with *G. plicata* (Borrill, 1956a). This suggests that the hybrid consists of different genotypes which have separate origins.

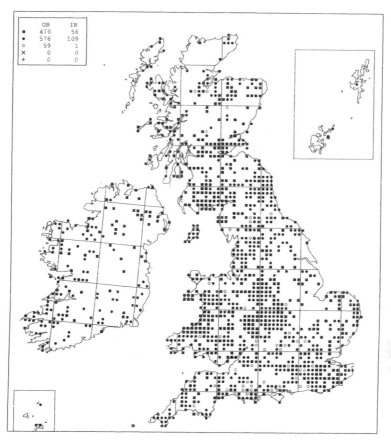

	GB	IR
■	470	56
●	576	109
○	59	1
×	0	0
+	0	0

Glyceria declinata Bréb.   **Small Sweet-grass**

This species may be found in shallow water at the edges
of ponds and in streams, but it is less frequent in permanently waterlogged habitats than the other members of the
genus. Typical habitats include trampled or waterlogged
mud by ponds, the muddy banks of streams and ditches,
flushes, winter-flooded ruts and depressions on tracks or
in pastures, and moist, trampled or cattle-poached grassland. It is even recorded from damp arable fields in
Berkshire (Bowen, 1968). It is found over a wide range of
acidic or circumneutral substrates, including peat, clay,
sand, gravel and river shingle, but it does not grow in the
highly calcareous sites in which *G. notata* can be found.
Characteristic associates include *Agrostis stolonifera*,
Isolepis setacea, *Juncus articulatus*, *J. bufonius*, *Lotus uligi-
nosus*, *Ranunculus flammula*, *R. repens* and *Stellaria uligi-
nosa* (Sinker *et al.*, 1985). In some parts of Scotland it is
confined to sites near the coast, and it may grow at the
upper edge of salt-marshes (Duncan, 1980). It is recorded
from sea-level to 500m at Llyn Crugnant.

This diploid perennial is usually self-pollinated

(Borrill, 1958b). The anthers are borne on very short
filaments and only just emerge from the florets. The
stigmas protrude only very slightly, and they are receptive at the time when the pollen is shed. The species is
highly self-fertile. Unlike *G. notata* and *G.* × *pedicellata*,
it lacks stolons.

In common with the other species in *Glyceria* section
Glyceria, this species is under-recorded in some areas. In
Northern Ireland, for example, the species was recorded
in only two sites before 1980, but was found in over 20
sites during fieldwork in 1986 (Leach *et al.*, 1987).
There is insufficient evidence to assess whether its distribution has changed in recent years.

This species is confined as a native to W. & C.
Europe, where it is found from the Azores and S. Spain
northwards to southernmost Scandinavia and eastwards
to Russia. It is recorded as an introduction in N. America (California, Nevada).

Although this species is less variable than *G. fluitans*,
populations differ genetically. Some have a particularly
luxuriant habit, and others have large vegetative and
floral parts (Borrill, 1958a).

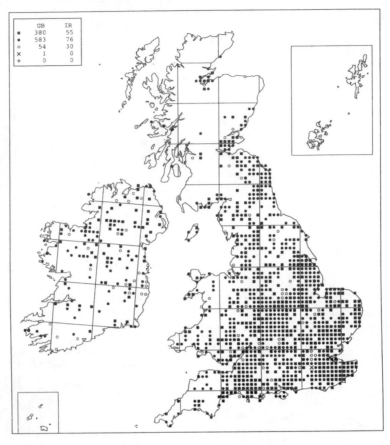

Glyceria notata Chevall.   Plicate Sweet-grass

Like *G. fluitans*, this is a species of shallow, nutrient-rich
water. Typical habitats include the gravelly beds of
streams, or their slow-flowing margins, shallow runnels
and ditches, and the muddy edges of ponds and marl-pits.
It can also be found in marshes and, less frequently, in
seasonally flooded areas. It is more calcicolous than the
other *Glyceria* species. It is locally frequent at the edge of
streams arising from limestone springs, and is scarce in
areas of poor soils. It has a predominantly coastal distribution in some areas, such as Cornwall and the Isle of Man,
where calcareous soils are infrequent. It is a predominantly lowland species, which ascends to 380m at Malham
Tarn. Old records from 460m in Derbyshire (Baker,
1884) may be correct but require confirmation.

G. *notata* is a stoloniferous perennial. It is the only
one of our species in section *Glyceria* in which flowering
is often delayed until the second year (Borrill, 1958a).
Like *G. declinata*, it is usually self-pollinating and highly
self-fertile (Borrill, 1958b). Crosses between populations

usually result in good seed-set. However, sterile clones
of *G. notata* are sometimes found, and these can be difficult to distinguish from *G.* × *pedicellata*. When fresh
seeds are sown few germinate immediately; most are
incorporated into a persistent seed-bank and they can
remain viable for at least five years (Roberts, 1986).
Vegetative reproduction, by detached stolons, is probably frequent.

There is little evidence for any change in the distribution of this species at a national scale.

G. *notata* is widespread in Europe from S.
Scandinavia southwards. It is also found in N. Africa
and extends east to C. Asia.

This species appears to be closely related to *G. declinata*, and it may even have arisen from that species by
autoploidy (Borrill, 1958b). Populations differ genetically in the extent to which they allocate resources to vegetative matter. Some produce a large vegetative biomass
in their first year but do not flower until the second year,
whereas others are smaller plants which flower in the
first year (Borrill, 1958a).

Phalaris L.

This is a genus of 15 species, which includes both annuals and perennials. Most of them are native to the Mediterranean region, or to arid areas in the New World, although many have been deliberately or accidentally introduced to other areas (Anderson, 1961). The single aquatic species in the genus, *P. arundinacea*, is unusual in having a more northerly distribution and a broad geographical range.

Phalaris arundinacea L.

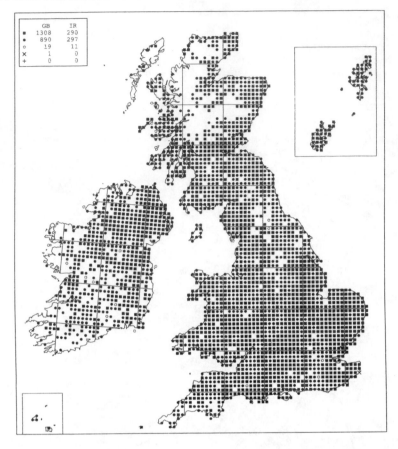

	GB	IR
■	1308	290
●	890	297
○	19	11
×	1	0
+	0	0

Phalaris arundinacea L.   Reed Canary-grass

This tall grass may form dense and virtually pure stands at the edge of standing and flowing waters, including lakes, reservoirs, rivers, streams, canals, ditches and flooded clay- and gravel-pits. It is most luxuriant at sites which are subject to large fluctuations in the water-table, such as the edges of spatey rivers, the margins of reservoirs and the Breckland meres. It is tolerant of a wide range of mineral substrates, from fine mud to coarse riverside shingle, but is rarely found over peat. It may also colonize artificial embankments. It grows at a range of nutrient levels from eutrophic to oligotrophic, although it rarely occurs in the most nutrient-poor sites. Few other species can grow in dense stands of *P. arundinacea* (Rodwell, 1995), and even if the stems are flattened by floodwater they tend to form a deep and tangled mass which prevents other species colonizing. Small patches are occasionally found in flushed terrestrial habitats such as cliffs and motorway embankments, or even in sites which appear to be dry. The superficially similar reed *Phragmites australis* differs from this species in its greater tolerance of permanently flooded sites, and its restriction to more nutrient-rich waters. *P. arundinacea* extends from sea-level to 475m (Llyn y Figyn).

   P. arundinacea is a rhizomatous perennial. The shoots emerge in early spring and die back it autumn. Unlike *Phragmites australis*, the dead stems are not persistent. The main method of reproduction of established stands is by growth of the rhizome, but many stands are clearly composed of more than one clone, the clones often differing in the pigmentation of their inflorescences. Plants usually flower prolifically, and are wind-pollinated. The species is self-incompatible but often sets abundant seed, although the flowers may become infected by the ergot *Claviceps purpurea* (Fr.) Tul. (Grime *et al.*, 1988). The seeds float, and may be dispersed in water. The fresh seeds show poor and irregular germination, as there is a water-soluble inhibitor in the embryo or caryopsis. The action of the germination inhibitor is less effective in well aerated conditions (Vose, 1962). In N. America, most seedlings are found in spring (Hoffman *et al.*, 1980). Detached vegetative shoots can root at the nodes and may float away to colonize new sites.

   This plant is frequent throughout much of Britain and Ireland. There is little evidence for any decline. It remains common along watercourses in the London area (Burton, 1983) and persists by canals in inner Dublin (Wyse Jackson & Sheehy Skeffington, 1984). It colonizes newly available habitats such as farm reservoirs and disused gravel-pits. Plants with striped leaves (var. *picta* L., or Gardeners' Garters) have been known since the 17th century, and are often cultivated. They are sometimes planted by village ponds or persist as garden throw-outs in terrestrial habitats.

   P. arundinacea is found throughout much of Europe, although it is rare in the Mediterranean region and is absent from some of the Mediterranean islands. It is widely distributed in the northern hemisphere, and has been introduced to C. & S. Africa, S. America, Java and Australia. The species is cytologically variable, with diploids, tetraploids, hexaploids and aneuploids all recorded.

Phragmites Adans.

This is a small genus which comprises three or
four species of perennial reeds, one of which
grows in Europe. For brief accounts of the taxonomy of the genus, see Clayton (1967, 1968).

Phragmites australis (Cav.) Trin. ex Steud.

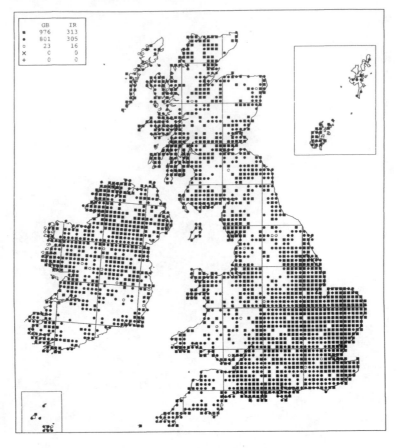

	GB	IR
■	976	313
●	801	305
○	23	16
×	0	0
+	0	0

Phragmites australis (Cav.) Trin. ex Steud.
Common Reed

This familiar emergent often forms dense and vigorous stands around the permanently flooded edges of still or slowly flowing waters, in low-lying areas which are intermittently flooded, or at the upper edge of salt-marshes, where it is tolerant of slight tidal scour. It is usually firmly rooted in the substrate, but it may also grow as lightly anchored, floating rafts of 'hover-reed'. Its competitive ability is reduced in drier sites and it tends to be replaced by other emergents at the upper end of swamps or as reed-beds become silted up. It is most frequent over nutrient-rich substrates, becomes confined to flushed areas in regions of poor soils and is absent from the most oligotrophic habitats. Under optimal conditions, stands of *P. australis* are dense and may exclude most competitors, but they are thinner and shorter towards the northern end of its range, or where ecological conditions are less favourable. *P. australis* is rarely present in sites where there are large and erratic fluctuations in water-level, often being replaced in such habitats by *Phalaris arundinacea*. Its vigour may also be reduced by grazing by stock, wildfowl or, formerly, coypu; when grazed it may be replaced by other emergents such as *Glyceria maxima* or *Phalaris arundinacea*. Small colonies grow in a range of terrestrial habitats including moorland, sea-cliffs, railway embankments, the central reservation of motorways and the edge of arable fields. This is primarily a lowland species, reaching 510m on Brown Clee Hill.

P. australis is a perennial with a relatively deep rhizome. All the shoots emerge in spring, and the species is therefore sensitive to summer mowing. The shoots are killed by severe frosts. The dead, hollow stems persist through the winter and are important as oxygen diffuses down them to the rhizome. The flowers are wind-pollinated. Seed-set and viability are very variable in Britain and Ireland. Seeds may germinate on damp terrestrial substrates or in very shallow water. Open habitats are colonized by wind-dispersed seed, but reproduction in closed stands is almost entirely by vegetative spread. Shoots detached as a result of disturbance float readily and are capable of regeneration (Grime *et al.*, 1988).

There is no conclusive fossil evidence that this species persisted in our area during the last glacial period, but it is recorded in the late glacial and is abundant in many later deposits (Godwin, 1975). It is a frequent species throughout most of its British and Irish range. However, there has been a progresssive loss of *P. australis* swamps in Broadland in the last 50 years, perhaps caused by the weakening of the stems when they grow in nitrogen-rich water and a consequent inability to withstand wind and waves (Boar *et al.*, 1989). Similar die-back has been noted at other sites, and it is a widespread phenomenon in Europe (Ostendorp, 1989). The reed has been described as 'a common species in decline' (Den Hartog *et al.*, 1989).

This species occurs throughout Europe north to 70°N., and has a virtually cosmopolitan world distribution. It is phenotypically, genetically and cytologically variable, and there are chromosome races at several ploidy levels from triploid to octoploid, as well as frequent aneuploids (Gorenflot, 1976; Gervais *et al.*, 1993). There are no detailed cytological data on plants in our area, although tetraploids (2n=48) occur in Co. Dublin (Curran, 1969).

This is an important species both to man and to other animals. For summaries of its ecology see Haslam (1972) and Rodwell (1995). An account of the growth of reed for thatching is provided by Haslam & McDougall (1972). Several birds require large stands of *P. australis* for breeding (Andrews & Ward, 1991).

SPARGANIACEAE

This family contains a single genus, *Sparganium*.

Sparganium L.

There are 14 species in this genus, all of them aquatic, perennial herbs. Most species are found in the northern hemisphere, especially in N. America and E. Asia, although the genus extends to the mountains of S.E. Asia and temperate regions of Australia and New Zealand. The species are cytologically uniform, all having a chromosome number of 2n=30.

The leafy stems of *Sparganium* species arise from a rather woody corm. Slender stolons or rhizomes with scale-like leaves grow out from the corm. The leaves die down in the winter and the corm itself dies, being replaced by new corms which develop on the upturned tips of some rhizomes. The species are variable in vegetative characters, but have a rather similar reproductive biology. The seeds germinate under water, and produce submerged juvenile leaves. At this stage the species are indistinguishable, but the mature vegetative leaves may remain submerged (as in *S.*

natans), or become floating (*S. angustifolium*) or erect (*S. erectum*). The leaves die down in winter. The flowers are borne on emergent shoots. Individual flowers are unisexual, and they are aggregated into compact, globose male or female heads. Plants are bisexual, with the male head above the female head on the same inflorescence. The female heads become receptive before the pollen is shed, with all the flowers in a head maturing simultaneously. Plants are self-compatible and wind-pollinated. The fruits have a hydrophobic surface, and may float for several months. If the fruit is frozen in ice the exocarp ruptures and the endocarp sinks. The seeds may be dispersed by moving water or may perhaps be carried externally by boats and large birds, or by birds which eat the seeds. They have a distinct micropylar plug, which has to be removed before they will germinate.

Like many aquatics, the species of *Sparganium* have a good fossil record. This indicates that all four species currently present in our area were present before the last glacial period and persisted through it (Godwin, 1975).

Sparganium erectum is an easily recognized species. Four subspecies are recognized in Britain

Sparganium erectum L.; *S. angustifolium* Michx.

and Ireland, differing from each other in the shape
of the fruit. These differences are maintained in
cultivation. The subspecies are very under-
recorded, as they can be recognized only late in
the season, when the fruits are mature. The differences between the taxa are not easily grasped, and
some are apparent only on dried material. The
compilation of records is also hampered by the
fact that some authors of county floras who have
critically examined the species do not provide
localized records or maps for the commoner subspecies in their area.The subspecies do not differ
in ecology, and all four can sometimes be found
growing together. They are mapped in the following pages, but the accompanying notes deal only
with their world distribution and any other individual characteristics of the taxon. Partially sterile
plants of *S. erectum* may be hybrids between the
subspecies.

S. emersum and *S. angustifolium* are sometimes
difficult to separate, and the fertile hybrid
between them, *S.* × *diversifolium*, is common in

northern Europe (Cook & Nicholls, 1986). The
hybrid is known from a few sites in Scotland, and
has recently been found in Northumberland. It
occurs in disturbed sites or in lakes and streams
which were once oligotrophic but are now
eutrophic (Cook, 1975b), and it may be more
frequent in our area than existing records suggest.
Dwarf forms of *S. angustifolium* are often recorded
as *S. natans*. These errors are presumably made by
botanists who are not familiar with the true
species, as the small, yellow-green leaves of *S.
natans* with small fruiting heads which emerge
directly from the leaf axils are distinctive. There
are no reliable records of hybrids between *S.
natans* and the other species in our area.

Much of the information in the species accounts is taken from the publications of Cook
(1961, 1962, 1975b, 1985) and Cook & Nicholls
(1986, 1987). Maps of the world distribution of
the species and subspecies are provided by Cook
& Nicholls (1986, 1987).

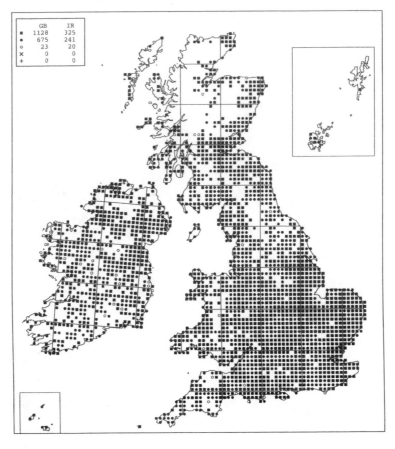

	GB	IR
■	1128	325
●	675	241
○	23	20
×	0	0
+	0	0

Sparganium erectum L.   **Branched Bur-reed**

This relatively short emergent is usually found in narrow bands in permanent water at the edge of lakes, rivers, streams, canals and ditches. Large expanses are infrequent but may occur in swamps or in ponds which are silting up, or as stands floating on liquid mud. It grows by still or slowly flowing water but is easily uprooted and is therefore absent from the edge of rapid waters or exposed shores. It is most vigorous in water 10–20cm deep, but may flower and fruit at depths up to 1m. It cannot withstand prolonged emersion. It is often replaced by stands of the leafy reeds *Glyceria maxima*, *Phalaris arundinacea* or *Phragmites australis* on the landward side, and may give way to the smooth-stemmed *Eleocharis palustris* or *Schoenoplectus lacustris* where the water deepens. It is found over a range of fine, and often anaerobic, mineral substrates in both mesotrophic and eutrophic habitats, and is very tolerant of eutrophication. It is palatable to stock and is often absent from lake shores to which cattle have access, but frequent on adjacent, ungrazed stretches. It may also be eliminated by summer cutting. *Alisma plantago-aquatica* is a frequent associate. It is primarily a lowland species, but reaches 425m at Nant Groes (Salter, 1935).

The leaves of *S. erectum* begin to grow in April; plants flower from July to September and the fruits may not ripen until November. The lowest female heads flower first and are no longer receptive when the pollen is shed by the upper inflorescences. This ensures some cross-pollination, but these lower heads rarely set much seed.

Emasculation experiments have shown that there is a much reduced seed-set in flowers which are 1m from a male inflorescence, but some pollen may travel up to 2m. Insects may visit the male flowers, but rarely visit female flowers and are therefore insignificant as pollen vectors (Pinkess, 1980; Cook & Nicholls, 1986). Some plants are partially sterile, and remain so in cultivation. In the wild the mortality of seedlings is high (especially in cloudy water) and, unlike the other *Sparganium* species, *S. erectum* rarely becomes established from seed. Plants spread vegetatively by growth of the rhizome, by dispersal of detached rhizomes and perhaps of uprooted plants. Plants stranded above the water are dwarfed but send out numerous rhizomes. In deep or rapidly flowing water *S. erectum* may occasionally be found with submerged or floating leaves, but such individuals do not flower.

This is a native member of our flora which may have persisted through the last glacial period (Godwin, 1975). It is frequent in much of its British and Irish range, and its tolerance of eutrophication doubtless assists its survival in many lowland habitats. It is able to colonize new habitats: it was present in only two of six Nottinghamshire ponds which were surveyed in 1923, but had reached them all 40 years later (Cook, 1962).

S. erectum is widespread in temperate regions of Europe, N. Africa and Asia, east to Japan; in western N. America from British Columbia to California; and in Australia, where it is probably introduced and appears to be spreading slowly. It is replaced by a similar species, *S. eurycarpa* Engelm., in much of N. America.

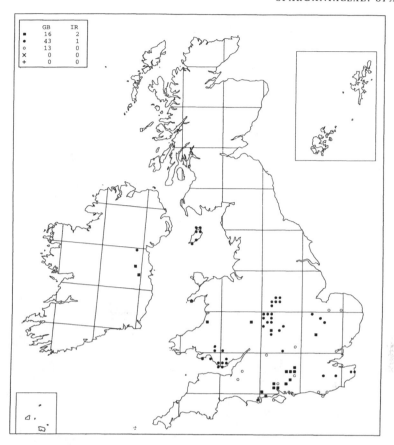

	GB	IR
■	16	2
●	43	1
○	13	0
×	0	0
+	0	0

Sparganium erectum L. subsp. *erectum*

S. erectum subsp. *erectum* is restricted to southern Britain and is rare in Ireland. It is, rather surprisingly, the prevalent subspecies in the Isle of Man (Allen, 1984). This taxon is particularly likely to be under-recorded as many authors of county floras assume that it is the commonest subspecies and therefore fail to provide critically determined, localized records.

This subspecies is common in C. & S. Europe, extending north to southernmost Scandinavia. It extends south and east to Turkey and the Caucasus.

In eastern Asia this subspecies is replaced by subsp. *stoloniferum* (Buch.-Ham. ex Graebn.) C. D. K. Cook & M. S. Nicholls. Subsp. *stoloniferum* is the only one of the five subspecies of *S. erectum* which is absent from Europe, but it has the widest distribution of them all, being the plant found in N. America and S.E. Australia.

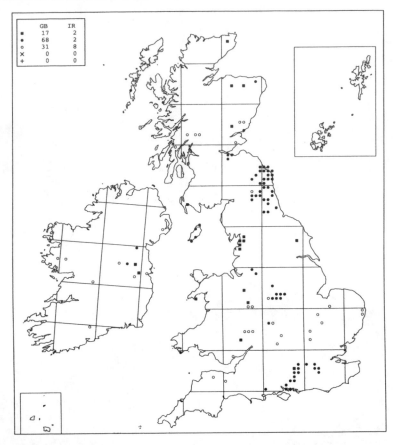

	GB	IR
■	17	2
●	68	2
○	31	8
×	0	0
+	0	0

Sparganium erectum subsp. *microcarpum* (Neuman) Domin

This is the most widespread of the four subspecies in the British Isles. In S. England and S. Wales it is rarer than subsp. *neglectum*, but it is the commonest subspecies in N. England (Graham, 1988) and the only one recorded from large areas of Scotland (McCallum Webster, 1978). It is widespread in Ireland.

This has the most northerly European distribution of the *S. erectum* subspecies, being particularly common in Scandinavia. It extends south to N. Africa, Turkey and the Caucasus.

Partially sterile hybrids between this subspecies and subsp. *neglectum* are occasionally found growing with the parents (Cook & Nicholls, 1987). We do not know if they occur in the British Isles.

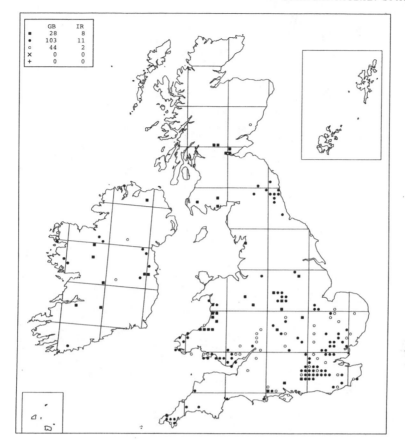

	GB	IR
■	28	8
●	103	11
○	44	2
×	0	0
+	0	0

Sparganium erectum subsp. *neglectum* (Beeby)
Schinz & Thell.

This is the commonest subspecies in some southern
counties, including Hampshire (Brewis *et al.*, 1996) and
Surrey (Lousley, 1976). However, it decreases in
frequency further north, being rare, for example, in Co.
Durham (Graham, 1988). It is widespread in Ireland.

Subsp. *neglectum* is widespread in Europe, including
the Mediterranean area, and extends north to southern
Scandinavia. It is also recorded from W. Asia.

This was the only segregate of *S. erectum* recognized
by many British botanists before the publication of
Cook's (1961) study of the genus. It was described by
Beeby (1885a,b), who first found it at Albury Ponds near
Guildford in October 1883.

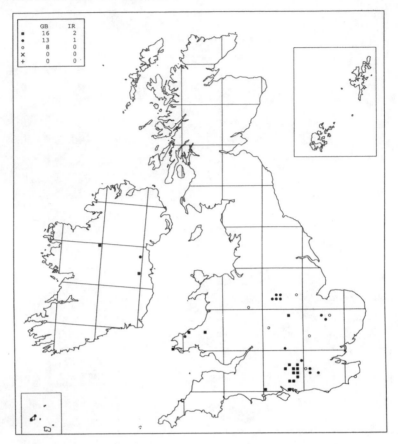

	GB	IR
■	16	2
●	13	1
○	8	0
×	0	0
+	0	0

Sparganium erectum subsp. *oocarpum* (Čelak.)
Domin

Subsp. *oocarpum* appears to be the rarest of the four subspecies of *S. erectum* in the British Isles. Like subsp. *erectum*, it has a predominantly southern distribution in Britain and is rare in Ireland.

This taxon is recorded from scattered localities in Europe, north to the British Isles. It appears to be most frequent in S.E. Europe, and is also found in N. Africa, Turkey and Iraq.

This subspecies is partially sterile, and Cook (1961, 1975b) and others have suggested that it may have originated as a hybrid between subsp. *erectum* and subsp. *neglectum*. However, it is unlikely to be a simple hybrid between these two taxa.

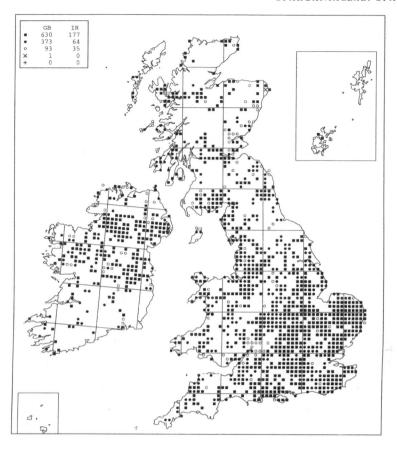

	GB	IR
■	630	177
●	373	64
○	93	35
×	1	0
+	0	0

Sparganium emersum Rehmann
Unbranched Bur-reed

S. emersum is found in still or slowly flowing, mesotrophic or eutrophic water in lakes, ponds, rivers, streams, canals and ditches. It is usually found in water 0.2–1.0m deep. It grows in deeper water than *S. erectum*, and in many sites the partially emergent or floating leaves of *S. emersum* are replaced towards the edge of the water by a band of the emergent *S. erectum*. It is usually found over fine, mineral substrates, and is a particularly characteristic species of rivers over clay (Holmes, 1983). In meandering rivers over coarser substrates it is often found at each curve, in sites where fine silt accumulates (Cook & Nicholls, 1986). It is not easily dislodged by flowing water, and so it may grow in rivers subject to spate. Although it is often frequent, it rarely forms dominant stands in which other species are unable to grow. *Nuphar lutea* is a characteristic associate. *S. emersum* is tolerant of disturbance by dredging or boat traffic, and it persists in heavily managed rivers from which most other aquatics have disappeared. Although it is primarily a lowland species, it ascends to over 500m in Crook Burn.

This is a deeply rooted perennial which spreads by growth of the slender rhizomes. The leaves are short-lived, dying after 31–39 days, but are rapidly replaced (Wiggers Nielsen *et al.*, 1985). *S. emersum* often persists as vegetative plants, but it flowers and fruits in well-illuminated, relatively shallow, still or sluggish water. It will withstand periods of emersion, but not prolonged drought.

This species may be under-recorded in some areas, as many botanists overlook non-flowering plants or will not attempt to identify them. It has decreased in some areas, and (like *Sagittaria sagittifolia*) it has sometimes disappeared from ponds and ditches in the river plain, but survives in the river channel.

S. emersum is found in temperate and boreal regions throughout the northern hemisphere. The European plant is subsp. *emersum*, which occurs throughout the range of the species. In western N. America populations have some characteristics of *S. angustifolium*, and may represent stabilized hybrids between these two species. Subsp. *acaule* (Beeby ex Macoun) C. D. K. Cook & M. S. Nicholls is confined to eastern N. America, where it is more frequent than subsp. *emersum*.

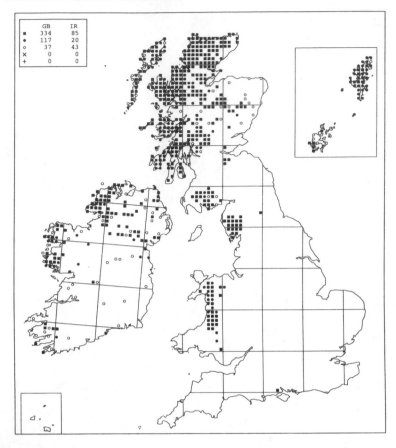

	GB	IR
■	334	85
●	117	20
○	37	43
×	0	0
+	0	0

Sparganium angustifolium Michx.

Floating Bur-reed

This is one of the most characteristic species of clear, oligotrophic lakes, and it is also found in pools, rivers, streams and ditches. It is more tolerant of exposure than the other species in the genus, but large stands characteristically occur away from the exposed edges of lakes, growing just offshore in water 0.3–1.5m deep. It sometimes grows in more mesotrophic waters. The species may grow amongst open stands of *Equisetum fluviatile*, or with submerged plants such as *Juncus bulbosus*, *Myriophyllum alternifolium* and *Utricularia vulgaris* sensu lato. It is recorded from sea-level (e.g. in the Outer Hebrides) to over 1000m near the summit of Beinn Heasgarnich.

This species has narrow floating leaves with an unwettable upper surface. It spreads vegetatively by the slender rhizomes. Many colonies flower only sparingly. The species will grow terrestrially on damp ground exposed above the water-level for periods of the summer.

The distribution map of *S. angustifolium* in the *Atlas of the British Flora* greatly under-estimated its distribution, and it is probably still under-recorded in northern and western areas where aquatic habitats have not been surveyed systematically. The pencil-thin leaves of some populations are distinctive, and are more likely to be confused with *Glyceria fluitans* than other *Sparganium* species, but plants with broader leaves are more difficult to separate from *S. emersum*. Even if plants are flowering and fruiting, they are often difficult to identify as they often grow several metres offshore and are not easily collected by grapnelling. Records of *S. angustifolium* outside the main range of the species have been accepted only if they have been verified by experts. The poor historical records make it difficult to assess changes in distribution. There is no reason to suspect that it has decreased in its main strongholds in the north and west, but the map suggests that, like other species of oligotrophic water, it may have decreased towards the eastern edge of its British and Irish ranges.

S. angustifolium is a plant of the boreal regions of Eurasia and N. America, although it is absent from the continental interior of Asia. In Europe it is frequent in the north, and extends south in the mountains to the Iberian peninsula, the Alps and the Balkans.

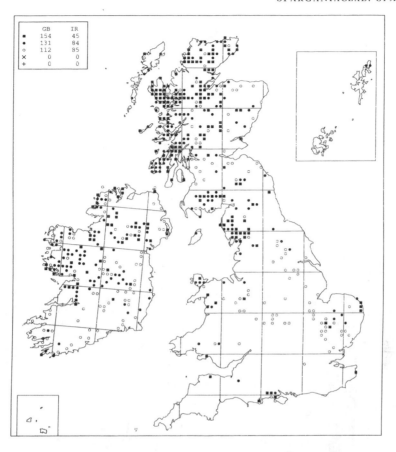

	GB	IR
■	154	45
●	131	84
○	112	85
×	0	0
+	0	0

Sparganium natans L.   **Least Bur-reed**

This small species grows in sheltered bays at the edge of lakes, in slowly flowing streams and drainage ditches, and in fenland pools. It usually grows in shallow water over peaty substrates, most frequently growing at depths of 10–50cm. It may grow terrestrially in damp moss carpets or as submerged, non-flowering plants in deeper water. It is found in peatlands over chalk and limestone, where it grows in clear, highly calcareous waters. In more acidic areas it is found in sites where the water or the substrate is enriched by local outcrops of base-rich rocks, such as basalt, limestone or mica-schist. *S. natans* is a plant of mesotrophic habitats, being absent from eutrophic or turbid waters. Characteristic associates include *Apium inundatum, Baldellia ranunculoides, Eleogiton fluitans, Juncus bulbosus, Menyanthes trifoliata, Potamogeton natans, P. polygonifolius* and *Utricularia* spp. It is recorded from sea-level up to 650m at Lochan Achlarich.

In shallow water this species has submerged and floating leaves; the floating leaves, unlike those of *S. angustifolium*, are wettable on both sides. Plants in shallow water flower and fruit freely, but those in terrestrial habitats are often less prolific (Cook & Nicholls, 1986). The usual mode of reproduction is probably by seed, and swards of seedlings can sometimes be seen in shallow water. The rhizomes are short and few in number.

There is very clear evidence that the species has decreased in England in historic times. Many populations have been exterminated by habitat destruction, with the drainage of fenland sites and the destruction of small ponds and ditches. The distribution of the species elsewhere is less clear, and there is insufficient evidence to comment on trends in its distribution. Terrestrial forms of *S. angustifolium* are often identified as this species, and *S. natans* is perhaps over-recorded in descriptions of aquatic vegetation (e.g. Spence, 1964; Rodwell, 1995). The diffuse distribution of *S. natans* compared to *S. angustifolium* makes it difficult to eliminate errors of identification, and the map doubtless contains some errors, especially in Scotland.

This species is found in boreal and temperate regions of the northern hemisphere; it is absent from much of eastern Asia but is recorded from Japan. In Europe it is absent from the Arctic and the Mediterranean regions, but widespread between these zones.

TYPHACEAE

This is a small family containing a single genus, *Typha*. It is closely related to the Sparganiaceae, and Müller-Doblies (1970) has even suggested that the two families should be united. Reasons for maintaining the taxa as separate families are given by Cook & Nicholls (1986). The affinities of both families to the rest of the monocotyledons are obscure.

Typha L.

The eight species in this genus are rhizomatous perennials which usually grow on wet ground or as emergents. The genus has a cosmopolitan distribution. The two species found in our area have been well studied, especially in N. America (where they are known as 'cattails'), and there is a substantial body of literature which deals with them. This is summarized in the ecological accounts by Grace & Harrison (1986) and Grime *et al.* (1988).

Both of our species are tall emergents with conspicuous inflorescences. The male and female flowers are aggregated into separate club-shaped heads, the female head below the male on the same inflorescence. Pollination is by wind. The female flowers become receptive shortly before the pollen is released, but remain receptive for up to four weeks so that most of them are self-pollinated. The fertile female flowers are mixed with sterile flowers in the inflorescence. The sterile flowers have club-shaped apices, and both fertile and sterile flowers have long hairs on the pedicels. The heads of the sterile flowers swell in wet weather, protecting the seeds from water. In dry weather the hairs on the pedicels spread and the inflorescence 'bursts'. The fruit drifts away, still attached to the hairy pedicel. If the diaspore lands on dry ground the hairs remain spread, and it is likely to be blown away. If it lands on water it initially rests on the tips of the hairs, and it may be blown across the surface. After a few minutes the hairs on the pedicel fold back, bringing the fruit into contact with the water. When wetted the pericarp springs open, releasing the seed which immediately sinks. For further details of this sophisticated dispersal mechanism, see Krattinger (1975) and Cook (1987).

A hybrid between *Typha angustifolia* and *T. latifolia*, *T.* × *glauca*, is known from scattered sites in England. Although accounts of it were given by Lousley (1947) and Tutin (1947), it is still unfamiliar to most British botanists and is almost

Typha latifolia L.; *T. angustifolia* L.

certainly under-recorded. The taxonomy and genetic variation of these two species and their putative hybrid has been studied in detail, especially in N. America (Krattinger *et al.*, 1979; Lee, 1975; Sharitz *et al.*, 1980; Smith, 1967). The hybrid nature of *T.* × *glauca* has been demonstrated by crossing experiments, and by isozyme studies. *T.* × *glauca* is partially sterile. It has been suggested that populations are not simple F1 hybrids, and that the hybrid backcrosses with

T. angustifolia, leading to introgression between the two species. Krattinger (1975) suggests that as the inflorescences are usually self-pollinated, the apparent backcrosses are more likely to be the result of segregation in the F2 generation.

In addition to the two native species of *Typha* and their hybrid, the alien *T. laxmanii* Lepech. is recorded as an introduction in a lake near Hextable (Clements & Foster, 1994). It is found as a native in S.E. Europe.

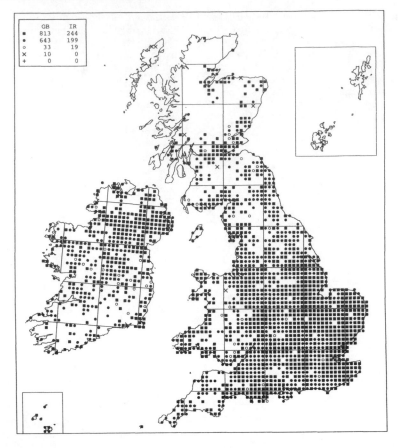

	GB	IR
■	813	244
●	643	199
○	33	19
×	10	0
+	0	0

Typha latifolia L.   Bulrush

T. latifolia grows in shallow water or on exposed mud at the edge of lakes, ponds, canals and ditches; it is less frequent by streams and rivers. It is a plant of nutrient-rich sites, and it is most frequent on mineral soils but also grows over peat. It is most vigorous in very shallow water, and it often fills shallow ditches and ponds. It will also colonize abandoned peat-diggings, especially where the peat has been extracted down to the underlying clay. Vigorous stands exclude most other species. It rarely grows at depths greater than 0.5m, frequently being replaced by *Phragmites australis* in saline sites or in deeper water. It is very tolerant of eutrophication. It is recorded from sea-level to 500m on Brown Clee Hill.

T. latifolia is a shallow-rooted perennial. The shoots may be produced throughout the growing season. They die in autumn, but the dead stems and infructescences often persist through the winter. Ramets (consisting of a rhizome branch and associated shoots) are fairly short-lived, lasting for up to three years. Colonies may expand by growth of the rhizome at rates of 4m per year, and detached portions of rhizome may float and establish new colonies. The pollen of *T. latifolia* is shed in tetrads, but it may be dispersed for at least 1km by strong winds (Krattinger, 1975). The flowers at the top of the female inflorescence set more seed, as they receive more pollen from the male inflorescence than those below. The inflorescences may produce over 200,000 seeds (Yeo, 1964).

The seeds may germinate when they are shed, but require moisture, light and relatively high temperatures (Sifton, 1959). At low light intensities, fluctuating temperatures will stimulate germination. Germination may take place in anaerobic conditions. If conditions are unsuitable, the seed may remain dormant for long periods. The species effectively colonizes ditches, sand- and gravel-pits, and even static water tanks and flooded building sites, by wind-blown seed.

The tetrads of *T. latifolia* pollen are distinctive fossils, and demonstrate that the species is a long-established native (Godwin, 1975). It appears to be more abundant now than it was last century. This increase has been noted in areas as dissimilar as Guernsey (McClintock, 1975), Cornwall (Margetts & David, 1981), the London area (Burton, 1983) and Northumberland (Swan, 1993). Burton suggests that it increased in response to the relaxation of grazing pressure around ponds on commons, and that these then provided foci from which the species spread. In other areas, such as the Isle of Man, it has apparently spread from ponds where it was initially planted as an ornamental species (Allen, 1984).

T. latifolia is widespread in temperate regions of the northern hemisphere, although it is rare in eastern Asia. It extends south to S. Africa, Madagascar, C. America and the West Indies, and is naturalized in Australia.

In N. America this has been shown to be a very variable species, with ecotypes which differ in many morphological and physiological characters.

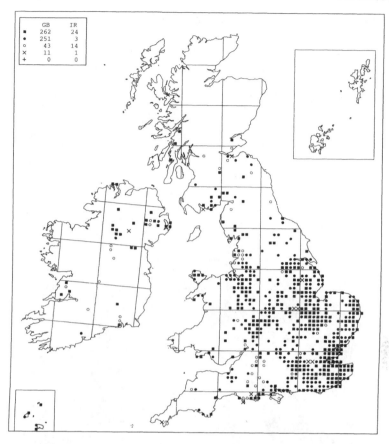

	GB	IR
■	262	24
●	251	3
○	43	14
×	11	1
+	0	0

Typha angustifolia L.   **Lesser Bulrush**

This species is similar in its ecology to *T. latifolia*, and like that species is often found in virtually pure stands. It is usually found in water at least 0.5m deep, and around some lakes a zone of *T. latifolia* may be replaced by one of *T. angustifolia* as the water deepens. *T. angustifolia* can be an important colonist of open water, sometimes growing in such sites with *Schoenoplectus lacustris*, and it can even extend over the surface of the water as a floating mat (Lambert, 1951; Lambert & Jennings, 1951; Sinker, 1962). In addition to lakes, its habitats include ponds, ditches, flooded clay- and gravel-pits and, occasionally, the banks of rivers and streams. *T. angustifolia* is less restricted to eutrophic sites than *T. latifolia*, and may grow in mesotrophic waters. It is palatable to stock, although the sites where it grows are often inaccessible to them, and it is also eaten by geese. This species was greatly reduced in numbers at some sites in Broadland when coypu numbers were high, as both the leaves and the rhizomes were eaten (Gosling, 1974). *T. angustifolia* is a lowland species.

T. angustifolia is a rhizomatous perennial. It has fewer but larger rhizomes than *T. latifolia*, and these form a denser mass but spread less rapidly (Grace & Wetzel, 1982). Its ramets live longer than those of that species. Colonies spread by the growth of the rhizome, by detached rhizome fragments or sometimes by large fragments of floating mat which may break off and blow across the surface of a lake. Rather surprisingly, a greater proportion of the biomass of *T. angustifolia* than of

T. latifolia is devoted to the reproductive structures. The pollen is released singly rather than in tetrads, and therefore blows away more readily. There is more cross- and less self-pollination in this species than in *T. latifolia* (Krattinger, 1975).

It is not easy to generalize about trends in the distribution of this species. In Broadland it spread after 1800 as nutrient levels and sedimentation increased, often acting as a pioneer colonist of open water with *Phragmites australis* following in its wake. It has decreased since 1940, probably because it is intolerant of very eutrophic conditions and of wave-scour from boats (George, 1992). *T. angustifolia* is often planted as an ornamental and has been introduced to new sites both within and outside its native range. It is much less frequent as a colonist of new sites than *T. latifolia*.

This species is widespread in temperate regions of the northern hemisphere. It is replaced in warmer climates by the closely related *T. domingensis* (Pers.) Steud. In N. America, as in Britain, it does not extend as far north as *T. latifolia*.

Experimental studies have shown that if *T. angustifolia* is transplanted into shallow water it can survive in the absence of *T. latifolia*, but is outcompeted by that species if both are present. *T. latifolia* is at an advantage because of its broader leaves with greater surface area but *T. angustifolia* is able to survive in deeper water because here the narrow leaves and greater food stores in the rhizome enable it to send leafy shoots above the water-level (Grace & Wetzel, 1981, 1982).

IRIDACEAE

This family contains 92 genera and 1850 species, most of which are geophytes with rhizomes, corms or bulbs. The family has a cosmopolitan distribution, but the species are particularly numerous in the seasonally dry climates of the eastern Mediterranean regions, S. Africa and C. & S. America. Many of the species have attractive flowers, and the family is important to horticulturists. It is of less interest to the aquatic botanist: only two genera contain aquatics, the S. American *Cypella* and *Iris*, which occurs in our area.

Iris L.

There are 300 species of *Iris*, most of which grow in north temperate regions. They have large flowers and include some favourite garden plants. For a survey of the genus, with particular emphasis on the cultivated species, see Mathew (1981). Mathew recognizes six subgenera, and places the two species native to our area in subgenus *Limniris* (Tausch) Spach. *I. pseudacorus*, an aquatic, belongs to series *Laevigatae* (Diels) G. H. M. Lawr., as do the naturalized taxa *I. ensata*, *I.* × *robusta* and *I. versicolor*. *I. foetidissima*, a plant of dry habitats, is the only member of series *Foetidissimae* (Diels) B. Mathew.

Hybridization is well known in the genus *Iris*. Many of the most popular cultivars are hybrids of ancient origin, and wild species in N. America provided a classic example of introgressive hybridization (Anderson, 1949). However, no hybrids of *I. pseudacorus* are known, and attempts to synthesize them artificially have been unsuccessful (Dykes, 1913).

The ecology of *I. pseudacorus* is summarized by Sutherland (1990), and a detailed study of the demography of populations colonizing abandoned meadows has been made by Falínska (1986).

Iris pseudacorus L.

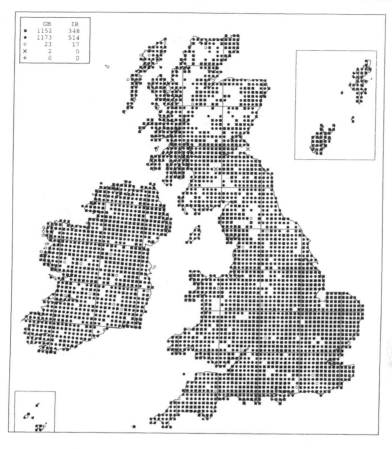

	GB	IR
■	1152	348
●	1173	514
○	23	17
×	2	0
+	0	0

Iris pseudacorus L.   Yellow Flag

This species grows as an emergent in shallow water at the edge of lakes, ponds, streams and rivers, in swamps, fens and the wettest parts of dune-slacks, by springs and flushes and in damp pastures, meadows and fen carr. In the south and east it is most frequent in water less than 0.3m deep at the edge of eutrophic lakes and rivers. It has also been found as a floating mat at the edge of lakes. In some northern and western areas (including the Hebrides) it is less frequent by lakes, but it grows in dense stands by coastal streams, on coastal shingle, upper salt-marshes, raised beaches and in lowland pastures (Adam *et al.*, 1977). It is found over a wide range of organic and inorganic, acidic or basic soils, and it may even grow on bare, irrigated rocks. However, it requires sites which are moderately fertile and are moist or flooded for most of the year. It tolerates occasional flooding by brackish water at its coastal sites, and periodic droughts. It is a lowland species, although there is an unlocalized record from 335m in Co. Sligo (Colgan & Scully, 1898).

    I. pseudacorus is a perennial with a thick rhizome. Its shoots may remain green in mild winters, or die back in more severe weather. Plants flower and set seed in open conditions, but flowering is reduced in shade. It is one of the few aquatics to have complex adaptations to insect-pollination. The flower consists of three units, which have to be visited separately by insects. Each of the large outer tepals provides a platform on which the insects

alight. A dark yellow patch surrounded by a brown zigzag line acts as a nectar guide. The pollinating insects follow it and crawl into a tube; the lower half of this tube is formed from the stalk of the tepal and the upper half is made of the flattened style and stigma. As the insect enters the tube, any pollen it carries is brushed from its back on to the stigma, and new pollen is collected from a single anther. Only long-tongued insects can obtain nectar from *I. pseudacorus*, which is usually pollinated by bumble-bees (Proctor & Yeo, 1973). The seeds float and are dispersed by water. Germina-tion usually occurs in spring and early summer, and both floating seeds and seeds on mud or in shallow water will germinate (Dymes, 1920). The seedling has a remarkably high proportion of root to shoot, an adaptation which anchors it in the water (Whitehead, 1971). Seedlings rarely become established in closed habitats but will do so where conditions are more open. Vegetative spread is by direct growth of the rhizome (clones 20m in diameter have been recorded in Ireland), or by rhizome fragments which are washed downstream. In a coastal population in S.W. Ireland, plants at the top of the salt-marsh were long-lived and seedlings were rare, whereas at lower levels plants were short-lived and seedlings numerous (Sutherland & Walton, 1990). *I. pseudacorus* contains poisonous glycosides, and it is avoided by almost all herbivorous vertebrates and by many invertebrates, although the brittle leaves are susceptible to trampling.

    This species grows in suitable habitats throughout its British and Irish range, and it is particularly common in lowland habitats in the north and west. In S.E. England habitat destruction has reduced its frequency, but plants have been deliberately planted by lakes and ponds or have escaped from cultivation, thus obscuring its native range. Both wild-type plants and cultivars are grown in gardens (Leslie, 1992).

I. pseudacorus is widespread in Europe, although absent from northern Scandinavia and Iceland. It also occurs as a native in N. Africa and in Asia east to the Caucasus. In N. America and New Zealand it has spread along water courses, and is now widely naturalized (Cody, 1961; Healy & Edgar, 1980; Raven & Thomas, 1970). There is some cytological variation, but all material from Scotland examined by Dyer *et al.* (1976) was uniform, with 2n=34.

Crawford (1989) suggests that emergents such as *Iris pseudacorus*, *Phragmites australis* and *Schoenoplectus lacustris* are the most likely plants to survive a nuclear winter, as they can withstand a period of enforced dormancy in anaerobic mud under ice.

CONSERVATION DESIGNATION OF THE RARER SPECIES

THE conservation designation of the rarer aquatic species in Britain and Ireland is outlined in this chapter.

RARE AND THREATENED PLANTS IN BRITAIN

The rarest species of vascular plants in Britain are included in the British Red Data Book: 1, Vascular Plants (Perring & Farrell, 1983). The species qualifying for inclusion are those which, at the time of publication, were known from 15 or fewer 10-km squares in Britain. The aquatic species are listed below under the International Union for the Conservation of Nature (IUCN) Red Data Book categories assigned to them:

Extinct:
Hydrilla verticillata

Endangered:
Alisma gramineum
Damasonium alisma
Schoenoplectus triqueter

Vulnerable:
Crassula aquatica
Najas marina
Sagittaria rigida

Rare:
Carex recta
Elatine hydropiper
Eriocaulon aquaticum
Ludwigia palustris
Najas flexilis
Potamogeton epihydrus
Potamogeton nodosus
Potamogeton rutilus
Rumex aquaticus

In 1994 the IUCN approved a new set of Red List categories to replace those given above.

The precise way in which these categories are applied to national lists is still under discussion, although the production of Red Lists for countries or regions, using the same principles, has been approved by IUCN. M. J. Wigginton (JNCC) has provisionally assigned the rare aquatics in Britain to the following categories:

Extinct:
Hydrilla verticillata

Critically endangered:
Alisma gramineum
Schoenoplectus pungens
Schoenoplectus triqueter

Endangered:
Damasonium alisma

Vulnerable:
Crassula aquatica
Najas marina
Potamogeton acutifolius
Ranunculus reptans
Ranunculus tripartitus
Rumex aquaticus

Lower risk (Near threatened):
Carex recta
Eriocaulon aquaticum
Ludwigia palustris
Potamogeton epihydrus
Potamogeton nodosus
Potamogeton rutilus

Sagittaria rigida is clearly an introduction and is no longer included on the Red List, and *Elatine hydropiper* and *Najas flexilis* are no longer considered to be threatened as they are now known from over 15 10-km squares. *Potamogeton acutifolius* has been added to the list as it occurs in fewer than 15 10-km squares, and *Ranunculus tripartitus* appears on the list in view of its rapid decline in recent years, even though it is still present in more than 15 10-km squares.

Seven aquatic plants are included amongst the species listed in Schedule 8 of the Wildlife & Countryside Act (1981, with subsequent amendments). They are *Alisma gramineum, Crassula aquatica, Damasonium alisma, Luronium natans, Najas flexilis, N. marina* and *Schoenoplectus triqueter*. These species are specially protected, and it is illegal to pick or uproot them unless a licence has been obtained from the appropriate conservation agency.

SCARCE PLANTS IN BRITAIN

The species which are present as native plants in 16–100 10-km squares in Britain are known as scarce, and are included in the book *Scarce Plants in Britain* (Stewart *et al.*, 1994). The following list of scarce species excludes *Ranunculus tripartitus*, recently reassigned to the Red List, but includes *Najas flexilis*, now known to occur in over 15 10–km squares. It includes *Myriophyllum verticillatum*, known from 100 squares when Stewart *et al.* (1994) went to press but subsequently recorded in several new squares.

> *Callitriche truncata*
> *Elatine hexandra*
> *Elatine hydropiper*
> *Isoetes echinospora*
> *Luronium natans*
> *Myriophyllum verticillatum*
> *Najas flexilis*
> *Nuphar pumila*
> *Nymphoides peltata*
> *Pilularia globulifera*
> *Potamogeton coloratus*
> *Potamogeton compressus*
> *Potamogeton filiformis*
> *Potamogeton trichoides*
> *Ruppia cirrhosa*
> *Stratiotes aloides*
> *Wolffia arrhiza*

RARE AND THREATENED PLANTS IN IRELAND

The rarest species of vascular plants in Ireland are included in *The Irish Red Data Book: 1, Vascular Plants* (Curtis & McGough, 1988). The species qualifying for inclusion are those which, at the time of publication, were known from ten or fewer 10-km squares in Ireland, or had apparently been lost from more than 66% of their 10-km squares since 1970. The aquatic species are listed below under the former IUCN Red Data Book categories assigned to them.

Vulnerable:
Groenlandia densa
Hydrilla verticillata
Najas flexilis
Pilularia globulifera
Schoenoplectus triqueter

Rare:
Callitriche truncata
Elatine hydropiper
Hottonia palustris
Ranunculus fluitans
Ranunculus tripartitus

Since 1988 the distribution of the Irish rare species has been surveyed in many counties, and new localities have been discovered for a number of these species (e.g. *Groenlandia densa, Najas flexilis*). However, *Ranunculus tripartitus* is now known to be extinct in Ireland and recent research has strengthened the suspicion that *Hottonia palustris* may be an introduction at all its Irish sites. In addition, *Luronium natans* has been rediscovered in Ireland, where some botanists consider it is native but others do not, and *Ceratophyllum submersum* has been added to the Irish flora.

The list of legally protected species in Ireland is currently under review.

PROTECTED SPECIES IN EUROPE

Two aquatic species covered in this book are specially protected under Appendix 1 of the Bern Convention and Annex IV of the European Community Habitats & Species Directive. They are *Luronium natans* and *Najas flexilis*. The governments of the member states of the Council of Europe and the EC have an obligation to protect these species. In addition, these species are listed on Annex II of the Habitats and Species Directive, which means that a series of protected areas must be designated for their conservation.

BRITISH AND IRISH SPECIES WITH A RESTRICTED WORLD DISTRIBUTION

Many aquatic plants have a wide distribution and occur throughout the northern hemisphere or (in some cases) virtually anywhere in the world where there are suitable habitats. However, some have a much more restricted distribution. The following species are endemic to Europe, or have primarily European distributions with populations in neighbouring regions or a few outlying areas elsewhere.

Apium inundatum
Baldellia ranunculoides
Callitriche obtusangula
Callitriche platycarpa
Callitriche truncata
Carex paniculata
Carex recta
Damasonium alisma
Elatine hexandra
Glyceria declinata
Groenlandia densa
Hottonia palustris
Juncus bulbosus
Littorella uniflora
Luronium natans
Oenanthe crocata
Oenanthe fistulosa
Oenanthe fluviatilis
Pilularia globulifera
Potamogeton acutifolius
Potamogeton coloratus
Potamogeton compressus
Potamogeton polygonifolius
Potamogeton rutilus
Ranunculus baudotii
Ranunculus fluitans
Ranunculus hederaceus
Ranunculus omiophyllus
Ranunculus peltatus
Rumex hydrolapathum
Ranunculus penicillatus
Ranunculus tripartitus
Stratiotes aloides

LIST OF LOCALITIES CITED IN THE TEXT

Localities mentioned in the text are given below, with their co–ordinates. Where possible, the 10-km square or squares are given but for larger areas or long rivers the 100-km codes are given. The accompanying map gives the alphabetical codes for the 100-km squares.

A

Adel Dam   SE24
Afon Teifi   SN
Albury Ponds   TQ04
Alderfen Broad   TG31
Alderminster   SP24
Alderney   WA50,60
Allangrange House, Munlochy   NH65
Allanton   NT85
Alston Moor   NY74
Amberley Wild Brooks   TQ01
An Fhaodhail   NM04
Angle Tarn, Place Fell   NY44
Arlesey   TL13
Ashdown Forest   TQ42,43,52,53
Ashton Canal   SJ89,99
Auchnagairn House, Kirkhill   NH54
Avery Hill   TQ47
Avielochan   NH91

B

Baldwin Reservoir   SC38
Balgavies Loch   NO55
Ballinamore & Ballyconnell Canal   H10,11
Ballydehob   V93
Balmadies Loch   NO54
Basingstoke Canal   SU,TQ
Bealanabrack River   L85,95
Beard Mill, Stanton Harcourt   SP30
Beinn Heasgarnich   NN43
Beinn nan Eachan   NN53
Berkeley Canal   ST
Birgham   NT73
Black Pond, Esher   TQ16
Blackwater River   SU
Blea Tarn   NY21
Bleddfa   SO26
Blind Tarn, Coniston   SD29
Bolder Mere   TQ05
Bomere Pool   SJ40

Braid Hills   NT26
Breadalbane   NN
Breckland   TF,TL
Bristol   ST56,57,66,67
Broadland   TG,TM
Brockenhurst   SU20,30
Brown Clee Hill   SO68
Bude Canal   SS20
Bugeilyn   SN89
Burren   M,R
Buxted   TQ42

C

Caenlochan Glen   NO17
Caerphilly Castle   ST18
Cairn of Claise   NO17
Cairngorm Mountains   NH,NJ,NN,NO
Calder & Hebble Navigation   SE12
Camborne   SW63
Cambridge   TL45
Cape Wrath   NC27
Capel Bangor   SN67
Captains Pit, Wallasey   SJ39
Carlcroft Burn   NT81
Carlingwark Loch   NX76
Carnedd Llewelyn   SH66
Castlederg   H28
Catfield Fen   TG32
Cauldshiels Loch   NT53
Chamberlayne's Mill   SY89
Chartners Lough   NZ05
Chelmer & Blackwater Navigation   TL70
Chelmsford   TL70
Chertsey   TQ06
Chesterfield Canal   SK58,68,78,79
Chichester Canal   SU80
Clowance Lake   SW63
Cogra Moss   NY01
Coire Dhubhchlair   NN43
Coll   NM15,25,26
Coniston Water   SD29,39
Connemara   L
Coole Lough   M40
Cow Green Reservoir   NY73,82,83
Cow Pond, Windsor Great Park   SU97
Craig Cerrig-gleisiad   SN92
Creag Meagaidh   NN48

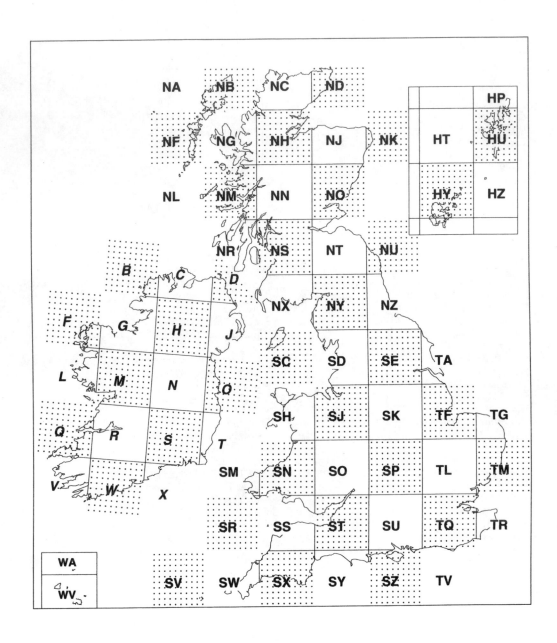

REFERENCES

Adam, P., 1981. The vegetation of British salt-marshes. *New Phytologist* **88**, 143–196.

Adam, P., Birks, H. J. B. & Huntley, B., 1977. Plant communities of the island of Arran, Scotland. *New Phytologist* **79**, 689–712.

Adams, K. [J.], 1988. *Callitriche truncata* Guss (Short leaved Water-starwort) new to Essex. *Bulletin of the Essex Field Club* no. **38**, 39–41.

Agami, M. & Waisel, Y., 1984. Germination of *Najas marina* L. *Aquatic Botany* **19**, 37–44.

Agami, M. & Waisel, Y., 1986. The role of mallard ducks (*Anas platyrhynchos*) in distribution and germination of seeds of the submerged hydrophyte *Najas marina* L. *Oecologia* **68**, 473–475.

Agami, M. & Waisel, Y., 1988. The role of fish in distribution and germination of seeds of the submerged macrophytes *Najas marina* L. and *Ruppia maritima* L. *Oecologia* **76**, 83–88.

Agami, M., Beer, S. & Waisel, Y., 1986. The morphology and physiology of turions in *Najas marina* L. in Israel. *Aquatic Botany* **26**, 371–376.

Aiken, S. G., Newroth, P. R. & Wile, I., 1979. The biology of Canadian weeds. 34. *Myriophyllum spicatum* L. *Canadian Journal of Plant Science* **59**, 201–215.

Airy Shaw, H. K., 1947. The botanical name of the wild tetraploid watercress. *Kew Bulletin* [2], 39–46.

Airy Shaw, H. K., 1949. Variation and ecology in the British watercresses. *In*: A. J. Wilmott (ed.), *British Flowering Plants and Modern Systematic Methods*, pp. 75–76. London: Botanical Society of the British Isles.

Airy Shaw, H. K., 1951. A binary name for the hybrid watercress. *Watsonia* **2**, 73–75.

Al-Bermani, A.-K. K. A., Al-Shammary, K. I. A., Bailey, J. P. & Gornall, R. J., 1993. Contributions to a cytological catalogue of the British and Irish flora, 3. *Watsonia* **19**, 269–271.

Alder, J., 1986. *A Survey of the Flora of the Chesterfield Canal*. NCC Contract Surveys no. 2. Peterborough: Nature Conservancy Council.

Allen, D. E., 1954. Variation in *Peplis portula* L. *Watsonia* **3**, 85–91.

Allen, D. E., 1984. *Flora of the Isle of Man*. Douglas: Manx Museum.

Amphlett, J. & Rea, C., 1909. *The Botany of Worcestershire*. Birmingham: Cornish Brothers.

Anderson, D. E., 1961. Taxonomy and distribution of the genus *Phalaris*. *Iowa State Journal of Science* **36**, 1–96.

Anderson, E., 1949. *Introgressive Hybridization*. New York: John Wiley & Sons.

Anderson, K. & Murphy, K. J., 1987. *The Aquatic Vegetation of the Union Canal (Lothian and Central Regions, Scotland)*. Unpublished report to Nature Conservancy Council, Edinburgh.

Anderson, L. C., Zeis, C. D. & Alam, S. F., 1974. Phytogeography and possible origins of *Butomus* in North America. *Bulletin of the Torrey Botanical Club* **101**, 292–296.

Andrews, J. & Ward, D., 1991. The management and creation of reedbeds – especially for rare birds. *British Wildlife* **3**, 81–91.

Arber, A., 1920. *Water Plants*. Cambridge: Cambridge University Press.

Archer, D., 1987. *An Investigation into the Effect of Ammonia Addition and Water Aeration on the Growth of* Potamogeton nodosus *Poir.* Unpublished report to Nature Conservancy Council.

Armstrong, W. P., 1989. Noteworthy collections. California. *Wolffia arrhiza* (L.) Horkel ex Wimmer [Lemnaceae]. *Madroño* **36**, 283.

Arts, G. H. P. & Heijden, R. A. J. M. van der, 1990. Germination ecology of *Littorella uniflora* (L.) Aschers. *Aquatic Botany* **37**, 139–151.

Aston, H. I., 1973. *Aquatic Plants of Australia*. Carlton: Melbourne University Press.

Aulio, K., 1987. Rapid decline of mass occurrences of *Juncus bulbosus* in a deacidified freshwater reservoir. *Memoranda Societatis pro Fauna et Flora Fennica* **63**, 41–44.

Babington, C. C., 1843. *Manual of British Botany*. London: John van Voorst.

Babington, C. C., 1848. On *Anacharis alsinastrum*, a supposed new British plant. *Annals and Magazine of Natural History*, 2nd series, **1**, 81–85.

Babington, C. C., 1850. A notice of *Potamogeton trichoides* of Chamisso as a native of Britain.

Botanical Gazette (edited by A. Henfrey) **2**, 285–288.

Babington, C. C., 1856. *Manual of British Botany*, edn 4. London: John van Voorst.

Babington, C. C., 1860. *Flora of Cambridgeshire*. London: John van Voorst.

Babington, C. C., 1862. *Manual of British Botany*, edn 5. London: John van Voorst.

Babington, C. C., 1863. On British species of *Isoëtes. Journal of Botany* **1**, 1–5.

Bailey, C., 1884. Notes on the structure, the occurrence in Lancashire, and the source of origin, of *Naias graminea* Delile, var. *Delilei* Magnus. *Journal of Botany* **22**, 305–333.

Baker, J. G. (ed.), 1879. Report of the plants gathered in 1878. *Report of the Botanical Exchange Club 1877–78*, 13–20.

Baker, J. G., 1884. On the upland botany of Derbyshire. *Journal of Botany* **22**, 6–15.

Baker, J. G. & Tate, G. R., 1868. A new Flora of Northumberland and Durham. *Natural History Transactions of Northumberland and Durham* **2**, 1–316.

Baker, J. G. & Trimen, H., 1867. Report of the London Botanical Exchange Club for the year 1866. *Journal of Botany* **5**, 65–73.

Bakker, D., 1954. Miscellaneous notes on *Scirpus lacustris* L. *sensu lat.* in the Netherlands. *Acta Botanica Neerlandica* **3**, 425–445.

Bance, H. M., 1946. A comparative account of the structure of *Potamogeton filiformis* Pers. and *P. pectinatus* L. in relation to the identity of a supposed hybrid of these species. *Transactions of the Botanical Society of Edinburgh* **34**, 361–367.

Barclay, A. M. & Crawford, R. M. M., 1982. Plant growth and survival under strict anaerobiosis. *Journal of Experimental Botany* **33**, 541–549.

Barrett, W. B. 1885. A contribution towards a Flora of Breconshire. *Journal of Botany* **23**, 39–44, 83–89, 107–112, 145–149.

Barry, D. H. & Jermy, A. C., 1953. Observations on *Najas marina* – 1. *Transactions of the Norfolk and Norwich Naturalists' Society* **17**, 294–297.

Barry, R. & Wade, P. M., 1986. Biological Flora of the British Isles no. 162. *Callitriche truncata* Guss. *Journal of Ecology* **74**, 289–294.

Barton, N. J., McOrist, S., McQueen, D. S. & O'Connor, P. F., 1983. Poisoning of cattle by *Glyceria maxima. Australian Veterinary Journal* **60**, 220–221.

Bass, W. J. & Holverda, W. J., 1966. *Hydrocotyle ranunculoides* L. f.: infiltrant in waterland? *Gorteria* **21**, 193–198.

Beal, E. O., 1956. Taxonomic revision of the genus *Nuphar* Sm. of North America and Europe. *Journal of the Elisha Mitchell Scientific Society* **72**, 317–346.

Beal, E. O., 1977. *A Manual of Marsh and Aquatic Vascular Plants of North Carolina with Habitat Data*. North Carolina Agricultural Experiment Station Technical Bulletin no. 247.

Bedford, B. L., Rappaport, N. R. & Bernard, J. M., 1988. A life history of *Carex lasiocarpa* Ehrh. ramets. *Aquatic Botany* **30**, 63–80.

Beeby, W. H., 1885a. A new *Sparganium. Journal of Botany* **23**, 26.

Beeby, W. H., 1885b. On *Sparganium neglectum. Journal of Botany* **23**, 193–194.

Beeby, W. H., 1886. *Equisetum litorale* Kühlewein, in Britain. *Journal of Botany* **24**, 54–55.

Belcher, H. & Swale, E., 1990. The South African water plant *Lagarosiphon major* (Ridley) Moss and its occurrence in Cambridgeshire. *Nature in Cambridgeshire* no. **32**, 63–66.

Bennett, A., 1883. On *Najas marina* L. as a British plant. *Journal of Botany* **21**, 353–354.

Bennett, A., 1884. Plants new to Norfolk, with notes on other species. *Transactions of the Norfolk and Norwich Naturalists' Society* **3**, 633–636.

Bennett, A., 1904. Contributions toward a Flora of Caithness. No. IV. *Annals of Scottish Natural History* **1904**, 224–233.

Bennett, A., 1908. *Potamogeton pensylvanicus* in England. *Naturalist, Hull* **1908**, 10–11.

Bennett, A., 1914. *Hydrilla verticillata* Casp. in England. *Journal of Botany* **52**, 257–258.

Benoit, P. M., 1954. The distinguishing characters of *Juncus bulbosus* and *J. kochii. Proceedings of the Botanical Society of the British Isles* **1**, 84.

Benoit, P. M., 1975. *Myosotis* L. *In*: C. A. Stace (ed.), *Hybridization and the Flora of the British Isles*, pp. 356–357. London: Academic Press.

Benoit, P. M., 1984. Tal-y-llyn Lake, Merioneth. 14th August. *Watsonia* **15**, 66.

Benoit, P. M. & Allen, D. E., 1968. Synthesised *Juncus bulbosus* × *kochii. Proceedings of the Botanical Society of the British Isles* **7**, 504.

Bernard, J. M., 1976. The life history and population dynamics of shoots of *Carex rostrata. Journal of Ecology* **64**, 1045–1048.

Biagi, J. A., Chamberlain, D. F., Hollands, R. C., King, R. A. & McKean, D. R., 1985. *Freshwater Macrophyte Survey of Selected Lochs in Lewis and Harris*. Unpublished report, Royal Botanic Garden, Edinburgh.

Birkinshaw, C. R., 1994. Aspects of the ecology and conservation of *Damasonium alisma* Miller in western Europe. *Watsonia* **20**, 33–39.

Björkqvist, I., 1967. Studies in *Alisma* L. I. Distribution, variation and germination. *Opera Botanica* **17**, 1–128.

Björkqvist, I., 1968. Studies in *Alisma* L. II. Chromosome studies, crossing experiments and taxonomy. *Opera Botanica* **19**, 1–138.

Blackburn, R. D. & Weldon, L. W., 1967. Eurasian watermilfoil – Florida's new underwater menace. *Hyacinth Control Journal* **1967**(6), 15–18.

Blake, S. T., 1939. A monograph of the genus *Eleocharis* in Australia and New Zealand. *Proceedings of the Royal Society of Queensland* **50**, 88–132.

Boar, R. R., Crook, C. E. & Moss, B., 1989. Regression of *Phragmites australis* reedswamps and recent changes of water chemistry in the Norfolk Broadland, England. *Aquatic Botany* **35**, 41–55.

Bodkin, P. C., Spence, D. H. N. & Weeks, D. C., 1980. Photoreversible control of heterophylly in *Hippuris vulgaris* L. *New Phytologist* **84**, 533–542.

Bogin, C., 1955. Revision of the genus *Sagittaria* (Alismataceae). *Memoirs of the New York Botanical Garden* **9**, 179–233.

Borrill, M., 1956a. A biosystematic study of some *Glyceria* species in Britain. 1. Taxonomy. *Watsonia* **3**, 291–298.

Borrill, M., 1956b. A biosystematic study of some *Glyceria* species in Britain. 2. Cytology. *Watsonia* **3**, 299–306.

Borrill, M., 1958a. A biosystematic study of some *Glyceria* species in Britain. 3. Biometrical studies. *Watsonia* **4**, 77–88.

Borrill, M., 1958b. A biosystematic study of some *Glyceria* species in Britain. 4. Breeding systems, fertility relationships and general discussion. *Watsonia* **4**, 89–100.

Boston, H. L., 1986. A discussion of the adaptations for carbon acquisition in relation to the growth strategy of aquatic isoetids. *Aquatic Botany* **26**, 259–270.

Bowen, H. J. M., 1968. *The Flora of Berkshire*. Oxford: privately published.

Bown, D., 1988. *Aroids*. London: Century Hutchinson.

Braithwaite, M. E. & Long, D. G., 1990. *The Botanist in Berwickshire*. Berwickshire Naturalists Club.

Brenan, J. P. M. & Chapple, J. F. G., 1949. The Australian *Myriophyllum verrucosum* Lindley in Britain. *Watsonia* **1**, 63–70.

Brewis, A., 1975. *Sagittaria subulata* (L.) Buch. in the British Isles. *Watsonia* **10**, 411.

Brewis, A., Bowman, R. P. & Rose, F., 1996. *The Flora of Hampshire*. Colchester: Harley Books.

Briggs, J., 1992. *Lemna minuta* and *Azolla filiculoides* in canals. *BSBI News* no. 60, 20.

Briggs, M., 1990. *Sussex Plant Atlas: Selected Supplement*. Brighton: Booth Museum of Natural History.

Briggs, T. R. A., 1880. *Flora of Plymouth*. London: John van Voorst.

Britton, C. E., 1909. A *Radicula*-hybrid. *Journal of Botany* **47**, 430.

Britton, C. E., 1915. Surrey plants. *Journal of Botany* **53**, 177.

Britton, C. E., 1928. *Veronica anagallis* L. and *V. aquatica* Bernh. *Report of the Botanical Society and Exchange Club of the British Isles* **8**, 548–550.

Brock, T. C. M., Mielo, H. & Oostermeijer, G., 1989. On the life cycle and germination of *Hottonia palustris* L. in a wetland forest. *Aquatic Botany* **35**, 153–166.

Brooks, R. E., 1976. A new *Veronica* (Scrophulariaceae) hybrid from Nebraska. *Rhodora* **78**, 773–775.

Bruhl, J. J., 1995. Sedge genera of the world: relationships and a new classification of the Cyperaceae. *Australian Systematic Botany* **8**, 125–305.

Bruinsma, J. H. P., 1996. Het voorkomen van Haarfonteinkruid (*Potamogeton trichoides* Cham. ex Schld.) op het Plistoceen van Zuid-Nederland. *Gorteria* **22**, 6–13.

Brummitt, R. K. & Powell, C. E., 1992. *Authors of Plant Names*. Kew: Royal Botanic Gardens.

Brux, H., Todeskino, D. & Wiegleb, G., 1987. Growth and reproduction of *Potamogeton alpinus* Balbis growing in disturbed habitats. *Archiv für Hydrobiologie Beihefte* **27**, 115–127.

Brux, H., Herr, W., Todeskino, D. & Wiegleb, G., 1988. A study on floristic structure and dynamics of communities with *Potamogeton alpinus* Balbis in water bodies in the northern part of the Federal Republic of Germany. *Aquatic Botany* **32**, 23–44.

Burk, C. J., 1977. A four year analysis of vegetation following an oil spill in a freshwater marsh. *Journal of Applied Ecology* **14**, 515–522.

Burnett, J. H., 1950. The correct name for *Veronica aquatica* Bernhardi. *Watsonia* **1**, 349–353.

Burton, R. M., 1983. *Flora of the London Area*. London: London Natural History Society.

Butcher, R. W., 1921. A new British flowering plant *Tillaea aquatica* L. *Naturalist, Hull* **1921**, 369–370.

Butcher, R. W., 1922. *Tillaea aquatica* L. *Report of the Botanical Society and Exchange Club of the British Isles* **6**, 281.

Buttery, B. R. & Lambert, J. M., 1965. Competition between *Glyceria maxima* and *Phragmites communis* in the region of Surlingham Broad I. The competition mechanism. *Journal of Ecology* **53**, 163–181.

Byfield, A. J., 1984. *Crassula helmsii* – again. *BSBI News* no. 37, 28.

Byfield, A. [J.] & Pearman, D. [A.], 1994. *Dorset's Disappearing Heathland Flora. Changes in the Distribution of Dorset's Rarer Heathland Species 1931 to 1993*. Unpublished report to Plantlife and the Royal Society for the Protection of Birds.

Caffrey, J., 1988. *The Status of Aquatic Plant Communities in the Royal and Grand Canals*. Central Fisheries Board unpublished report.

Camus, J. M., Jermy, A. C., Paul, A. M., Preston, C. D. & Taylor, W. C., 1988. Hybridization and speciation in North Temperate *Isoetes*. *BSBI News* no. 49, 41–42.

Candlish, P. A., 1975. *Grantham Canal Survey*. Unpublished report, Nature Conservancy Council.

Caspar, S. J. & Manitz, H., 1975. Beiträge zur Taxonomie und Chorologie der mitteleuropäischen *Utricularia*-Arten 2. Androsporogenese, Chromosomenzahlen und Pollenmorphologie. *Feddes Repertorium* **86**, 211–232.

Catling, P. M. & Dobson, I., 1985. The biology of Canadian weeds. 69. *Potamogeton crispus* L. *Canadian Journal of Plant Science* **65**, 655–668.

Catling, P. M. & Dore, W. G., 1982. Status and identification of *Hydrocharis morsus-ranae* and *Limnobium spongia* (Hydrocharitaceae) in north-eastern North America. *Rhodora* **84**, 523–545.

Caulton, E., 1961. Fresh-water studies in Derbyshire – 2. *Calla palustris* Linn. from Allestree Park Lake, 1960 – a county record. *Transactions of the Derby Natural History Society* **2**, 16–17.

Ceska, A. & Ceska, O., 1986. Notes on *Myriophyllum* (Haloragaceae) in the Far East: the identity of *Myriophyllum sibiricum* Komarov. *Taxon* **35**, 95–100.

Chamberlain, D. F., King, R. A., McKean, D. R., Miller, A. G. & Nyberg, J. A., 1984. *Freshwater Macrophyte Survey of Selected Lochs in the Uists.* Unpublished report, Royal Botanic Garden, Edinburgh.

Chambers, P. A., Spence, D. H. N. & Weeks, D. C., 1985. Photocontrol of turion formation by *Potamogeton crispus* L. in the laboratory and natural water. *New Phytologist* **99**, 183–194.

Chase, M. W., Soltis, D. E., Olmstead, R. G., Morgan, D. *et al.* [38 others], 1993. Phylogenetics of seed plants: an analysis of nucleotide sequences from the plastid gene *rbcL*. *Annals of the Missouri Botanical Garden* **80**, 528–580.

Chippindale, H. G. & Milton, W. E. J., 1934. On the viable seeds present in the soil beneath pastures. *Journal of Ecology* **22**, 508–531.

Church, A. H., 1925. *Introduction to the Plant-life of the Oxford district II. The Annual Succession. (Jan.-June).* Botanical Memoirs no. 14. Oxford: Oxford University Press.

Clapham, A. R. (ed.), 1969. *Flora of Derbyshire.* Derby: County Borough of Derby Museum and Art Gallery.

Clapham, A. R., Tutin, T. G. & Moore, D. M., 1987. *Flora of the British Isles*, edn 3. Cambridge: Cambridge University Press.

Clapham, A. R., Tutin, T. G. & Warburg, E. F., 1952. *Flora of the British Isles.* Cambridge: Cambridge University Press.

Clapham, A. R., Tutin, T. G. & Warburg, E. F., 1962. *Flora of the British Isles*, edn 2. Cambridge: Cambridge University Press.

Clark, W. A., 1943. Pondweeds from North Uist (V.-C.110), with a special consideration of *Potamogeton rutilus* Wolfg. and a new hybrid. *Proceedings of the University of Durham Philosophical Society* **10**, 368–373.

Clarke, W. G. & Gurney, R., 1921. Notes on the genus *Utricularia* and its distribution in Norfolk. *Transactions of the Norfolk and Norwich Naturalists' Society* **11**, 128–161.

Clatworthy, J. N. & Harper, J. L., 1962. The comparative biology of closely related species living in the same area. V. Inter- and intra-specific interference within cultures of *Lemna* spp. and *Salvinia natans*. *Journal of Experimental Botany* **13**, 307–324.

Clayton, W. D., 1967. Studies in the Gramineae: XIV. *Kew Bulletin* **21**, 111–117.

Clayton, W. D., 1968. The correct name of the common reed. *Taxon* **17**, 168–169.

Clayton, W. D. & Renvoize, S. A., 1986. *Genera Graminum*. Kew Bulletin Additional Series XIII. London: Her Majesty's Stationery Office.

Clement, E. J. & Foster, M. C. 1994. *Alien Plants of the British Isles*. London: Botanical Society of the British Isles.

Cody, W. J., 1961. *Iris pseudacorus* L. escaped from cultivation in Canada. *Canadian Field-Naturalist* **75**, 139–142.

Coleman, W. H., 1844. Observations on a new species of *Oenanthe*. *Annals and Magazine of Natural History* **13**, 188–191.

Coleman, W. H., 1849. Note on *Oenanthe fluviatilis*. *In*: R. H. Webb & W. H. Coleman, *Flora Hertfordiensis*, pp. 368–371. London: W. Pamplin.

Colgan, N. & Scully, R. W., 1898. *Contributions towards a Cybele Hibernica*, edn 2. Dublin: Edward Ponsonby.

Combes, R., 1965. Contribution a l'étude de *Veronica anagallis-aquatica* L.: existence d'une pseudo-cléistogamie expérimentale. *Revue Générale de Botanique* **72**, 323–330.

Conway, V. M., 1936a. Studies in the autecology of *Cladium mariscus* R. Br. I. Structure and development. *New Phytologist* **35**, 177–204.

Conway, V. M., 1936b. Studies in the autecology of *Cladium mariscus* R. Br. II. Environmental conditions at Wicken Fen, with special reference to soil temperatures and the soil atmosphere. *New Phytologist* **35**, 359–380.

Conway, V. M., 1937. Studies in the autecology of *Cladium mariscus* R. Br. Part III. The aeration of the subterranean parts of the plant. *New Phytologist* **36**, 64–96.

Conway, V. M., 1938a. Studies in the autecology of *Cladium mariscus* R. Br. IV. Growth rates of the leaves. *New Phytologist* **37**, 254–278.

Conway, V. M., 1938b. Studies in the autecology of *Cladium mariscus* R. Br. V. The distribution of the species. *New Phytologist* **37**, 312–328.

Conway, V. M., 1942. Biological Flora of the British Isles. *Cladium mariscus* (L.) R. Br. *Journal of Ecology* **30**, 211–216.

Cook, C. D. K., 1961. *Sparganium* in Britain. *Watsonia* **5**, 1–10.

Cook, C. D. K., 1962. Biological Flora of the British Isles. *Sparganium erectum* L. *Journal of Ecology* **50**, 247–255.

Cook, C. D. K., 1966a. A monographic study of *Ranunculus* subgenus *Batrachium* (DC.) A. Gray. *Mitteilungen der Botanischen Staatssammlung München* **6**, 47–237.

Cook, C. D. K., 1966b. Studies in *Ranunculus* subgenus *Batrachium* (DC.) A. Gray. III. *Ranunculus hederaceus* L. and *R. omiophyllus* Ten. *Watsonia* **6**, 246–259.

Cook, C. D. K., 1969. On the determination of leaf form in *Ranunculus aquatilis*. *New Phytologist* **68**, 469–480.

Cook, C. D. K., 1970. Hybridization in the evolution of *Batrachium*. *Taxon* **19**, 161–166.

Cook, C. D. K., 1973. New and noteworthy plants from the northern Italian ricefields. *Berichte der Schweizerischen Botanischen Gesellschaft* **83**, 54–65.

Cook, C. D. K., 1975a. *Ranunculus* L. subg. *Batrachium* (DC.) A. Gray. *In*: C. A. Stace (ed.), *Hybridization and the Flora of the British Isles*, pp. 125–129. London: Academic Press.

Cook, C. D. K., 1975b. *Sparganium* L. *In*: C. A. Stace (ed.), *Hybridization and the Flora of the British Isles*, pp. 508–509. London: Academic Press.

Cook, C. D. K., 1976. Autecology. *In*: J. B. Simmons, R. I. Beyer, P. E. Brandham, G. L. Lucas & V. T. H. Parry (eds), *Conservation of Threatened Plants*, pp. 207–210. New York: Plenum Press.

Cook, C. D. K., 1978. The *Hippuris* syndrome. *In*: H. E. Street (ed.), *Essays in Plant Taxonomy*, pp. 163–176. London: Academic Press.

Cook, C. D. K., 1983. Aquatic plants endemic to Europe and the Mediterranean. *Botanische Jahrbücher* **103**, 539–582.

Cook, C. D. K., 1984. One new taxon and two new combinations in *Ranunculus* subgenus *Batrachium* (DC.) A. Gray. *Anales del Jardin Botanico de Madrid* **40**, 473.

Cook, C. D. K., 1985. *Sparganium*: some old names and their types. *Botanische Jahrbücher* **107**, 269–276.

Cook, C. D. K., 1987. Dispersion in aquatic and amphibious vascular plants. *In*: R. M. M. Crawford (ed.), *Plant Life in Aquatic and Amphibious Habitats*, pp. 179–190. British Ecological Society Special Publication no. 5. Oxford: Blackwell Scientific Publications.

Cook, C. D. K., 1988. Wind pollination in aquatic angiosperms. *Annals of the Missouri Botanical Garden* **75**, 768–777.

Cook, C. D. K., 1990a. *Aquatic Plant Book*. The Hague: SPB Academic Publishing.

Cook, C. D. K., 1990b. Seed dispersal of *Nymphoides peltata* (S. G. Gmelin) O. Kuntze (Menyanthaceae). *Aquatic Botany* **37**, 325–340.

Cook, C. D. K. & Lüönd, R., 1982a. A revision of the genus *Hydrilla* (Hydrocharitaceae). *Aquatic Botany* **13**, 485–504.

Cook, C. D. K. & Lüönd, R., 1982b. A revision of the genus *Hydrocharis* (Hydrocharitaceae). *Aquatic Botany* **14**, 177–204.

Cook, C. D. K. & Nicholls, M. S., 1986. A monographic study of the genus *Sparganium* (Sparganiaceae). Part 1. Subgenus *Xantho-*

sparganium Holmberg. *Botanica Helvetica* **96**, 213–267.

Cook, C. D. K. & Nicholls, M. S., 1987. A monographic study of the genus *Sparganium* (Sparganiaceae). Part 2. Subgenus *Sparganium*. *Botanica Helvetica* **97**, 1–44.

Cook, C. D. K. & Urmi-König, K., 1983. A revision of the genus *Stratiotes* (Hydrocharitaceae). *Aquatic Botany* **16**, 213–249.

Cook, C. D. K. & Urmi-König, K., 1984. A revision of the genus *Egeria* (Hydrocharitaceae). *Aquatic Botany* **19**, 73–96.

Cook, C. D. K. & Urmi-König, K., 1985. A revision of the genus *Elodea* (Hydrocharitaceae). *Aquatic Botany* **21**, 111–156.

Cook, S. A. & Johnson, M. P., 1968. Adaptation to heterogeneous environments. I. Variation in heterophylly in *Ranunculus flammula* L. *Evolution* **22**, 496–516.

Coombe, D. E., Douse, A. F. G. & Preston, C. D., 1981. The vegetation of Ringmere in August 1974. *Transactions of the Norfolk and Norwich Naturalists' Society* **25**, 206–217.

Cooper, M. R. & Johnson, A. W., 1984. *Poisonous Plants in Britain and their Effects on Animals and Man*. Ministry of Agriculture Fisheries and Food Reference Book 161. London: Her Majesty's Stationery Office.

Corporaal, A., 1987. Rijpe vruchten van *Carex aquatilis* Wahlenb. *Gorteria* **13**, 207–210.

Countryman, W. D., 1968. *Alisma gramineum* in Vermont. *Rhodora* **70**, 577–579.

Crackles, F. E., 1974. A rush called the Dumbles. *Local Historian* **11**, 63–67.

Crackles, F. E., 1982. *Stratiotes aloides* L. in the East Riding of Yorkshire. *Naturalist, Hull* **107**, 99–101.

Crackles, F. E., 1983a. Dumbles or bumbles. *Lore and Language* **3(8)**, 53–55.

Crackles, F. E., 1983b. *Ruppia spiralis* L. ex Dumort. and *R. maritima* L. in S.E. Yorkshire. *Watsonia* **14**, 274–275.

Crackles, F. E., 1984. *Carex acuta* L. × *C. acutiformis* Ehrh. in S.E. Yorkshire. *Watsonia* **15**, 33.

Crackles, F. E., 1990. *Flora of the East Riding of Yorkshire*. Hull: Hull University Press & Humberside County Council.

Cramp, S. (ed.), 1977. *Handbook of the Birds of Europe, the Middle East and North Africa: The Birds of the Western Palearctic*, **1**. *Ostrich to Ducks*. Oxford: Oxford University Press.

Crawford, R. M. M., 1989. *Studies in Plant Survival*. Studies in Ecology 11. Oxford: Blackwell Scientific Publications.

Croft, J., 1994. *Lythrum portula* (L.) D. Webb (*Peplis portula* L.) Water-purslane re-found in Monks Wood. *Annual Report of the Huntingdonshire Fauna and Flora Society* **1993**, 8.

Crompton, G. & Whitehouse, H. L. K., 1983. *A Checklist of the Flora of Cambridgeshire*. Cambridge: privately published.

Cronquist, A., 1981. *An Integrated System of Classification of Flowering Plants.* New York: Columbia University Press.

Curran, P. L., 1969. Fertility of *Phragmites communis* Trin. *Irish Naturalists' Journal* 16, 242.

Curtis, T. G. F. & McGough, H. N., 1988. *The Irish Red Data Book. 1 Vascular Plants.* Dublin: Stationery Office.

Czapik, R., 1978. The karyology of *Hydrilla* (Hydrocharitaceae) from Ireland and Poland. *Proceedings of the Royal Irish Academy* 78B, 267–272.

Dahlgren, G., 1991. Karyological investigations in *Ranunculus* subg. *Batrachium* (Ranunculaceae) on the Aegean islands. *Plant Systematics and Evolution* 177, 193–211.

Dahlgren, R. M. T., Clifford, H. T. & Yeo, P. F., 1985. *The Families of the Monocotyledons.* Berlin: Springer-Verlag.

Dalgliesh, J. G., 1926. Notes on duckweeds. *Journal of Botany* 64, 272–274.

Dandy, J. E., 1958. *List of British Vascular Plants.* London: British Museum (Natural History) & Botanical Society of the British Isles.

Dandy, J. E., 1975. *Potamogeton* L. *In:* C. A. Stace (ed.), *Hybridization and the Flora of the British Isles*, pp. 444–459. London: Academic Press.

Dandy, J. E. & Taylor, G., 1938a. Studies of British Potamogetons. – I. *Journal of Botany* 76, 89–92.

Dandy, J. E. & Taylor, G., 1938b. Studies of British Potamogetons. – II. Some British records of *Potamogeton trichoides. Journal of Botany* 76, 166–171.

Dandy, J. E. & Taylor, G., 1938c. Studies of British Potamogetons. – III. *Potamogeton rutilus* in Britain. *Journal of Botany* 76, 239–241.

Dandy, J. E. & Taylor, G., 1939a. Studies of British Potamogetons. – IV. The identity of *Potamogeton Drucei. Journal of Botany* 77, 56–62.

Dandy, J. E. & Taylor, G., 1939b. Studies of British Potamogetons. – VII. Some new county records. *Journal of Botany* 77, 253–259.

Dandy, J. E. & Taylor, G., 1939c. Studies of British Potamogetons. – IX. × *Potamogeton Bennettii* and × *P. Lintonii. Journal of Botany* 77, 304–311.

Dandy, J. E. & Taylor, G., 1940a. Studies of British Potamogetons. – XII. *Potamogeton pusillus* in Great Britain. *Journal of Botany* 78, 1–11.

Dandy, J. E. & Taylor, G., 1940b. Studies of British Potamogetons. – XIII. *Potamogeton Berchtoldii* in Great Britain. *Journal of Botany* 78, 49–66.

Dandy, J. E. & Taylor, G., 1940c. Studies of British Potamogetons. – XIV. *Potamogeton* in the Hebrides (Vice-county 110). *Journal of Botany* 78, 139–147.

Dandy, J. E. & Taylor, G., 1942. Studies of British Potamogetons. – XVI. × *Potamogeton*

olivaceus (*P. alpinus* × *crispus*). *Journal of Botany* 80, 117–120.

Dandy, J. E. & Taylor, G., 1946. An account of × *Potamogeton suecicus* Richt. in Yorkshire and the Tweed. *Transactions of the Botanical Society of Edinburgh* 34, 348–360.

Darwin, C., 1875. *Insectivorous Plants.* London: John Murray.

Daumann, E., 1965. Insekten- und Windbestäubung bei *Alisma plantago-aquatica* L. *Österreichische Botanische Zeitschrift* 112, 295–310.

David, R. W., 1958. An introduction to the British species of *Callitriche. Proceedings of the Botanical Society of the British Isles* 3, 28–32.

Davies, E. W., 1956. Cytology, evolution and origin of the aneuploid series in the genus *Carex. Hereditas* 42, 349–365.

Davis, G. J., Jones, M. N. & Davis, D., 1973. Seed germination in *Myriophyllum spicatum* L. *Journal of the Elisha Mitchell Scientific Society* 89, 246.

Dawson, F. H., 1980. Flowering of *Ranunculus penicillatus* (Dum.) Bab. var. *calcareus* (R. W. Butcher) C. D. K. Cook in the River Piddle (Dorset, England). *Aquatic Botany* 9, 145–157.

Dawson, F. H., 1993. Comparison of the rates of naturalisation of the invasive alien aquatics, *Crassula helmsii* and *Myriophyllum aquaticum. BSBI News* no. 63, 47–48.

Dawson, F. H., 1994. Spread of *Crassula helmsii* in Britain. *In:* L. C. de Waal, L. E. Child & P. M. Wade (eds), *Ecology and Management of Invasive Riverside Plants*, pp. 1–14. Chichester: John Wiley & Sons.

Dawson, F. H. & Warman, E. A., 1987. *Crassula helmsii* (T. Kirk) Cockayne: is it an aggressive alien aquatic plant in Britain? *Biological Conservation* 42, 247–272.

De Sloover, J. R., Iserentant, R. & Lebrun, J., 1977. La renoncule à feuilles de lierre (*Ranunculus hederaceus* L.) au Plateau des Tailles. *Bulletin de la Société Royale de Botanique de Belgique* 110, 49–64.

De Tabley, Lord, 1899. *The Flora of Cheshire.* London: Longmans, Green & Co.

Den Hartog, C., Kvet, J. & Sukopp, H., 1989. Reed. A common species in decline. *Aquatic Botany* 35, 1–4.

Dennis, R. W. G. & Ellis, E. A., 1956. The ergot on *Eleocharis* in Norfolk. *Transactions of the Norfolk and Norwich Naturalists' Society* 18(3), 18–20.

Department of Botany, British Museum (Natural History), 1970. Still more about Mull. *Watsonia* 8, 89.

Dietrich, H., 1971. Blütenmorphologische und palynologische Untersuchungen an *Littorella. Feddes Repertorium* 82, 155–165.

Doarks, C., 1990. *Changes in the Flora of Grazing Marsh Dykes in Broadland, between 1972–74 and*

1988–89. Peterborough: England Field Unit, Nature Conservancy Council.

Dony, J. G., 1967. *Flora of Hertfordshire*. Hitchin: Hitchin Urban District Council.

Dony, J. G., 1976. *Bedfordshire Plant Atlas*. Luton: Borough of Luton Museum and Art Gallery.

Dore, W. G., 1968. Progress of the European frog-bit in Canada. *Canadian Field-Naturalist* 82, 76–84.

Douglas, C. & Lockhart, N., 1983. *Pre-drainage Survey, Finn-Lacky River Catchment, Counties Monaghan and Cavan*. Forest and Wildlife Service unpublished report.

Douglas, D., 1880. Notes on the water thyme (*Anacharis alsinastrum*, Bab.): first occurrence of its male flowers in Britain. *Hardwicke's Science-Gossip* 16, 227–229.

Driscoll, R. J., 1975. *Distribution Maps of Aquatic Macrophytes and Bryophytes Recorded in Drainage Dykes in Broadland, 1972–1974*. Unpublished report.

Driscoll, R. J., 1983. Improvements in land management: their effects on aquatic plants in Broadland. *Watsonia* 14, 276–277.

Driscoll, R. J., 1985a. The effect of changes in land management on the dyke flora at Somerton and Winterton. *Transactions of the Norfolk and Norwich Naturalists' Society* 27, 33–41.

Driscoll, R. J., 1985b. Floating Water-plantain, *Luronium natans*, in Norfolk. *Transactions of the Norfolk and Norwich Naturalists' Society* 27, 43–44.

Dromey, M., Johnston, B. & Keane, S., 1992. *Ecological Survey of the Grand Canal*. Unpublished report.

Dromey, M., Johnston, B. & Nairn, R., 1991. *Ecological Survey of the Royal Canal*. Report of the Wildlife Service and Waterways Section, Office of Public Works.

Druce, G. C., 1911. The International Phytogeographical Excursion in the British Isles. III. – The floristic results. *New Phytologist* 10, 306–328.

Druce, G. C., 1912. *Veronica anagallis-aquatica*, L. *Report of the Botanical Exchange Club and Society of the British Isles* 3, 26–27.

Druce, G. C., 1916. *Hydrilla verticillata* in Britain. *Journal of Botany* 54, 172–173.

Druce, G. C., 1922. *Tillaea aquatica* L. *Report of the Botanical Society and Exchange Club of the British Isles* 6, 281–282.

Druce, G. C., 1929. *Scirpus americanus* Pers. *Report of the Botanical Society and Exchange Club of the British Isles* 8, 762–763.

Druce, G. C., 1932. *The Comital Flora of the British Isles*. Arbroath: T. Buncle & Co.

Duckett, J. G., 1970a. The coning behaviour of the genus *Equisetum* in Britain. *British Fern Gazette* 10, 107–112.

Duckett, J. G., 1970b. Spore size in the genus *Equisetum*. *New Phytologist* 69, 333–346.

Duckett, J. G., 1970c. Sexual behaviour of the genus *Equisetum*, subgenus *Equisetum*. *Botanical Journal of the Linnean Society* 63, 327–352.

Duckett, J. G., 1979. An experimental study of the reproductive biology and hybridization in the European and North American species of *Equisetum*. *Botanical Journal of the Linnean Society* 79, 205–229.

Duckett, J. G. & Duckett, A. R., 1980. Reproductive biology and population dynamics of wild gametophytes of *Equisetum*. *Botanical Journal of the Linnean Society* 80, 1–40.

Duckett, J. G. & Page, C. N., 1975. *Equisetum* L. *In*: C. A. Stace (ed.), *Hybridization and the Flora of the British Isles*, pp. 99–103. London: Academic Press.

Duffey, E., 1968. Ecological studies on the Large Copper butterfly *Lycaena dispar* Haw. *batavus* Obth. at Woodwalton Fen National Nature Reserve, Huntingdonshire. *Journal of Applied Ecology* 5, 69–96.

Duffey, E., 1977. The re-establishment of the Large Copper butterfly *Lycaena dispar batava* Obth. on Woodwalton Fen National Nature Reserve, Cambridgeshire, England, 1969–73. *Biological Conservation* 12, 143–158.

Duncan, U. K., 1980. *Flora of East Ross-shire*. Edinburgh: Botanical Society of Edinburgh.

Duvigneaud, J. & Schotsman, H. D., 1977. Le genre *Callitriche* en Belgique et dans les régions avoisinantes. *Natura Mosana* 30, 1–22.

Dyer, A. F., Ellis, T. H. N., Lithgow, E., Lowther, S., Mason, I. & Williams, D., 1976. The karyotype of *Iris pseudacorus* L. *Transactions of the Botanical Society of Edinburgh* 42, 421–429.

Dykes, W. R., 1913. *The Genus* Iris. Cambridge: Cambridge University Press.

Dymes, T. A., 1920. Note on the life-history of *Iris Pseudacorus*, Linn., with special reference to its seeds and seedlings. *Proceedings of the Linnean Society of London* 132, 59–63.

Easy, G. 1990. A note on Cambridgeshire bulrushes. *Nature in Cambridgeshire* no. 32, 58.

Edmondson, S. E., Edmondson, M. R., Gateley, P. S., Rooney, P. J. & Smith, P. H., 1993. Flowering plants. *In*: D. Atkinson & K. J. Houston (eds), *The Sand Dunes of the Sefton Coast*, pp. 109–118. Liverpool: National Museums and Galleries on Merseyside.

Ehrle, E. B., 1960. *Eleocharis acicularis* in acid mine drainage. *Rhodora* 62, 95–97.

Ellis, E. A., 1965. *The Broads*. London: Collins.

Ellis, R. G., 1983. *Flowering Plants of Wales*. Cardiff: National Museum of Wales.

Environmental Consultancy Services, 1989. *Restoration of the Ballinamore and Ballyconnell Canal. Environmental Impact Statement*. Unpublished report to ESB International.

Faegri, K., 1982. The *Myriophyllum spicatum* group in north Europe. *Taxon* 31, 467–471.

Faegri, K. & Pijl, L. van der, 1971. *The Principles of Pollination Ecology*, edn 2. Oxford: Pergamon Press.

Falínska, K., 1986. Demography of *Iris pseudo-acorus* L. populations in abandoned meadows. *Ekologia Polska* 34, 583–613.

Farmer, A. M., 1987. Terrestrial and aquatic variants of *Lobelia dortmanna* L. *Watsonia* 16, 432–435.

Farmer, A. M., 1989. Biological Flora of the British Isles no. 165. *Lobelia dortmanna* L. *Journal of Ecology* 77, 1161–1173.

Farmer, A. M. & Spence, D. H. N., 1985. Studies of diurnal acid fluctuations in British isoetid-type submerged aquatic macrophytes. *Annals of Botany* new series 56, 347–350.

Farmer, A. M. & Spence, D. H. N., 1986. The growth strategies and distribution of isoetids in Scottish freshwater lochs. *Aquatic Botany* 26, 247–258.

Farmer, A. M. & Spence, D. H. N., 1987. Flowering, germination and zonation of the submerged aquatic plant *Lobelia dortmanna* L. *Journal of Ecology* 75, 1065–1076.

Farrell, L., 1983. *The Status of* Eriocaulon aquaticum *in Western Scotland*. Unpublished report, Nature Conservancy Council, South West (Scotland) region.

Fassett, N. C., 1940. *A Manual of Aquatic Plants*. New York: McGraw-Hill Book Company.

Faulkner, J. S., 1972. Chromosome studies on *Carex* section *Acutae* in north-west Europe. *Botanical Journal of the Linnean Society* 65, 271–301.

Faulkner, J. S., 1973. Experimental hybridization of north-west European species in *Carex* section *Acutae*. *Botanical Journal of the Linnean Society* 67, 233–253.

Ferguson, I. K. & Ferguson, L. F., 1971. *Eriocaulon aquaticum* (Hill) Druce – on Ardnamurchan (Westerness, v.c. 97). *Watsonia* 8, 400.

Fernald, M. L., 1932. The linear-leaved North American species of *Potamogeton*, section *Axillares*. *Memoirs of the American Academy of Arts and Sciences* 17, 1–183.

Fitter, A. & Smith, C., 1979. *A Wood in Ascam*. York: William Sessions.

Friday, L. E., 1988. *Utricularia vulgaris*, an aquatic carnivore at Wicken Fen. *Nature in Cambridgeshire* no. 30, 50–54.

Friday, L. E., 1989. Rapid turnover of traps in *Utricularia vulgaris* L. *Oecologia* 80, 272–277.

Friday, L. E., 1991. The size and shape of traps of *Utricularia vulgaris* L. *Functional Ecology* 5, 602–607.

Friday, L. E., 1992. Measuring investment in carnivory: seasonal and individual variation in trap number and biomass in *Utricularia vulgaris* L. *New Phytologist* 121, 439–445.

Fryer, A. 1886. Notes on pondweeds. 1. *Potamogeton natans* L. *Journal of Botany* 24, 337–338.

Fryer, A., 1887a. Notes on pondweeds. 3. *Potamogeton lucens* L. *Journal of Botany* 25, 50–52.

Fryer, A., 1887b. Notes on pondweeds. 6. On land-forms of *Potamogeton*. *Journal of Botany* 25, 306–310.

Fryer, A., 1890. Supposed hybridity in *Potamogeton*. *Journal of Botany* 28, 173–179.

Fryer, A., 1892. On the specific rank of *Potamogeton Zizii*. *Journal of Botany* 30, 114–118.

Fryer, A. & Bennett, A., 1915. *The Potamogetons (Pond Weeds) of the British Isles*. London: L. Reeve & Co.

Geldart, A. M., 1906. *Stratiotes aloides* L. *Transactions of the Norfolk and Norwich Naturalists' Society* 8, 181–200.

George, M., 1992. *The Land Use, Ecology and Conservation of Broadland*. Chichester: Packard Publishing.

Gerarde, J., 1597. *The Herball or Generall Historie of Plantes*. London.

Gerarde, J., 1633. *The Herball or Generall Historie of Plantes*, edn 2, revised by T. Johnson. London.

Gervais, C., Trahan, R., Moreno, D. & Drolet, A.-M., 1993. Le *Phragmites australis* au Québec: distribution géographique, nombres chromo-somiques et reproduction. *Canadian Journal of Botany* 71, 1386–1393.

Gibbs, P. E. & Gornall, R. J., 1976. A biosystem-atic study of the creeping spearworts at Loch Leven, Kinross. *New Phytologist* 77, 777–785.

Gilbert-Carter, H. 1964. *Glossary of the British Flora*, edn 3. Cambridge: Cambridge University Press.

Godfrey, R. K. & Wooten, J. W., 1979. *Aquatic and Wetland Plants of Southeastern United States. Monocotyledons*. Athens: University of Georgia Press.

Godfrey, R. K. & Wooten, J. W., 1981. *Aquatic and Wetland Plants of Southeastern United States. Dicotyledons*. Athens: University of Georgia Press.

Godwin, H., 1929. The "sedge" and "litter" of Wicken Fen. *Journal of Ecology* 17, 148–160.

Godwin, H., 1941. Studies in the ecology of Wicken Fen IV. Crop-taking experiments. *Journal of Ecology* 29, 83–106.

Godwin, H., 1975. *The History of the British Flora*, edn 2. Cambridge: Cambridge University Press.

Godwin, H. & Tansley, A. G., 1929. The vegeta-tion of Wicken Fen. *In:* J. S. Gardiner (ed.), *The Natural History of Wicken Fen* 5, pp. 385–446. Cambridge: Bowes & Bowes.

Goetghebeur, P. & Simpson, D. A., 1991. Critical notes on *Actinoscirpus, Bolboschoenus, Isolepis, Phylloscirpus* and *Amphiscirpus* (Cyperaceae). *Kew Bulletin* 46, 169–178.

Goodwillie, R., 1992. *Turloughs over Ten Hectares. Vegetation Survey and Evaluation*. National Parks and Wildlife Service unpublished report.

Goodwillie, R. N., Buckley, P. & Douglas, C., 1992. *Owenmore River: Proposed Arterial Drainage. Environmental Impact Assessment. Botanical and Ornithological Surveys.* Report to National Parks and Wildlife Service, Office of Public Works.

Gorenflot, R., 1976. Le complexe polyploïde du *Phragmites australis* (Cav.) Trin. ex Steud. (= *P. communis* Trin.). *Bulletin de la Société Botanique de France* 123, 261–271.

Gorham, E. & Somers, M. G., 1973. Seasonal changes in the standing crop of two montane sedges. *Canadian Journal of Botany* 51, 1097–1108.

Gornall, R. J., 1987. Notes on a hybrid spearwort, *Ranunculus flammula* L. × *R. reptans* L. *Watsonia* 16, 383–388.

Gosling, L. M., 1974. The coypu in East Anglia. *Transactions of the Norfolk and Norwich Naturalists' Society* 23, 49–59.

Grace, J. B. & Harrison, J. S., 1986. The biology of Canadian weeds. 73. *Typha latifolia* L., *Typha angustifolia* L. and *Typha* × *glauca* Godr. *Canadian Journal of Plant Science* 66, 361–379.

Grace, J. B. & Wetzel, R. G., 1978. The production biology of Eurasian Watermilfoil (*Myriophyllum spicatum* L.): a review. *Journal of Aquatic Plant Management* 16, 1–11.

Grace, J. B. & Wetzel, R. G., 1981. Habitat partitioning and competitive displacement in cattails (*Typha*): experimental field studies. *American Naturalist* 118, 463–474.

Grace, J. B. & Wetzel, R. G., 1982. Niche differentiation between two rhizomatous plant species: *Typha latifolia* and *Typha angustifolia*. *Canadian Journal of Botany* 60, 46–57.

Graham, G. G., 1988. *The Flora & Vegetation of County Durham.* Durham: Durham Flora Committee and Durham County Conservation Trust.

Graham, R., 1841. Account of botanical excursions from Edinburgh in Autumn 1839. *Transactions of the Botanical Society of Edinburgh* 1, 19–26.

Grayum, M. H., 1987. A summary of evidence and arguments supporting the removal of *Acorus* from the Araceae. *Taxon* 36, 723–729.

Green, P., 1989. *Haloragis micrantha* – creeping raspwort. *BSBI News* no. 51, 48.

Green, P. S., 1962. Watercress in the New World. *Rhodora* 64, 32–43.

Greenwood, E. F., 1974. Herbicide treatments on the Lancaster Canal. *Nature in Lancashire* 4, 24–36.

Grenfell, A. L. (comp.), 1984. Adventive News 27. *BSBI News* no. 36, 28–29.

Griffiths, B. M., 1932. The ecology of Butterby Marsh, Durham. *Journal of Ecology* 20, 105–127.

Grigg, H. C., 1951. *Elodea callitrichoides* and associated river plants. *Botanical Society of the British Isles Year Book* 1951, 80–81.

Grigson, G., 1958. *The Englishman's Flora.* London: Phoenix House.

Grime, J. P., Hodgson, J. G. & Hunt, R., 1988. *Comparative Plant Ecology.* London: Unwin Hyman.

Grinling, C. H., Ingram, T. A. & Polkinghorne, B. C. (eds), 1909. *A Survey and Record of Woolwich and West Kent.* Woolwich: South Eastern Union of Scientific Societies.

Guo, Y.-H. & Cook, C. D. K., 1989. Pollination efficiency of *Potamogeton pectinatus* L. *Aquatic Botany* 34, 381–384.

Guo, Y.-H. & Cook, C. D. K., 1990. The floral biology of *Groenlandia densa* (L.) Fourreau (Potamogetonaceae). *Aquatic Botany* 38, 283–288.

Guo, Y.-H., Sperry, R., Cook, C. D. K. & Cox, P. A., 1990. The pollination ecology of *Zannichellia palustris* L. (Zannichelliaceae). *Aquatic Botany* 38, 341–356.

Guppy, H. B., 1894. Water-plants and their ways. *Hardwicke's Science-Gossip* new series 1, 195–199.

Guppy, H. B., 1906. *Observations of a Naturalist in the Pacific between 1896 and 1899, 2. Plant-dispersal.* London: Macmillan & Co.

Gurney, R., 1922. *Utricularia* in Norfolk in 1921: the effects of drought and temperature. *Transactions of the Norfolk and Norwich Naturalists' Society* 11, 260–266.

Gurney, R., 1949. Notes on frog-bit (*Hydrocharis*) and hair-weed (*Potamogeton pectinatus*.). *Transactions of the Norfolk and Norwich Naturalists' Society* 16, 381–385

Hackney, P. (comp. & ed.), 1992. *Stewart & Corry's Flora of the North-east of Ireland*, edn 3. Belfast: Institute of Irish Studies, Queen's University of Belfast.

Haeupler, H. & Schönfelder, P., 1989. *Atlas der Farn- und Blütenpflanzen der Bundesrepublik Deutschland.* Stuttgart: Eugen Ulmer.

Hagstrom, J. O., 1916. Critical researches on the Potamogetons. *Kungliga Svenska Vetenskapsakademiens Handlingar* 55(5), 1–281.

Hall, C., 1988. *Survey of the Flora of the Basingstoke Canal 1986–87.* NCC Contract Surveys no. 14. Peterborough: Nature Conservancy Council.

Hall, P. C., 1980. *Sussex Plant Atlas.* Brighton: Booth Museum of Natural History.

Hall, P. M., 1939. The British species of *Utricularia. Report of the Botanical Society and Exchange Club of the British Isles* 12, 100–117.

Harper, J. L. & Chancellor, A. P., 1959. The comparative biology of closely related species living in the same area. IV. *Rumex*: interference between individuals in populations of one and two species. *Journal of Ecology* 47, 679–695.

Harris, S. & Lording, T. A., 1973. Distribution of *Vallisneria spiralis* L. in the River Lea Navigation Canal (Essex-Hertfordshire border). *Watsonia* 9, 253–256.

Harris, S. A., Maberly, S. C. & Abbott, R. J., 1992. Genetic variation within and between populations of *Myriophyllum alterniflorum* DC. *Aquatic Botany* **44**, 1–21.

Harron, J., 1986. *Flora of Lough Neagh*. Belfast: Irish Naturalists' Journal Committee & University of Ulster, Coleraine.

Hart, H. C., 1885. Botanical notes along the Rivers Nore, Blackwater, etc. *Journal of Botany* **23**, 228–233.

Hartleb, C. F., Madsen, J. D. & Boylen, C. W., 1993. Environmental factors affecting seed germination in *Myriophyllum spicatum* L. *Aquatic Botany* **45**, 15–25.

Haslam, S. M., 1965. Ecological studies in Breck fens I. Vegetation in relation to habitat. *Journal of Ecology* **53**, 599–619.

Haslam, S. M., 1972. Biological Flora of the British Isles. *Phragmites communis* Trin. *Journal of Ecology* **60**, 585–610.

Haslam, S. M., 1978. *River Plants*. Cambridge: Cambridge University Press.

Haslam, S. M. & McDougall, D. S. A., 1972. *The Reed ('Norfolk Reed')*, edn 2. Norfolk Reed Growers' Association.

Haslam, S. M., Sinker, C. A. & Wolseley, P. A., 1975. British Water Plants. *Field Studies* **4**, 243–351.

Hauke, R. [L.], 1963. A taxonomic monograph of the genus *Equisetum* subgenus *Hippochaete*. *Beihefte zur Nova Hedwigia* **8**, 1–123.

Hauke, R. L., 1974. The taxonomy of *Equisetum* – an overview. *New Botanist* **1**, 89–95.

Hauke, R. L., 1978. A taxonomic monograph of *Equisetum* subgenus *Equisetum*. *Nova Hedwigia* **30**, 385–455.

Haynes, R. R., 1974. A revision of North American *Potamogeton* subsection *Pusilli* (Potamogetonaceae). *Rhodora* **76**, 564–649.

Haynes, R. R., 1979. Revision of North and Central American *Najas* (Najadaceae). *Sida* **8**, 34–56.

Haynes, R. R., 1985. A revision of the clasping-leaved *Potamogeton* (Potamogetonaceae). *Sida* **11**, 173–188.

Healy, A. J. & Edgar, E., 1980. *Flora of New Zealand, 3. Adventive Cyperaceous, Petalous and Spathaceous Monocotyledons*. Wellington: P.D. Hasselberg.

Heckard, L. & Rubtzoff, P., 1977. Additional notes on *Veronica anagallis-aquatica* × *catenata* (Scrophulariaceae). *Rhodora* **79**, 579–582.

Hegelmaier, F., 1867. Zur Systematik von *Callitriche*. *Verhandlungen des Botanischen Vereins für die Provinz Brandenburg und die angrenzenden Länder* **9**, 1–41.

Henderson, A. C. B., 1982. *Biological Survey of the Pastures and Dykes of the Sandwich–Deal Area, 1982*. Unpublished report to Nature Conservancy Council.

Hennedy, R., 1891. *The Clydesdale Flora*, edn 5. Glasgow: Hugh Hopkins.

Heslop-Harrison, J., 1953. The North American and Lusitanian elements in the flora of the British Isles. *In*: J. E. Lousley (ed.), *The Changing Flora of Britain*, pp. 105–123. Oxford: Botanical Society of the British Isles.

Heslop Harrison, J. W., 1949. Potamogetons in the Scottish Western Isles, with some remarks on the general natural history of the species. *Transactions of the Botanical Society of Edinburgh* **35**, 1–25.

Heslop Harrison, J. W., 1950. A pondweed, new to the European flora, from the Scottish Western Isles, with some remarks on the phytogeography of the island group. *Phyton, Horn* **2**, 104–109.

Heslop-Harrison, Y., 1953a. Variation in *Nymphaea alba* in the British Isles. *Botanical Society of the British Isles Year Book* **1953**, 58–59.

Heslop-Harrison, Y., 1953b. *Nuphar intermedia* Ledeb., a presumed relict hybrid, in Britain. *Watsonia* **3**, 7–25.

Heslop-Harrison, Y., 1955a. Biological Flora of the British Isles. *Nuphar* Sm. *Journal of Ecology* **43**, 342–364.

Heslop-Harrison, Y., 1955b. Biological Flora of the British Isles. *Nymphaea* L. em. Sm. (nom. conserv.). *Journal of Ecology* **43**, 719–734.

Heslop-Harrison, Y., 1955c. British water lilies. *New Biology* **18**, 111–120.

Heslop-Harrison, Y., 1975. *Nuphar* Sm. *In*: C. A. Stace (ed.), *Hybridization and the Flora of the British Isles*, pp. 131–132. London: Academic Press.

Heuff, H., 1984. *The Vegetation of Irish lakes*. Unpublished report.

Heuff, H., 1987. *The Vegetation of Irish Rivers*. Unpublished report to Wildlife Service, Office of Public Works.

Hewett, D. G., 1964. Biological Flora of the British Isles. *Menyanthes trifoliata* L. *Journal of Ecology* **52**, 723–735.

Hickey, R. J., Taylor, W. C. & Luebke, N. T., 1989. The species concept in Pteridophyta with special reference to *Isoëtes*. *American Fern Journal* **79**, 78–89.

Hiern, W. P., 1908. *Sagittaria heterophylla* Pursh in Devon. *Journal of Botany* **46**, 273–278.

Hoffman, G. R., Hogan, M. B. & Stanley, L. D., 1980. Germination of plant species common to reservoir shores in the northern Great Plains. *Bulletin of the Torrey Botanical Club* **107**, 506–513.

Holdgate, M. W., 1955. The vegetation of some British upland fens. *Journal of Ecology* **43**, 389–403.

Holland, S. C. (ed.), 1986. *Supplement to the Flora of Gloucestershire*. Bristol: Grenfell Publications.

Hollingsworth, P. M., Gornall, R. J. & Preston, C. D., 1995. Genetic variability in British populations of *Potamogeton coloratus*

(Potamogetonaceae). *Plant Systematics and Evolution* 197, 71–85.

Hollingsworth, P. M., Preston, C. D. & Gornall, R. J., 1995. Isozyme evidence for hybridization between *Potamogeton natans* and *P. nodosus* (Potamogetonaceae) in Britain. *Botanical Journal of the Linnean Society* 117, 59–69.

Hollingsworth, P. M., Preston, C. D. & Gornall, R. J., 1996. Isozyme evidence for the parentage and multiple origins of *Potamogeton × suecicus* (*P. pectinatus × P. filiformis*, Potamogetonaceae). *Plant Systematics and Evolution* 202, 219–232.

Holmes, N. T. H., 1979. *A Guide to Identification of Batrachium Ranunculus species of Britain.* NCC CST Notes no. 14. London: Nature Conservancy Council.

Holmes, N. T. H., 1980. *Ranunculus penicillatus* (Dumort.) Bab. in the British Isles. *Watsonia* 13, 57–59.

Holmes, N. T. H., 1983. *Typing British Rivers according to their Flora.* Focus on Nature Conservation no. 4. Shrewsbury: Nature Conservancy Council.

Holmes, N. T. H. & Whitton, B. A., 1977a. Macrophytes of the River Wear: 1966–1976. *Naturalist, Hull* 102, 53–73.

Holmes, N. T. H. & Whitton, B. A., 1977b. The macrophytic vegetation of the River Tees in 1975: observed and predicted changes. *Freshwater Biology* 7, 43–60.

Holmes, N. T. H. & Whitton, B. A., 1977c. Macrophytic vegetation of the River Swale, Yorkshire. *Freshwater Biology* 7, 545–558.

Hong, D.-Y., 1991. A biosystematic study on *Ranunculus* subgenus *Batrachium* in S. Sweden. *Nordic Journal of Botany* 11, 41–59.

Hooker, J. D., 1884. *The Student's Flora of the British Islands*, edn 3. London: Macmillan & Co.

Hotta, M., 1971. Study of the family Araceae: general remarks. *Japanese Journal of Botany* 20, 269–310.

Howard, H. W. & Lyon, A. G., 1950. The identification and distribution of the British watercress species. *Watsonia* 1, 228–233.

Howard, H. W. & Lyon, A. G., 1951. Distribution of the British watercress species. *Watsonia* 2, 91–92.

Howard, H. W. & Lyon, A. G., 1952. Biological Flora of the British Isles. *Nasturtium* R. Br. *Journal of Ecology* 40, 228–245.

Howard, H. W. & Manton, I., 1946. Autopolyploid and allopolyploid watercress with the description of a new species. *Annals of Botany* new series 10, 1–13.

Hroudová, Z., 1989. Growth of *Butomus umbellatus* at a stable water level. *Folia Geobotanica et Phytotaxonomica* 24, 371–385.

Hroudová, Z., Hrouda, L., Zákravský, P. & Ostrý, I., 1988. Ecobiology and distribution of *Sagittaria sagittifolia* L. in Czechoslovakia.

Folia Geobotanica et Phytotaxonomica 23, 337–373.

Hroudová, Z. & Zákravský, P., 1993a. Ecology of two cytotypes of *Butomus umbellatus* II. Reproduction, growth and biomass production. *Folia Geobotanica et Phytotaxonomica* 28, 413–424.

Hroudová, Z. & Zákravský, P., 1993b. Ecology of two cytotypes of *Butomus umbellatus* III. Distribution and habitat differentiation in the Czech and Slovak Republics. *Folia Geobotanica et Phytotaxonomica* 28, 425–435.

Hroudová, Z., Zákravský, P., Hrouda, L. & Ostrý, I., 1992. *Oenanthe aquatica* (L.) Poir.: seed reproduction, population structure, habitat conditions and distribution in Czechoslovakia. *Folia Geobotanica et Phytotaxonomica* 27, 301–355.

Hubbard, C. E., 1942. The collective species *Glyceria fluitans. Journal of Ecology* 30, 233.

Hull, P. & Nicholl, M. J., 1982. Hybridization between *Rumex aquaticus* L. and *Rumex obtusifolius* L. in Britain. *Annals of Botany* new series 49, 127–129.

Hultén, E., 1958. The amphi-atlantic plants and their phytogeographical connections. *Kunglinga svenska Vetenskapsakademiens Handlingar, Fjärde Serien*, 7, 1–340.

Hultén, E. & Fries, M., 1986. *Atlas of North European Vascular Plants North of the Tropic of Cancer.* 3 vols. Königstein: Koeltz Scientific Books.

Hultgren, A. B. C., 1988. A demographic study of aerial shoots of *Carex rostrata* in relation to water level. *Aquatic Botany* 30, 81–93.

Idle, E. T., 1968. *Rumex aquaticus* L. at Loch Lomondside. *Transactions of the Botanical Society of Edinburgh* 40, 445–449.

Idle, E. T., Mitchell, J. & Stirling, A. McG., 1970. *Elatine hydropiper* L. – new to Scotland. *Watsonia* 8, 45–46.

Ingram, R. & Noltie, H. J., 1981. *The Flora of Angus (Forfar, v.c. 90).* Dundee: Dundee Museums & Art Galleries.

Jacobs, D. L., 1947. An ecological life-history of *Spirodela polyrhiza* (Greater Duckweed) with emphasis on the turion phase. *Ecological Monographs* 17, 437–469.

Jackson, A. B. & Domin, K., 1908. *S. lacustris × triqueter (S. carinatus, Sm.). Report of the Botanical Exchange Club of the British Isles* 2, 314–316.

Jalas, J. & Suominen, J. (eds), 1972. *Atlas Florae Europaeae*, 1. *Pteridophyta (Psilotaceae to Azollaceae).* Helsinki: Committee for Mapping the Flora of Europe & Societas Biologica Fennica Vanamo.

Jalas, J. & Suominen, J. (eds), 1979. *Atlas Florae Europaeae*, 4. *Polygonaceae.* Helsinki: Committee for Mapping the Flora of Europe & Societas Biologica Fennica Vanamo.

Jalas, J. & Suominen, J. (eds), 1989. *Atlas Florae Europaeae*, 8. *Nymphaeaceae to Ranunculaceae*. Helsinki: Committee for Mapping the Flora of Europe & Societas Biologica Fennica Vanamo.

Jalas, J. & Suominen, J. (eds), 1994. *Atlas Florae Europaeae*, 10. *Cruciferae (Sisymbrium to Aubrieta)*. Helsinki: Committee for Mapping the Flora of Europe & Societas Biologica Fennica Vanamo.

Jans, A., 1989. Broedknoppen in de bloeiwijze van *Butomus umbellatus* L. *Dumortiera* 45, 18–19.

Jermy, A. C., 1967. *Carex* section *Carex (=Acutae* Fr.). *Proceedings of the Botanical Society of the British Isles* 6, 375–379.

Jermy, A. C., 1984. *Pilularia* survey. *BSBI News* no. 38, 22.

Jermy, A. C. & Crabbe, J. A. (eds), 1978. *The Island of Mull: a Survey of its Flora and Environment*. London: British Museum (Natural History).

Jermy, A. C. & Tutin, T. G., 1968. *British Sedges*. [BSBI Handbook no. 1.] London: Botanical Society of the British Isles.

Jermy, A. C., Arnold, H .R., Farrell, L. & Perring, F. H., 1978. *Atlas of Ferns of the British Isles*. London: Botanical Society of the British Isles & British Pteridological Society.

Jermy, A. C., Chater, A. O. & David, R. W., 1982. *Sedges of the British Isles*. BSBI Handbook no. 1, edn 2. London: Botanical Society of the British Isles.

Jermyn, S. T., 1974. *Flora of Essex*. Fingringhoe: Essex Naturalists' Trust.

Jessen, K., Andersen, S. T. & Farrington, A., 1959. The interglacial deposit near Gort, Co. Galway, Ireland. *Proceedings of the Royal Irish Academy* 60B, 2–77.

Johnston, G., 1853. *The Botany of the Eastern Borders*. London: John van Voorst.

Jonsell, B., 1968. Studies in the north-west European species of *Rorippa* s. str. *Symbolae Botanicae Upsalienses* 19(2), 1–222.

Jonsell, B., 1975. Hybridization in yellow-flowered European *Rorippa* species. *In*: S.M. Walters (ed.), *European Floristic and Taxonomic Studies*, pp. 101–110. Faringdon: E.W. Classey.

Jovet, P. & Jovet-Ast, S., 1966. *Lemna valdiviana* Philippi, espèce signalée pour la première fois en Europe. *Bulletin du Centre d'Études et de Recherches Scientifiques, Biarritz* 6, 57–64.

Jovet, P. & Jovet-Ast, S., 1967. Floraison, fructification, germination du *Lemna valdiviana* au Lac Marion (B.P.). *Bulletin du Centre d'Études et de Recherches Scientifiques, Biarritz* 6, 729–734.

Jupp, B. P. & Spence, D. H. N., 1977. Limitations on macrophytes in a eutrophic lake, Loch Leven. I. Effects of phytoplankton. *Journal of Ecology* 65, 175–186.

Justice, O. L., 1941. *A Study of Dormancy in Seeds of* Polygonum. Cornell University Agricultural Experiment Station Memoir no. 235. Ithaca: Cornell University.

Karagatzides, J. D. & Hutchinson, I., 1991. Intraspecific comparisons of biomass dynamics in *Scirpus americanus* and *Scirpus maritimus* on the Fraser River Delta. *Journal of Ecology* 79, 459–476.

Kaul, R. B., 1985. Reproductive phenology and biology in annual and perennial Alismataceae. *Aquatic Botany* 22, 153–164.

Keble Martin, W. & Fraser, G. T., 1939. *Flora of Devon*. Arbroath: T. Buncle & Co.

Keeley, J. E., 1982. Distribution of diurnal acid metabolism in the genus *Isoetes*. *American Journal of Botany* 69, 254–257.

Keeley, J. E. & Morton, B. A., 1982. Distribution of diurnal acid metabolism in submerged aquatic plants outside the genus *Isoetes*. *Photosynthetica* 16, 546–553.

Kendall, S., 1987. *Botanical Survey of the Leven Canal (with Notes on Hydrology)*. Unpublished report, Nature Conservancy Council.

Kent, D. H., 1955a. *Egeria densa* Planch. *Proceedings of the Botanical Society of the British Isles* 1, 322.

Kent, D. H., 1955b. *Lagarosiphon major* (Ridley) C. E. Moss. *Proceedings of the Botanical Society of the British Isles* 1, 322–323.

Kent, D. H., 1975. *The Historical Flora of Middlesex*. London: Ray Society.

Kent, D. H., 1992. *List of Vascular Plants of the British Isles*. London: Botanical Society of the British Isles.

Kerner von Marilaun, A., 1894. *The Natural History of Plants*, translated by F. W. Oliver. 2 vols. London: Blackie & Son.

Keynes, G. (ed.), 1964. *The Works of Sir Thomas Browne*, 4. *Letters*. London: Faber & Faber.

Kirby, J. E., 1964. *Crassula helmsii* in Britain. *Cactus and Succulent Journal of Great Britain* 26, 15–16.

Klötzli, F., 1988. Conservational status and use of sedge wetlands. *Aquatic Botany* 30, 157–168.

Knuth, P., 1906–1909. *Handbook of Flower Pollination*, translated by J. R. Ainsworth Davis. 3 vols. Oxford: Clarendon Press.

Korschgen, C. E. & Green, W. L., 1988. *American Wildcelery* (Vallisneria americana): *Ecological Considerations for Restoration*. Fish and Wildlife Technical Report no. 19. Washington: US Department of the Interior.

Kott, L. S. & Britton, D. M., 1982. A comparative study of spore germination of some *Isoetes* species of northeastern North America. *Canadian Journal of Botany* 60, 1679–1687.

Koyama, T., 1958. Taxonomic study of the genus *Scirpus* Linné. *Journal of the Faculty of Science, University of Tokyo*, Section III (Botany), 7, 271–366.

Krahulcová, A. & Jarilímová, V., 1993. Ecology of two cytotypes of *Butomus umbellatus* I.

Karyology and breeding behaviour. *Folia Geobotanica et Phytotaxonomica* **28**, 385–411.

Krattinger, K., 1975. Genetic mobility in *Typha*. *Aquatic Botany* **1**, 57–70.

Krattinger, K., Rast, D. & Karesch, H., 1979. Analysis of pollen proteins of *Typha* species in relation to identification of hybrids. *Biochemical Systematics and Ecology* **7**, 125–128.

Kubitzki, K. (ed.), 1990. *The Families and Genera of Vascular Plants*, **1**. *Pteridophytes and Gymnosperms*. Berlin: Springer-Verlag.

Kubitzki, K. (ed.), 1993. *The Families and Genera of Vascular Plants*, **2**. *Flowering Plants: Dicotyledons: Magnoliid, Hamamelid and Caryophyllid Families*. Berlin: Springer-Verlag.

Kukkonen, I. & Toivonen, H., 1988. Taxonomy of wetland carices. *Aquatic Botany* **30**, 5–22.

Lambert, J. M., 1946. The distribution and status of *Glyceria maxima* (Hartm.) Holmb. in the region of Surlingham and Rockland Broads, Norfolk. *Journal of Ecology* **33**, 230–267.

Lambert, J. M., 1947a. A note on the physiognomy of *Glyceria maxima* reedswamps in Norfolk. *Transactions of the Norfolk and Norwich Naturalists' Society* **16**, 246–259.

Lambert, J. M., 1947b. *Glyceria maxima* (Hartm.) Holmberg – effect of drought on flowering. *Report of the Botanical Society and Exchange Club of the British Isles* **13**, 41.

Lambert, J. M., 1947c. Biological Flora of the British Isles. *Glyceria maxima* (Hartm.) Holmb. *Journal of Ecology* **34**, 310–344.

Lambert, J. M., 1948. A survey of the Rockland–Claxton level, Norfolk. *Journal of Ecology* **36**, 120–135.

Lambert, J. M., 1951. Alluvial stratigraphy and vegetational succession in the region of the Bure Valley Broads III. Classification, status and distribution of communities. *Journal of Ecology* **39**, 149–170.

Lambert, J. M. & Jennings, J. N., 1951. Alluvial stratigraphy and vegetational succession in the region of the Bure Valley Broads II. Detailed vegetational-stratigraphical relationships. *Journal of Ecology* **39**, 120–148.

Landolt, E., 1979. *Lemna minuscula* Herter (= *L. minima* Phil.), eine in Europa neu eingebürgerte amerikanische Wasserpflanze. *Berichte des Geobotanischen Instituts der Eidg. Techn. Hochschule, Stiftung Rübel, in Zurich* **46**, 86–89.

Landolt, E., ed. 1980. Biosystematic investigations in the family of duckweeds (Lemnaceae), **1**. *Veröffentlichungen des Geobotanischen Instituts der Eidg. Techn. Hochschule, Stiftung Rübel, in Zurich* **70**. 247pp.

Landolt, E., 1986. Biosystematic investigations in the family of duckweeds (Lemnaceae), **2**. The family of Lemnaceae – a monographic study. Volume 1. *Veröffentlichungen des Geobotanischen Instituts der Eidg. Techn. Hochschule, Stiftung Rübel, in Zürich* **71**. 566pp.

Landolt, E. & Kandeler, R., 1987. Biosystematic investigations in the family of duckweeds (Lemnaceae), **4**. The family of Lemnaceae – a monographic study. Volume 2. *Veröffentlichungen des Geobotanischen Instituts der Eidg. Techn. Hochschule, Stiftung Rübel, in Zürich* **95**. 638pp.

Lange, L. de, Pieterse, A. H. & Wetsteyn, L. P. M. J., 1984. On the occurrence of the flat form of *Lemna gibba* L. in nature. *Acta Botanica Neerlandica* **33**, 469–474.

Lange, L. de & Westinga, E., 1979. The distinction between *Lemna gibba* L. and *Lemna minor* L. on the basis of vegetative characters. *Acta Botanica Neerlandica* **28**, 169–176.

Last, B., 1990. *Lemna minuscula* in Wiltshire. *BSBI News* no. **56**, 9–10.

Laundon, J. R., 1961. An Australasian species of *Crassula* introduced into Britain. *Watsonia* **5**, 59–63.

Lavin, J. C. & Wilmore, G. T. D., 1994. *The West Yorkshire Plant Atlas*. Bradford: City of Bradford Metropolitan Council.

Le Sueur, F., 1984. *Flora of Jersey*. Jersey: Société Jersiaise.

Leach, S. J., Wolfe Murphy, S. A. & McMullin, A. S., 1987. *Glyceria declinata* Breb. in Northern Ireland. *Irish Naturalists' Journal* **22**, 261–262.

Lee, D. W., 1975. Population variation and introgression in North American *Typha*. *Taxon* **24**, 633–641.

Lee, J. A., 1977. The vegetation of British inland salt marshes. *Journal of Ecology* **65**, 673–698.

Leereveld, H., Meeuse, A. D. J. & Stelleman, P., 1981. Anthecological relations between reputedly anemophilous flowers and syrphid flies. IV. A note on the anthecology of *Scirpus maritimus* L. *Acta Botanica Neerlandica* **30**, 465–473.

Les, D. H., 1986. The evolution of achene morphology in *Ceratophyllum* (Ceratophyllaceae), I. Fruit-spine variation and relationships of *C. demersum*, *C. submersum*, and *C. apiculatum*. *Systematic Botany* **11**, 549–558.

Les, D. H., 1988a. The origin and affinities of the Ceratophyllaceae. *Taxon* **37**, 326–345.

Les, D. H., 1988b. The evolution of achene morphology in *Ceratophyllum* (Ceratophyllaceae), II. Fruit variation and systematics of the 'spiny-margined' group. *Systematic Botany* **13**, 73–86.

Les, D. H., 1988c. The evolution of achene morphology in *Ceratophyllum* (Ceratophyllaceae), III. Relationships of the 'facially-spined' group. *Systematic Botany* **13**, 509–518.

Les, D. H., 1989. The evolution of achene morphology in *Ceratophyllum* (Ceratophyllaceae), IV. Summary of proposed relationships and evolutionary trends. *Systematic Botany* **14**, 254–262.

Les, D. H., 1991. Genetic diversity in the monoecious hydrophile *Ceratophyllum* (Ceratophyllaceae). *American Journal of Botany* **78**, 1070–1082.

Les, D. H. & Stuckey, R. L., 1985. The introduction and spread of *Veronica beccabunga* (Scrophulariaceae) in Eastern North America. *Rhodora* 87, 503–515.

Les, D. H., Garvin, D. K. & Wimpee, C. F., 1993. Phylogenetic studies in the monocot subclass Alismatidae: evidence for a reappraisal of the aquatic order Najadales. *Molecular Phylogenetics and Evolution* 2, 304–314.

Leslie, A. C., 1987. *Flora of Surrey Supplement and Checklist*. Guildford: A.C. & P. Leslie.

Leslie, A. C., 1992. *Iris pseudacorus*. *The Garden* 1992 (May), 204–206.

Leslie, A. C. & Walters, S. M., 1983. The occurrence of *Lemna minuscula* Herter in the British Isles. *Watsonia* 14, 243–248.

Lester-Garland, L. V., 1903. *A Flora of the Island of Jersey*. London: West, Newman & Co.

Lewis, K. R. & John, B., 1961. Hybridisation in a wild population of *Eleocharis palustris*. *Chromosoma* 12, 433–448.

Lewis-Jones, L. J. & Kay, Q. O. N., 1977. The cytotaxonomy and distribution of water starworts (*Callitriche* spp) in West Glamorgan. *Nature in Wales* 15, 180–183.

Leyshon, O. J. & Moore, N. W., 1993. A note on the British Dragonfly Society's survey of *Anaciaeschna isosceles* at Castle Marshes, Barnby, Suffolk, 1991–1992. *Journal of the British Dragonfly Society* 9, 5–9.

Libbey, R. P. & Swann, E. L., 1973. *Alisma gramineum* Lej.: a new county record. *Nature in Cambridgeshire* no. 16, 39–41.

Lightcap, B. W. & Schuyler, A. E., 1984. *Scirpus triqueter* established along tidal portions of the Columbia River system. *Bartonia* 50, 23–24.

Lightfoot, J., 1777. *Flora Scotica*. 2 vols. London.

Lindblad, R. & Ståhl, B., 1990. Revsvalting, *Baldellia repens*, i Norden. [*Baldellia repens* (Alismataceae) in the Nordic countries.] *Svensk Botanisk Tidskrift* 84, 253–258.

Livermore, L. A. & Livermore, P. D., [1988]. *The Flowering Plants, Ferns & Rusts of the Lancaster Canal in the Lancaster District*. Unpublished report.

Lloyd, F. E., 1942. *The Carnivorous Plants*. Waltham: Chronica Botanica.

Lockton, A. J. & Whild, S. J., 1995. *Rare Plants of Shropshire*. Shrewsbury: Shrewsbury Museums Service.

Lohammar, G., 1954. Bulbils in the inflorescences of *Butomus umbellatus*. *Svensk Botanisk Tidskrift* 48, 485–488.

Lousley, J. E., 1931. The *Schoenoplectus* group of the genus *Scirpus* in Britain. *Journal of Botany* 69, 151–163.

Lousley, J. E., 1937. *Callitriche truncata* Gussone, var. *occidentalis* Rouy. *Report of the Botanical Society and Exchange Club of the Britsh Isles* 11, 400–401.

Lousley, J. E., 1939. *Rumex aquaticus* L. as a British plant. *Journal of Botany* 77, 149–152.

Lousley, J. E., 1944a. *Potamogeton nodosus* Poir. *Report of the Botanical Society and Exchange Club of the British Isles* 12, 507.

Lousley, J. E., 1944b. Notes on British Rumices: II. *Report of the Botanical Society and Exchange Club of the British Isles* 12, 547–585.

Lousley, J. E., 1947. *Typha angustifolia* L. × *latifolia* L. *Report of the Botanical Society and Exchange Club of the British Isles* 13, 174.

Lousley, J. E., 1957. *Alisma gramineum* in Britain. *Proceedings of the Botanical Society of the British Isles* 2, 346–353.

Lousley, J. E., 1971. *Flora of the Isles of Scilly*. Newton Abbot: David & Charles.

Lousley, J. E., 1975. *Scirpus* L. (*Schoenoplectus* (Reichb.) Palla). *In*: C. A. Stace (ed.), *Hybridization and the Flora of the British Isles*, pp. 510–512. London: Academic Press.

Lousley, J. E., 1976. *Flora of Surrey*. Newton Abbot: David & Charles.

Lousley, J. E. & Kent, D. H., 1981. *Docks and Knotweeds of the British Isles*. BSBI Handbook no. 3. London: Botanical Society of the British Isles.

Löve, Á. & Löve, D., 1958. The American element in the flora of the British Isles. *Botaniska Notiser* 111, 376–388.

Lowden, R. M., 1982. An approach to the taxonomy of *Vallisneria* L. (Hydrocharitaceae). *Aquatic Botany* 13, 269–298.

Lowden, R. M., 1986. Taxonomy of the genus *Najas* L. (Najadaceae) in the Neotropics. *Aquatic Botany* 24, 147–184.

Lund, J. W. G., 1979. The mystery of *Elodea* Michx in Britain. *Watsonia* 12, 388.

Lüönd, A., 1983. Biosystematic investigations in the family of duckweeds (Lemnaceae), 3. Das Wachstum von Wasserlinsen (Lemnaceae) in Abhängigkeit des Nährstoffangebots, insbesondere Phosphor und Stickstoff. *Veröffentlichungen des Geobotanischen Instituts der Eidg. Techn. Hochschule, Stiftung Rübel, in Zürich* 80. 116pp.

Mabberley, D. J., 1987. *The Plant-book*. Cambridge: Cambridge University Press.

Mabbott, D. C., 1920. Food habits of seven species of American shoal-water ducks. *United States Department of Agriculture Bulletin* no. 862.

Macan, T. T., 1977. Changes in the vegetation of a moorland fishpond in twenty-one years. *Journal of Ecology* 65, 95–106.

McCallum Webster, M., 1978. *Flora of Moray, Nairn & East Inverness*. Aberdeen: Aberdeen University Press.

McClintock, D., 1968. *Eriocaulon aquaticum* (*E. septangulare*) in Ardnamurchan. *Proceedings of the Botanical Society of the British Isles* 7, 509.

McClintock, D., 1975. *The Wild Flowers of Guernsey*. London: Collins.

McClintock, K. A. & Waterway, M. J., 1993. Patterns of allozyme variation and clonal diversity in *Carex lasiocarpa* and *C. pellita* (Cyperaceae). *American Journal of Botany* 80, 1251–1263.

McConchie, C. A. & Kadereit, J. W., 1987. Floral structure of *Vallisneria caulescens* Bailey & F. Mueller. *Aquatic Botany* 29, 101–110.

McGough, H. N., 1984. *A Report on the Vegetation of the Nanny Subcatchment of the Corrib/Clare Drainage Catchment*. Wildlife Service unpublished report.

McIlraith, A. L., Robinson, G. G. C. & Shay, J. M., 1989. A field study of competition and interaction between *Lemna minor* and *Lemna trisulca*. *Canadian Journal of Botany* 67, 2904–2911.

McVean, D. N. & Ratcliffe, D. A., 1962. *Plant Communities of the Scottish Highlands*. Monographs of the Nature Conservancy no. 1. London: Her Majesty's Stationery Office.

Maitland, P. S., 1971. A population of coloured goldfish, *Carassius auratus* in the Forth and Clyde Canal. *Glasgow Naturalist* 18, 565–568.

Mansel-Pleydell, J. C., 1895. *The Flora of Dorsetshire*, edn 2. Dorchester: privately published.

Manton, I., 1932. Introduction to the general cytology of the Cruciferae. *Annals of Botany* 46, 509–556.

Manton, I., 1935. The cytological history of Watercress (*Nasturtium officinale* R. Br.). *Zeitschrift für induktive Abstammungs- u. Vererbungslehre* 69, 132–157.

Manton, I., 1950. *Problems of Cytology and Evolution in the Pteridophyta*. Cambridge: Cambridge University Press.

Marchant, C. J., 1973. Chromosome variation in Araceae: V. Acoreae to Lasieae. *Kew Bulletin* 28, 199–210.

Marchant, N. G., 1970. *Experimental Taxonomy of Veronica section Beccabungae Griseb*. PhD thesis, University of Cambridge.

Margetts, L. J. & David, R. W., 1981. *A Review of the Cornish Flora 1980*. Redruth: Institute of Cornish Studies.

Marquand, E. D., 1901. *Flora of Guernsey and the Lesser Channel Islands*. London: Dulau & Co.

Marren, P. & Rich, T., 1993. Back from the brink – conserving our rarest flowering plants. *British Wildlife* 4, 296–304.

Marsh, A. S., 1914. *Azolla* in Britain and in Europe. *Journal of Botany* 52, 209–213.

Marshall, E. S., 1889. Notes on Highland plants. *Journal of Botany* 27, 229–236.

Marshall, E. S., 1892. On an apparently endemic British *Ranunculus*. *Journal of Botany* 30, 289–290.

Marshall, E. S., 1898. *Ranunculus petiolaris*. *Journal of Botany* 36, 103.

Marshall, W., 1852. Excessive and noxious increase of *Udora canadensis* (*Anacharis alsinastrum*). *Phytologist* 4, 705–715.

Marshall, W., 1857. The American water-weed *Anacharis alsinastrum*. *Phytologist* new series 2, 194–197.

Martin, A. C., Zim, H. S. & Nelson, A. L., 1961. *American Wildlife & Plants*. New York: Dover Publications.

Martinsson, K., 1984. Blomningen hos *Lemna minor*, andmat. [The flowering of *Lemna minor*.] *Svensk Botanisk Tidskrift* 78, 9–15.

Martinsson, K., 1991. Geographical variation in fruit morphology in Swedish *Callitriche hermaphroditica* (Callitrichaceae). *Nordic Journal of Botany* 11, 497–512.

Mason, H. L., 1957. *A Flora of the Marshes of California*. Berkeley & Los Angeles: University of California Press.

Mason, R., 1960. Three waterweeds of the family Hydrocharitaceae in New Zealand. *New Zealand Journal of Science* 3, 382–395.

Mathew, B., 1981. *The Iris*. London: B.T. Batsford.

Meikle, R. D. & Sandwith, N. Y., 1956. *Callitriche palustris* L. *Proceedings of the Botanical Society of the British Isles* 2, 135–136.

Mériaux, J.-L., 1981. Espèces rares ou menacées des biotopes lacustres et fluviatiles du nord-ouest de la France (Ptéridophytes et Spermatophytes). *Natura Mosana* 34, 177–194.

Meusel, H. & Jäger, E., 1992. *Vergleichende Chorologie der Zentraleuropäischen Flora*, 3. 2 vols. Jena: Gustav Fischer.

Meusel, H., Jäger, E. & Weinert, E., 1965. *Vergleichende Chorologie der Zentraleuropäischen Flora*, 1. 2 vols. Jena: Gustav Fischer.

Meusel, H., Jäger, E., Rauschert, S. & Weinert, E., 1978. *Vergleichende Chorologie der Zentraleuropäischen Flora*, 2. 2 vols. Jena: Gustav Fischer.

Meyers, D. G., 1982. Darwin's investigations of carnivorous aquatic plants of the genus *Utricularia*: misconception, contribution, and controversy. *Proceedings of the Academy of Natural Sciences of Philadelphia* 134, 1–11.

Milner, J. M., 1979. *Myriophyllum aquaticum* (Velloso) Verdc. in East Sussex. *Watsonia* 12, 259.

Mitchell, J., 1981. *Elatine hydropiper* at Kilmannan Reservoir. *Glasgow Naturalist* 20, 185–186.

Mitchell, J., 1982. A new colony of *Rumex aquaticus* L. *Glasgow Naturalist* 20, 260.

Mitchell, J., 1983. The dock of Loch Lomond. *Living Countryside* 11, 2488–2489.

Mitchell, J., 1993. *Loch Lomond National Nature Reserve. The Reserve Record – Pt II The Flora of the Reserve*. Scottish Natural Heritage Research, Survey and Monitoring Report. Scottish Natural Heritage.

Morin, N. R. (ed.), 1993. *Flora of North America*, 2. *Pteridophytes and Gymnosperms*. New York: Oxford University Press.

Morton, J. D., 1966. Water Plantain (*Alisma plantago-aquatica*): opening of flowers. *Glasgow Naturalist* 18, 454.

Mountford, J. O., 1994a. Floristic change in English grazing marshes: the impact of 150 years of drainage and land-use change. *Watsonia* 20, 3–24.

Mountford, J. O., 1994b. Problems in reconstructing floristic change: interpreting the sources for English grazing marshes. *Watsonia* 20, 25–31.

Mountford, J. O. & Sheail, J., 1989. *The Effects of Agricultural Land Use Change on the Flora of Three Grazing Marsh Areas*. Focus on Nature Conservation no. 20. Peterborough: Nature Conservancy Council.

Müller-Doblies, D., 1970. Über die Verwandtschaft von *Typha* und *Sparganium* im Infloreszenz- und Blütenbau. *Botanische Jahrbücher* 89, 451–562.

Murphy, K. J. & Eaton, J. W., 1983. Effects of pleasure-boat traffic on macrophyte growth in canals. *Journal of Applied Ecology* 20, 713–729.

Murray, C. W., 1980. *The Botanist in Skye*, edn 2. Botanical Society of the British Isles.

Murrell, G. & Sell, P., 1990. *Ranunculus calcareus* and *R. pseudofluitans* in Cambridgeshire. *Nature in Cambridgeshire* no. 32, 14–16.

Muscott, J., 1983. *Azolla filiculoides* in Edinburgh. *Fern Gazette* 12, 301.

Naustdal, J., 1974. *Callitriche pedunculata* i Noreg. *Blyttia* 32, 15–19.

Newton, A., 1971. *Flora of Cheshire*. Chester: Cheshire Community Council.

Newton, A., [1991]. *Supplement to Flora of Cheshire*. Leamington Spa: privately published.

Nic Lughadha, E. M. & Parnell, J. A. N., 1989. Heterostyly and gene-flow in *Menyanthes trifoliata* L. (Menyanthaceae). *Botanical Journal of the Linnean Society* 100, 337–354.

Nilsson, Ö. & Gustafsson, L.-Å., 1978. Projekt Linné rapporterar 64–79. *Svensk Botanisk Tidskrift* 72, 1–24.

Norman, E. G., 1978. Spineless-fruited *Ceratophyllum*. *Watsonia* 12, 50–51.

Obermeyer, A. A., 1964. The South African species of *Lagarosiphon*. *Bothalia* 8, 139–146.

Odell, T. W., 1886. *Azolla caroliniana*. *Hardwicke's Science-Gossip* 19, 279.

Olesen, I. & Warncke, E., 1992. Breeding system and seasonal variation in seed set in a population of *Potentilla palustris*. *Nordic Journal of Botany* 12, 373–380.

Oliver, D., 1851. Notes of a botanical ramble in Ireland last autumn. *Phytologist* 4, 125–128.

Oliver, J. E., 1991. Spread of *Lemna minuscula* in Wilts. –2. *BSBI News* no. 58, 10.

Olney, P. J. S., 1963. The food and feeding habits of teal *Anas crecca crecca* L. *Proceedings of the Zoological Society of London* 140, 169–210.

Olney, P. J. S., 1968. The food and feeding-habits of the pochard, *Aythya ferina*. *Biological Conservation* 1, 71–76.

Orchard, A. E., 1980. *Callitriche* (Callitrichaceae) in South Australia. *Journal of the Adelaide Botanic Garden* 2, 191–194.

Orchard, A. E., 1981. A revision of South American *Myriophyllum* (Haloragaceae), and its repercussions on some Australian and North American species. *Brunonia* 4, 27–65.

Orchard, A. E., 1986. *Myriophyllum* (Haloragaceae) in Australasia. II. The Australian species. *Brunonia* 8, 173–291.

Ørgaard, M., 1991. The genus *Cabomba* (Cabombaceae) – a taxonomic study. *Nordic Journal of Botany* 11, 179–203.

Ørnduff, R., 1966. The origin of dioecism from heterostyly in *Nymphoides* (Menyanthaceae). *Evolution* 20, 309–314.

Ostendorp, W., 1989. 'Die-back' of reeds in Europe – a critical review of literature. *Aquatic Botany* 35, 5–26.

Otzen, D., 1962. Chromosome studies in the genus *Scirpus* L., section *Schoenoplectus* Benth. et Hook., in the Netherlands. *Acta Botanica Neerlandica* 11, 37–46.

Ozimek, T. & Klekot, L., 1979. *Glyceria maxima* (Hartm.) Holmb. in ponds supplied with post-sewage water. *Aquatic Botany* 7, 231–239.

Öztürk, A. & Fischer, M. A., 1982. Karyosystematics of *Veronica* sect. *Beccabunga* (Scrophulariaceae) with special reference to the taxa in Turkey. *Plant Systematics and Evolution* 140, 307–319.

Packer, J. G. & Ringius, G. S., 1984. The distribution and status of *Acorus* (Araceae) in Canada. *Canadian Journal of Botany* 62, 2248–2252.

Packham, J. R. & Liddle, M. J., 1970. The Cefni salt marsh, Anglesey, and its recent development. *Field Studies* 3, 331–356.

Padmore, P. A., 1957. The varieties of *Ranunculus flammula* L. and the status of *R. scoticus* E. S. Marshall and of *R. reptans* L. *Watsonia* 4, 19–27.

Page, C. N., 1982. *The Ferns of Britain and Ireland*. Cambridge: Cambridge University Press.

Page, C. N., 1988. *Ferns: their Habitats in the British and Irish Landscape*. London: Collins.

Page, C. N. & Barker, M. A., 1985. Ecology and geography of hybridisation in British and Irish horsetails. *Proceedings of the Royal Society of Edinburgh* 86B, 265–272.

Paley, W., 1802. *Natural Theology*. London.

Palmer, M. [A.], 1986. The impact of a change from permanent pasture to cereal farming on the flora and invertebrate fauna of watercourses in the Pevensey Levels, Sussex. *Proceedings EWRS/AAB 7th symposium on aquatic weeds, 1986*.

Palmer, M. [A.], 1991. *A summary of the work of the England Field Unit 1979 to 1991*. Peterborough: Nature Conservancy Council.

Palmer, M. A., Bell, S. L. & Butterfield, I., 1992. A botanical classification of standing waters in

Britain: applications for conservation and monitoring. *Aquatic Conservation: Marine and Freshwater Ecosystems* **2**, 125–143.

Palmer, M. A. & Bratton, J. H. (eds), 1995. *A Sample Survey of the Flora of Britain and Ireland.* UK Nature Conservation no. 8. Peterborough: Joint Nature Conservation Committee.

Patten, B. C., 1955. Germination of the seed of *Myriophyllum spicatum* L. *Bulletin of the Torrey Botanical Club* **82**, 50–56.

Payne, R., 1989. Frogbit *Hydrocharis morsus-ranae* at the Nene Washes in 1988. *Nature in Cambridgeshire* no. **31**, 31.

Payne, R. G., 1994. A new waterweed in Essex. *Essex Field Club Newsletter* no. **9**, 2–3.

Pearman, D., 1994. *Sedges and Their Allies in Dorset.* Dorchester: Dorset Environmental Records Centre.

Pearman, D., 1995. Plant records & notes for 1994. *Recording Dorset* no. **5**, 24–29.

Pearsall, W. H., 1915. *Naias flexilis* R. & S. *Report of the Botanical Society and Exchange Club of the British Isles* **4**, 167.

Pearsall, W. H., 1921. *Hydrilla verticillata* Casp. *Annual Report of the Watson Botanical Exchange Club* **3**, 149.

Pearsall, W. H., 1931. Notes on *Potamogeton*. II. The larger British species. *Report of the Botanical Society and Exchange Club of the British Isles* **9**, 380–415.

Pearsall, W. H., 1933. *Zannichellia*. *Report of the Botanical Society and Exchange Club of the British Isles* **10**, 235–241.

Pearsall, W. H., 1935. The British species of *Callitriche*. *Report of the Botanical Society and Exchange Club of the British Isles* **10**, 861–871.

Pearsall, W. H. [fil.], 1936. *Hydrilla verticillata* Presl, a plant new to Ireland. *Irish Naturalists' Journal* **6**, 20–21.

Peck, J. H., 1980. *Equisetum* × *litorale* in Illinois, Iowa, Minnesota, and Wisconsin. *American Fern Journal* **70**, 33–38.

Pedersen, A., 1976. Najadaceernes, Potamogeton-aceernes, Ruppiaceernes, Zannichelliaceernes og Zosteraceernes udbredelse i Danmark. *Botanisk Tidsskrift* **70**, 205–262.

Percy, A. A., 1964. "Narrow Water-Plantain" (*Alisma lanceolatum*) in Lanarkshire. *Glasgow Naturalist* **18**, 383.

Perring, F. H., 1963. The Irish problem. *Proceedings of the Bournemouth Natural Science Society* **52**, 36–48.

Perring, F. H. & Farrell, L., 1983. *British Red Data Books: 1. Vascular Plants,* edn 2. Lincoln: Royal Society for Nature Conservation.

Perring, F. H. & Sell, P. D., 1967. *Catabrosa aquatica* (L.) Beauv. *Watsonia* **6**, 317–318.

Perring, F. H. & Sell, P. D., 1968. *Critical Supplement to the Atlas of the British Flora.* London: Thomas Nelson & Sons.

Perring, F. H. & Walters, S. M. (eds), 1962. *Atlas of the British Flora.* London: Thomas Nelson & Sons.

Perring, F. H., Sell, P. D. & Walters, S. M., 1964. *A Flora of Cambridgeshire.* Cambridge: Cambridge University Press.

Philbrick, C. T., 1983. Aspects of floral biology in three species of *Potamogeton* (pondweeds). *Michigan Botanist* **23**, 35–38.

Philbrick, C. T. & Anderson, G. J., 1992. Pollination biology in the Callitrichaceae. *Systematic Botany* **17**, 282–292.

Philp, E. G., 1982. *Atlas of the Kent Flora.* Kent Field Club.

Philp, E. G., 1992. Notes from Maidstone Museum. *Bulletin of the Kent Field Club* no. **37**, 37–38.

Pieterse, A. H., de Lange, L. & Van Vliet, J. P., 1977. A comparative study of *Azolla* in the Netherlands. *Acta Botanica Neerlandica* **26**, 433–449.

Pietsch, W., 1981. Zur Bioindikation *Najas marina* s.l. – und *Hydrilla verticillata* (L. fil.) Royle – reicher Gewässer Mitteleuropas. *Feddes Repertorium* **92**, 125–174.

Pinkess, L. H., 1980. The possibility of pollination of *Sparganium erectum* by insects. *Proceedings of the Birmingham Natural History Society* **24**, 101–102.

Pizarro, J., 1995. Contribución al estudio taxonómico de *Ranunculus* L. subgen. *Batrachium* (DC.) A. Gray (Ranunculaceae). *Lazaroa* **15**, 21–113.

Pogan, E., 1971. Karyological studies in a natural hybrid of *Alisma lanceolatum* With. × *Alisma plantago-aquatica* L. and its progeny. *Genetica Polonica* **12**, 219–222.

Poore, M. E. D. & Walker, D., 1959. Wybunbury Moss, Cheshire. *Memoirs and Proceedings of the Manchester Literary and Philosophical Society* **101**, 72–95.

Praeger, R. L., 1901. Irish Topographical Botany. *Proceedings of the Royal Irish Academy,* 3rd series, **7**, 1–410.

Praeger, R. L., 1913. On the buoyancy of the seeds of some Britannic plants. *Scientific Proceedings of the Royal Dublin Society* new series **14**, 13–62.

Praeger, R. L., 1917. *Equisetum litorale* in Ireland. *Irish Naturalist* **26**, 141–147.

Praeger, R. L., 1934a. *The Botanist in Ireland.* Dublin: Hodges, Figgis & Co.

Praeger, R. L., 1934b. Propagation from aerial shoots in *Equisetum*. *Journal of Botany* **72**, 175–176.

Praeger, R. L., 1938a. *A Flora of the North-east of Ireland by Samuel Alexander Stewart, F.B.S.E., and Thomas Hughes Corry, M.A., F.L.S.,* edn. 2. *Flowering Plants, Vascular Cryptogams and Charophytes.* Belfast: Quota Press.

Praeger, R. L., 1938b. A note on Mr. Pugsley's *Myriophyllum alternifolium* [sic] var. *americanum*. *Journal of Botany* 76, 53–54.

Prankerd, T. L., 1911. On the structure and biology of the genus *Hottonia*. *Annals of Botany* 25, 253–267.

Preston, C. D., 1986. Is *Groenlandia densa* extinct in Scotland? *BSBI Scottish Newsletter* 8, 13–14.

Preston, C. D., 1988. The *Potamogeton* L. taxa described by Alfred Fryer. *Watsonia* 17, 23–35.

Preston, C. D., 1991. *Lemna minuscula* in Cambridge. *Nature in Cambridgeshire* no. 33, 52–54.

Preston, C. D., 1995. *Pondweeds of Great Britain and Ireland*. BSBI Handbook no. 8. London: Botanical Society of the British Isles.

Preston, C. D. & March, M. D., 1996. *Hydrocharis morsus-ranae* L. (Hydrocharitaceae) fruited in Britain in 1995. *Watsonia* 21, 206–208.

Preston, C. D. & Stewart, N. F., 1992. Irish Pondweeds III. *Potamogeton* × *griffithii* A. Benn. in Co Donegal, new to Ireland. *Irish Naturalists' Journal* 24, 143–147.

Preston, C. D. & Stewart, N. F., 1994. Irish Pondweeds V. *Potamogeton* × *suecicus* K. Richter in Co Donegal, new to Ireland. *Irish Naturalists' Journal* 24, 485–489.

Priestley, C. A., 1953. *Alisma plantago-aquatica* and *A. lanceolatum* in the Cambridge district. *Botanical Society of the British Isles Year Book* 1953, 63.

Primavesi, A. L. & Evans, P. A. (eds), 1988. *Flora of Leicestershire*. Leicester: Leicestershire Museums, Art Galleries and Records Service.

Proctor, H. G., 1978. Changes in the large-aquatic flora of the Cow Green basin. *In*: A. R. Clapham (ed.), *Upper Teesdale*, pp. 191–194. London: Collins.

Proctor, M. C. F., 1974. The vegetation of the Malham Tarn fens. *Field Studies* 4, 1–38.

Proctor, M. [C. F.] & Yeo, P. [F.], 1973. *The Pollination of Flowers*. London: Collins.

Pugsley, H. W., 1938. A new variety of *Myriophyllum alterniflorum* DC. *Journal of Botany* 76, 51–53.

Pulteney, R., 1813. *Catalogues of the Birds, Shells and some of the more rare Plants, of Dorsetshire*. London.

Ranwell, D. S., Bird, E. C. F., Hubbard, J. C. E. & Stebbings, R. E., 1964. *Spartina* salt marshes in southern England V. Tidal submergence and chlorinity in Poole Harbour. *Journal of Ecology* 52, 627–641.

Rataj, K., 1972a. Revision of the genus *Sagittaria*. Part I. (Old World species). *Annotationes Zoologicae et Botanicae, Bratislava*, no. 76, 1–31.

Rataj, K., 1972b. Revision of the genus *Sagittaria*. Part II. (The species of West Indies, Central and South America). *Annotationes Zoologicae et Botanicae, Bratislava*, no. 78, 1–61.

Raven, P. H., 1963. The Old World species of *Ludwigia* (including *Jussiaea*), with a synopsis of the genus (Onagraceae). *Reinwardtia* 6, 327–427.

Raven, P. H. & Thomas, J. H., 1970. *Iris pseudacorus* in western North America. *Madroño* 20, 390–391.

Ray, J., 1660. *Catalogus Plantarum circa Cantabrigiam nascentium*. Cambridge.

Ray, J., 1724. *Synopsis Methodica Stirpium Britannicarum*, edn 3. London.

Raymond, M. & Kucyniak, J., 1948. Six additions to the adventitious flora of Quebec. *Rhodora* 50, 176–180.

Reese, G., 1962. Zur intragenerischen Taxonomie der Gattung *Ruppia* L. *Zeitschrift für Botanik* 50, 237–264.

Reese, G., 1963. Über die deutschen *Ruppia*- und *Zannichellia*- Kategorien und ihre Verbreitung in Schleswig-Holstein. *Schriften des Naturwissenschaftlichen Vereins für Schleswig-Holstein* 34, 44–70.

Rich, T. C. G., 1991. *Crucifers of Great Britain and Ireland*. BSBI Handbook no. 6. London: Botanical Society of the British Isles.

Rich, T. C. G., Kay, G. M. & Kirschner, J., 1995. Floating water plantain *Luronium natans* (L.) Raf. (Alismataceae) present in Ireland. *Irish Naturalists' Journal* 25, 140–145.

Rich, T. C. G. & Woodruff, E. R., 1996. Changes in the vascular plant floras of England and Scotland between 1930–1960 and 1987–1988: the BSBI Monitoring Scheme. *Biological Conservation* 75, 217–229.

Richards, A. J. & Blakemore, J., 1975. Factors affecting the germination of turions in *Hydrocharis morsus-ranae* L. *Watsonia* 10, 273–275.

Ridley, H. N., 1885. Two new British plants. *Journal of Botany* 23, 289–291.

Ridley, H. N., 1930. *The Dispersal of Plants Throughout the World*. Ashford: L. Reeve & Co.

Riemer, D. N., 1985. Seed germination in spatterdock (*Nuphar advena* Ait). *Journal of Aquatic Plant Management* 23, 46–47.

Robe, W. E. & Griffiths, H., 1992. Seasonal variation in the ecophysiology of *Littorella uniflora* (L.) Ascherson in acidic and eutrophic habitats. *New Phytologist* 120, 289–304.

Roberts, F. J. & Halliday, G., 1979. The altitudinal range of *Catabrosa aquatica* (L.) Beauv. *Watsonia* 12, 342–343.

Roberts, H. A., 1986. Persistence of seeds of some grass species in cultivated soil. *Grass and Forage Science* 41, 273–276.

Rodriguez-Oubiña, J. & Ortiz, S., 1991. *Luronium natans* (Alismataceae) in the Iberian peninsula. *Willdenowia* 21, 77–80.

Rodwell, J. S. (ed.), 1991. *British Plant Communities*, 2. *Mires and Heaths*. Cambridge: Cambridge University Press.

Rodwell, J. S. (ed.), 1995. *British Plant Communities*, 4. *Aquatic Communities, Swamps and Tall-*

herb Fens. Cambridge, Cambridge University Press.

Roe, R. G. B., 1981. *The Flora of Somerset.* Taunton: Somerset Archaeological and Natural History Society.

Roelofs, J. G. M., 1983. Impact of acidification and eutrophication on macrophyte communities in soft waters in the Netherlands I. Field observations. *Aquatic Botany* 17, 139–155.

Roelofs, J. G. M. & Schuurkes, J. A. A. R., 1983. Impact of acidification and eutrophication on macrophyte communities in soft waters. *Proceedings of the International Symposium on Aquatic Macrophytes, Nijmegen, 18–23 September, 1983*, 197–202.

Roelofs, J. G. M., Schuurkes, J. A. A. R. & Smits, A. J. M., 1984. Impact of acidification and eutrophication on macrophyte communities in soft waters. II. Experimental studies. *Aquatic Botany* 18, 389–411.

Ronse Decraene, L.-P. & Akeroyd, J. R., 1988. Generic limits in *Polygonum* and related genera (Polygonaceae) on the basis of floral characters. *Botanical Journal of the Linnean Society* 98, 321–371.

Rørslett, B. & Brettum, P., 1989. The genus *Isoëtes* in Scandinavia: an ecological review and perspectives. *Aquatic Botany* 35, 223–261.

Röst, L. C. M., 1978. Biosystematic investigations with *Acorus* L. (Araceae). 1. Communication. Cytotaxonomy. *Proceedings of the Koninklijke Nederlandse Akademie van Wetenschappen* 81C, 428–441.

Röst, L. C. M., 1979. Biosystematic investigations with *Acorus* 4. Communication. A synthetic approach to the classification of the genus. *Planta Medica* 37, 289–307.

Rothrock, P. E. & Wagner, R. H., 1975. *Eleocharis acicularis* (L.) R. & S.: the autecology of an acid tolerant sedge. *Castanea* 40, 279–289.

Royal Botanic Garden, Edinburgh, [1983]. *Survey of Aquatic Vegetation on South Uist and Benbecula 25 July–5 August 1983.* Unpublished report.

Rumsey, F. J., Thompson, P. & Sheffield, E., 1993. Triploid *Isoetes echinospora* (Isoetaceae: Pteridophyta) in northern England. *Fern Gazette* 14, 215–221.

Salisbury, E. J., 1960. Variation in the flowers of *Ranunculus circinatus* Sibth. *Kew Bulletin* 14, 34–36.

Salisbury, E. J., 1967. On the reproduction and biology of *Elatine hexandra* (Lapierre) DC. (Elatinaceae); a typical species of exposed mud. *Kew Bulletin* 21, 139–149.

Salisbury, E. [J.], 1970. The pioneer vegetation of exposed muds and its biological features. *Philosophical Transactions of the Royal Society of London* 259B, 207–255.

Salisbury, E. J., 1972. *Ludwigia palustris* (L.) Ell. in England with special reference to its dispersal and germination. *Watsonia* 9, 33–37.

Salisbury, E. [J.], 1974. The floral morphology of *Ranunculus tripartitus* var. *terrestris* (*R. lutarius*) and comparison with related taxa. *Proceedings of the Royal Society of London* 186B, 89–97.

Salter, J. H., 1935. *The Flowering Plants and Ferns of Cardiganshire.* Cardiff: University Press Board.

Sandwith, C., 1927. The hornworts and their occurrence in Britain. *Proceedings of the Bristol Naturalists' Society* 6, 303–311.

Sauvageau, C., 1893–94. Notes biologiques sur les *Potamogeton. Journal de Botanique* 8, 1–9, 21–43, 45–58, 98–106, 112–123, 140–148, 165–172.

Savidge, J. P., 1958. Distribution of *Callitriche* in north-west Europe. *Proceedings of the Botanical Society of the British Isles* 3, 103.

Savidge, J. P., 1960. The experimental taxonomy of European *Callitriche. Proceedings of the Linnean Society of London* 171, 128–130.

Savidge, J. P. (ed.), 1963. *Travis's Flora of South Lancashire.* Liverpool: Liverpool Botanical Society.

Savidge, J. P., 1967. Recognition of *Callitriche* spp. in Britain. *Proceedings of the Botanical Society of the British Isles* 6, 380–383.

Scannell, M. J. P., 1971. *Ceratophyllum demersum* L. in County Dublin. *Irish Naturalists' Journal* 17, 61.

Scannell, M. J. P., 1976a. *Ceratophyllum demersum* L. and fruit performance. *Irish Naturalists' Journal* 18, 348–349.

Scannell, M. J. P., 1976b. *Hydrilla verticillata* in flower in the wild. *Irish Naturalists' Journal* 18, 350.

Scannell, M. J. P. & Webb, D. A., 1976. The identity of the Renvyle *Hydrilla. Irish Naturalists' Journal* 18, 327–331.

Schotsman, H. D., 1949. Korte mededeling betreffende het geslacht *Alisma* in Nederland. *Nederlandsch Kruidkundig Archief* 56, 199–203.

Schotsman, H. D., 1954. A taxonomic spectrum of the section *Eu-Callitriche* in the Netherlands. *Acta Botanica Neerlandica* 3, 313–384.

Schotsman, H. D., 1958a. Beitrag zur Kenntnis der *Callitriche*-Arten in Bayern. *Bericht der Bayerischen Botanischen Gesellschaft zur Erforschung der heimischen Flora* 32, 128–140.

Schotsman, H. D., 1958b. Notes on *Callitriche hermaphroditica* Jusl. *Acta Botanica Neerlandica* 7, 519–523.

Schotsman, H. D., 1961a. Notes on some Portuguese species of *Callitriche. Boletim da Sociedade Broteriana* 35, 95–127.

Schotsman, H. D., 1961b. Contribution a l'étude des *Callitriche* du Canton de Neuchatel. *Bulletin de la Société Neuchâteloise des Sciences Naturelles* 84, 89–101.

Schotsman, H. D., 1961c. Races chromosomiques chez *Callitriche stagnalis* Scop. et *Callitriche obtusangula* Legall. *Berichte der Schweizerischen Botanischen Gesellschaft* 71, 5–17.

Schotsman, H. D., 1967. *Les Callitriches*. Paris: Éditions Paul Lechevalier.

Schotsman, H. D., 1972. Note sur la répartition des Callitriches en Sologne et dans les régions limitrophes. *Bulletin du Centre d'Études et de Recherches Scientifiques, Biarritz*, **9**, 19–52.

Schotsman, H. D., 1977. Callitriches de la région méditerranéenne. *Bulletin du Centre d'Études et de Recherches Scientifiques, Biarritz*, **11**, 241–312.

Schotsman, H. D., 1982. Biologie florale des *Callitriche*: étude sur quelques espèces d'Espagne méridionale. *Bulletin du Muséum National d'Histoire Naturelle*, series 4, **4**, section B, *Adansonia*, 111–160.

Schotsman, H. D. & Haldimann, G., 1981. Callitriches inédites du Jura français: *C. cophocarpa* Sendtn., *C. platycarpa* Kütz. (Angiospermae) et l'hybride dans la partie septentrionale. *Bulletin de la Société Neuchâteloise des Sciences Naturelles* **104**, 131–143.

Schuster, R., 1967. Taxonomische Untersuchungen über die Serie *Palustres* M. Pop. der Gattung *Myosotis* L. *Feddes Repertorium* **74**, 39–98.

Schuurkes, J. A. A. R., Elbers, M. A., Gudden, J. J. F. & Roelofs, J. G. M., 1987. Effects of simulated ammonium sulphate and sulphuric acid rain on acidification, water quality and flora of small-scale soft water systems. *Aquatic Botany* **28**, 199–226.

Scott, W. & Palmer, R. C., 1987. *The Flowering Plants and Ferns of the Shetland Islands*. Lerwick: Shetland Times.

Scribailo, R. W. & Posluszny, U., 1984. The reproductive biology of *Hydrocharis morsus-ranae*. I. Floral biology. *Canadian Journal of Botany* **62**, 2779–2787.

Scribailo, R. W. & Posluszny, U., 1985. The reproductive biology of *Hydrocharis morsus-ranae*. II. Seed and seedling morphology. *Canadian Journal of Botany* **63**, 492–496.

Scully, R. W., 1916. *Flora of County Kerry*. Dublin: Hodges, Figgis & Co.

Sculthorpe, C. D., 1967. *The Biology of Aquatic Vascular Plants*. London: Edward Arnold.

Seddon, B., 1965. Occurrence of *Isoetes echinospora* in eutrophic lakes in Wales. *Ecology* **46**, 747–748.

Seddon, B., 1972. Aquatic macrophytes as limnological indicators. *Freshwater Biology* **2**, 107–130.

Segal, S., 1967. Some notes on the ecology of *Ranunculus hederaceus* L. *Vegetatio* **15**, 1–26.

Sharitz, R. R., Wineriter, S. A., Smith, M. H. & Lu, E. H., 1980. Comparison of isozymes among *Typha* species in the eastern United States. *American Journal of Botany* **67**, 1297–1303.

Shaver, G. R., Chapin, F. S. & Billings, W. D., 1979. Ecotypic differentiation in *Carex aquatilis* on ice-wedge polygons in the Alaskan coastal tundra. *Journal of Ecology* **67**, 1025–1045.

Shi, D.-J. & Hall, D. O., 1988. *Azolla* and immobilized cyanobacteria (blue-green algae): from traditional agriculture to biotechnology. *Plants Today* **1**, 5–12.

Sifton, H. B., 1959. The germination of light-sensitive seeds of *Typha latifolia* L. *Canadian Journal of Botany* **37**, 719–739.

Silverside, A. J., 1984. The subspecies of *Ranunculus flammula*. *BSBI Scottish Newsletter* no. **6**, 4–7.

Simpson, D. A., 1984. A short history of the introduction and spread of *Elodea* Michx in the British Isles. *Watsonia* **15**, 1–9.

Simpson, D. [A.], 1985. *Elodea nuttallii* (Planch.) St John in Ireland. *Irish Naturalists' Journal* **21**, 497–498.

Simpson, D. A., 1986. Taxonomy of *Elodea* Michx in the British Isles. *Watsonia* **16**, 1–14.

Simpson, D. A., 1988. Phenotypic plasticity of *Elodea nuttallii* (Planch.) H. St John and *Elodea canadensis* Michx in the British Isles. *Watsonia* **17**, 121–132.

Simpson, D. A., 1990. Displacement of *Elodea canadensis* Michx by *Elodea nuttallii* (Planch.) H. St John in the British Isles. *Watsonia* **18**, 173–177.

Simpson, F. W., 1982. *Simpson's Flora of Suffolk*. Ipswich: Suffolk Naturalists' Society.

Sinker, C. A., 1960. The vegetation of the Malham Tarn area. *Proceedings of the Leeds Philosophical and Literary Society Scientific Section* **8**, 139–175.

Sinker, C. A., 1962. The North Shropshire Meres and Mosses: a background for ecologists. *Field Studies* **1** (4), 101–138.

Sinker, C. A., Packham, J. R., Trueman, I. C., Oswald, P. H., Perring, F. H. & Prestwood, W. V., 1985. *Ecological Flora of the Shropshire Region*. Shrewsbury: Shropshire Trust for Nature Conservation.

Slack, A. A., 1964. "Floating Water-Plantain" (*Luronium natans*) in Argyll. *Glasgow Naturalist* **18**, 382.

Sledge, W. A., 1942. *Vallisneria spiralis* L. *Report of the Botanical Society and Exchange Club of the British Isles* **12**, 424.

Sledge, W. A., 1945. Disappearance of *Tillaea aquatica* L. at Adel. *Naturalist, Hull* **1945**, 149.

Smith, J. E., 1990. *Surrey Flora Committee Newsletter* **1990** (February), 3.

Smith, R. A. H., Stewart, N. F., Taylor, N. W. & Thomas, R. E., 1992. *Checklist of the Plants of Perthshire*. Perth: Perthshire Society of Natural Science.

Smith, S. G., 1967. Experimental and natural hybrids in North American *Typha* (Typhaceae). *American Midland Naturalist* **78**, 257–287.

Smith, S. J. & Wolfe-Murphy, S. A., 1991. *Ceratophyllum submersum* L. Soft Hornwort, a species new to Ireland. *Irish Naturalists' Journal* 23, 374–376.

Smith, W. G., 1905. Botanical survey of Scotland. III and IV. – Forfar and Fife. *Scottish Geographical Magazine* 21, 57–83.

Smith, W. W., 1907. Note on a peculiar tussock-formation. *Transactions of the Botanical Society of Edinburgh* 23, 234–235.

Smits, A. J. M., Van Avesaath, P. H. & Velde, G. van der, 1990. Germination requirements and seed banks of some nymphaeid macrophytes: *Nymphaea alba* L., *Nuphar lutea* (L.) Sm. and *Nymphoides peltata* (Gmel.) O. Kuntze. *Freshwater Biology* 24, 315–326.

Smits, A. J. M., Van Ruremonde, R. & Velde, G. van der, 1989. Seed dispersal of three nymphaeid macrophytes. *Aquatic Botany* 35, 167–180.

Smits, A. J. M. & Wetzels, A. M. M., 1986. Germination studies on three nymphaeid species (*Nymphaea alba* L., *Nuphar lutea* (L.) Sm. and *Nymphoides peltata* (Gmel.) O. Kuntze). *Proceedings EWRS/AAB 7th Symposium on Aquatic Weeds*, 315–320.

Soukupová, L., 1988. Short life-cycles in two wetland sedges. *Aquatic Botany* 30, 49–62.

Sowter, A. J., 1971. *Crassula aquatica* (L.) Schönl. – new to Scotland. *Watsonia* 8, 294.

Sowter, A. J., Sowter, M. M. & Webster, M. McC., 1972. *Crassula aquatica* (L.) Schönl. in v.c. 97 – further observations. *Watsonia* 9, 140.

Spence, D. H. N., 1964. The macrophytic vegetation of freshwater lochs, swamps and associated fens. *In:* J. H. Burnett (ed.) *The Vegetation of Scotland*, pp. 306–425. Edinburgh & London: Oliver & Boyd.

Spence, D. H. N., 1967. Factors controlling the distribution of freshwater macrophytes with particular reference to the lochs of Scotland. *Journal of Ecology* 55, 147–170.

Spence, D. H. N., 1982. The zonation of plants in freshwater lakes. *Advances in Ecological Research* 12, 37–125.

Spence, D. H. N., Barclay, A. M. & Allen, E. D., 1984. Limnology and macrophyte vegetation of a deep, clear limestone lake, Loch Borralie. *Transactions of the Botanical Society of Edinburgh* 44, 187–204.

Spencer, D. F. & Anderson, L. W. J., 1987. Influence of photoperiod on growth, pigment composition and vegetative propagule formation for *Potamogeton nodosus* Poir. and *Potamogeton pectinatus* L. *Aquatic Botany* 28, 103–112.

Spencer, D. F. & Ksander, G. G., 1990. Influence of planting depth on *Potamogeton gramineus* L. *Aquatic Botany* 36, 343–350.

Spencer-Jones, D., 1994. Some observations on the use of herbicides for control of *Crassula helmsii*. *In:* L. C. de Waal, L. E. Child & P. M. Wade (eds), *Ecology and Management of Invasive Riverside Plants*, pp. 15–18. Chichester: John Wiley & Sons.

Spicer, K. W. & Catling, P. M., 1988. The biology of Canadian weeds. 88. *Elodea canadensis* Michx. *Canadian Journal of Plant Science* 68, 1035–1051.

Stace, C. A., 1991. *New Flora of the British Isles.* Cambridge: Cambridge University Press.

Stewart, A., Pearman, D. A. & Preston, C. D. (comps & eds), 1994. *Scarce Plants in Britain.* Peterborough: Joint Nature Conservation Committee.

Stewart, O. [M.], 1988. *Pilularia globulifera* in Kirkcudbrightshire. *Transactions of the Dumfriesshire and Galloway Natural History and Antiquarian Society* 63, 1–4.

Stewart, O. M. & Preston, C. D., 1990. *Potamogeton × lintonii* – new to Scotland. *BSBI News* no. 54, 20–21.

Stewart, S. A. & Corry, T. H., 1888. *A Flora of the North-east of Ireland.* Belfast: Belfast Naturalists' Field Club.

Stoeva, M. P., 1992. Karyological study of *Bolboschoenus maritimus* (L.) Palla and *Holoschoenus vulgaris* Link (Cyperaceae) in Bulgaria. *Comptes rendus de l'Académie Bulgare des Sciences* 45, 61–63.

Stokoe, R., 1978. *Isoetes echinospora* Durieu new to northern England. *Watsonia* 12, 51–52.

Stokoe, R., 1983. *Aquatic Macrophytes in the Tarns and Lakes of Cumbria.* Freshwater Biological Association Occasional Publication no. 18.

Strandhede, S.-O., 1960. A note on *Scirpus palustris* L. *Botaniska Notiser* 113, 161–171.

Strandhede, S.-O., 1961. *Eleocharis Palustres* in Scandinavia and Finland. *Botaniska Notiser* 114, 417–434.

Strandhede, S.-O., 1965. Chromosome studies in *Eleocharis*, subser. *Palustres* III. Observations on western European taxa. *Opera Botanica* 9(2), 1–86.

Strandhede, S.-O., 1966. Morphologic variation and taxonomy in European *Eleocharis*, subser. *Palustres*. *Opera Botanica* 10(2), 1–187.

Strandhede, S.-O., 1967. *Eleocharis*, subser. *Eleocharis* in North America: taxonomical comments and chromosome numbers. *Botaniska Notiser* 120, 355–368.

Stuckey, R. L., 1972. Taxonomy and distribution of the genus *Rorippa* (Cruciferae) in North America. *Sida* 4, 279–430.

Stuckey, R. L., 1974. The introduction and distribution of *Nymphoides peltatum* (Menyanthaceae) in North America. *Bartonia* 42, 14–23.

Stuckey, R. L., 1979. Distributional history of *Potamogeton crispus* (Curly Pondweed) in North America. *Bartonia* 46, 22–42.

Sturrock, A., 1876. *Naias flexilis* in Perthshire. *Scottish Naturalist* 3, 198–199.

Sutherland, W. J., 1990. Biological Flora of the British Isles no. 169. *Iris pseudacorus* L. *Journal of Ecology* 78, 833–848.

Sutherland, W. J. & Walton, D., 1990. The changes in morphology and demography of *Iris pseudacorus* L. at different heights on a salt-marsh. *Functional Ecology* 4, 655–659.

Svendäng, M. U., 1990. The growth dynamics of *Juncus bulbosus* L. – a strategy to avoid competition? *Aquatic Botany* 37, 123–138.

Swan, G. A., 1993. *Flora of Northumberland.* Newcastle upon Tyne: Natural History Society of Northumbria.

Swann, E. L., 1975. *Supplement to the Flora of Norfolk.* Norwich: F. Crowe & Sons.

Swindells, P., 1983. *Waterlilies.* Beckenham: Croom Helm.

Sykes, N., 1993. *Wild Plants and Their Habitats in the North York Moors.* Helmsley: North York Moors National Park.

Symoens, J. J. & Triest, L., 1983. Monograph of the African genus *Lagarosiphon* Harvey (Hydrocharitaceae). *Bulletin du Jardin Botanique National de Belgique* 53, 441–488.

Talavera, S., García-Murillo, P. & Herrera, J., 1993. Chromosome numbers and a new model for karyotype evolution in *Ruppia* L. (Ruppiaceae). *Aquatic Botany* 45, 1–13.

Talavera, S., García Murillo, P. & Smit, H., 1986. Sobre el genero *Zannichellia* L. (Zannichelliaceae). *Lagascalia* 14, 241–271.

Tan, B. C., Payawal, P., Watanabe, I., Lacdan, N. & Ramirez, C., 1986. Modern taxonomy of *Azolla*: a review. *Philippine Agriculturist* 69, 491–512.

Tansley, A. G., 1939. *The British Islands and their Vegetation.* Cambridge: Cambridge University Press.

Tarpey, T. & Heath, J., 1990. *Wild Flowers of North East Essex.* Colchester: Colchester Natural History Society.

Tarver, D. P. & Sanders, D. R., 1977. Selected life cycle features of fanwort. *Journal of Aquatic Plant Management* 15, 18–22.

Taylor, P., 1989. *The genus* Utricularia *– a taxonomic monograph.* Kew Bulletin Additional Series XIV. London: Her Majesty's Stationery Office.

Thommen, G. H. & Westlake, D. F., 1981. Factors affecting the distribution of populations of *Apium nodiflorum* and *Nasturtium officinale* in small chalk streams. *Aquatic Botany* 11, 21–36.

Thor, G., 1979. *Utricularia* i Sverige, speciellt de förbisedda arterna *U. australis* och *U. ochroleuca.* [*Utricularia* in Sweden, especially the overlooked species *U. australis* and *U. ochroleuca.*] *Svensk Botanisk Tidskrift* 73, 381–395.

Thor, G., 1987. Sumpbläddra, *Utricularia stygia*, en ny svensk art. [*Utricularia stygia* Thor, a new *Utricularia* species in Sweden.] *Svensk Botanisk Tidskrift* 81, 273–280.

Thor, G., 1988. The genus *Utricularia* in the Nordic countries, with special emphasis on *U. stygia* and *U. ochroleuca. Nordic Journal of Botany* 8, 213–225.

Tolhurst, S. A., 1987. *A Survey of the Aquatic Flora of the Pocklington Canal, Yorkshire, 1986,* edited by H. E. Stace & M. A. Palmer. NCC Contract Surveys no. 4. Peterborough: Nature Conservancy Council.

Tomlinson, P. B., 1982. *Anatomy of the Monocotyledons, 7. Helobiae (Alismataceae) (including the Sea Grasses).* Oxford: Clarendon Press.

Townsend, F., 1850. On a supposed new species of *Glyceria. Annals and Magazine of Natural History,* 2nd series, 5, 104–108.

Townsend, F., 1883. *Flora of Hampshire.* London: L. Reeve & Co.

Trail, J. W. H., 1923. Flora of the city parish of Aberdeen. *In: James William Helenus Trail, a Memorial Volume,* pp. 57–331. Aberdeen: Aberdeen University Press.

Tralau, H., 1959. Extinct aquatic plants of Europe. *Botaniska Notiser* 112, 385–406.

Travis, W. G., 1929. *Scirpus americanus* Pers. & *Weingaertneria canescens* Bernh. in Lancashire. *North Western Naturalist* 4, 175–177.

Triest, L., 1987. A revision of the genus *Najas* L. (Najadaceae) in Africa and surrounding islands. *Mémoires, Academie Royale des Sciences d'Outre-Mer, Classe des Sciences Naturelles et Médicales,* new series, 21 (4), 1–88.

Triest, L., 1988. A revision of the genus *Najas* L. (Najadaceae) in the Old World. *Mémoires, Academie Royale des Sciences d'Outre-Mer, Classe des Sciences Naturelles et Médicales,* new series, 22 (1), 1–172.

Trimen, H., 1866. *Wolffia arrhiza*, Wimmer, in England. *Journal of Botany* 4, 219–223.

Tutin, T. G., 1947. *Typha angustifolia* L. × *latifolia* L. *Report of the Botanical Society and Exchange Club of the British Isles* 13, 173–174.

Tutin, T. G., 1975. *Apium* L. *In*: C. A. Stace (ed.), *Hybridization and the Flora of the British Isles,* pp. 268–269. London: Academic Press.

Tutin, T. G., 1980. *Umbellifers of the British Isles.* BSBI Handbook no. 2. London: Botanical Society of the British Isles.

Tutin, T. G., Burges, N. A., Chater, A. O., Edmondson, J. R., Heywood, V. H., Moore, D. M., Valentine, D. H., Walters, S. M. & Webb, D. A. (eds), 1993. *Flora Europaea,* 1, edn 2. *Psilotaceae to Platanaceae.* Cambridge: Cambridge University Press.

Tutin, T. G., Heywood, V. H., Burges, N. A., Moore, D. M., Valentine, D. H., Walters, S. M. & Webb, D. A. (eds), 1968. *Flora Europaea,* 2. *Rosaceae to Umbelliferae.* Cambridge: Cambridge University Press.

Tutin, T. G., Heywood, V. H., Burges, N. A., Moore, D. M., Valentine, D. H., Walters, S. M. & Webb, D. A. (eds), 1972. *Flora Europaea,* 3. *Diapensiaceae to Myoporaceae.* Cambridge: Cambridge University Press.

Tutin, T. G., Heywood, V. H., Burges, N. A., Moore, D. M., Valentine, D. H., Walters, S. M. & Webb, D. A. (eds), 1980. *Flora Europaea*, 5. *Alismataceae to Orchidaceae (Monocotyledones)*. Cambridge: Cambridge University Press.

Uotila, P., 1974. *Elatine hydropiper* L. aggr. in northern Europe. *Memoranda Societatis pro Fauna et Flora Fennica* 50, 113–123.

Uotila, P., Van Vierssen, W. & Van Wijk, R. J., 1983. Notes on the morphology and taxonomy of *Zannichellia* in Turkey. *Annales Botanici Fennici* 20, 351–356.

Urbanska-Worytkiewicz, K., 1975. Cytological variation within *Lemna* L. *Aquatic Botany* 1, 377–394.

Urbanska-Worytkiewicz, K., 1980. Cytological variation within the family of Lemnaceae. *Veröffentlichungen des Geobotanischen Instituts der Eidg. Techn. Hochschule, Stiftung Rübel, in Zurich* 70, 30–101.

Van Bruggen, H. W. E., 1985. Monograph of the genus *Aponogeton* (Aponogetonaceae). *Bibliotheca Botanica* 137, 1–76.

Van Vierssen, W., 1982a. The ecology of communities dominated by *Zannichellia* taxa in western Europe. I. Characterization and autecology of the *Zannichellia* taxa. *Aquatic Botany* 12, 103–155.

Van Vierssen, W., 1982b. The ecology of communities dominated by *Zannichellia* taxa in western Europe. II. Distribution, synecology and productivity aspects in relation to environmental factors. *Aquatic Botany* 13, 385–483.

Van Vierssen, W., 1982c. Reproductive strategies of *Zannichellia* taxa in western Europe. *In*: J. J. Symoens, S. S. Hooper & P. Compère (eds), *Studies on Aquatic Vascular Plants*, pp. 144–149. Brussels: Royal Botanical Society of Belgium.

Van Vierssen, W., Van Kessel, C. M. & Zee, J. R. van der, 1984. On the germination of *Ruppia* taxa in western Europe. *Aquatic Botany* 19, 381–393.

Van Vierssen, W., Van Wijk, R. J. & Zee, J. R. van der, 1981. Some additional notes on the cytotaxonomy of *Ruppia* taxa in western Europe. *Aquatic Botany* 11, 297–301.

Van Wijk, R. J., 1988. Ecological studies on *Potamogeton pectinatus* L. I. General characteristics, biomass production and life cycles under field conditions. *Aquatic Botany* 31, 211–258.

Van Wijk, R. J., 1989. Ecological studies on *Potamogeton pectinatus* L. III. Reproductive strategies and germination ecology. *Aquatic Botany* 33, 271–299.

Van Wijk, R. J. & Trompenaars, H. J. A. J., 1985. On the germination of turions and the life cycle of *Potamogeton trichoides* Cham. & Schld. *Aquatic Botany* 22, 165–172.

Van Wijk, R. J., Van Goor, E. M. J. & Verkley, J. A. C., 1988. Ecological studies on *Potamogeton pectinatus* L. II. Autecological characteristics, with emphasis on salt tolerance, intraspecific

variation and isoenzyme patterns. *Aquatic Botany* 32, 239–260.

Vanhecke, L., 1986. *Scirpus* × *carinatus* Smith, *S.* × *scheuchzeri* Brügger en *S. triqueter* L. in België. *Dumortiera* 34–35, 94–100.

Varopoulos, A., 1979. Breeding systems in *Myosotis scorpioides* L. (Boraginaceae) I. Self–incompatibility. *Heredity* 42, 149–157.

Veken, P. van der, 1965. Contribution à l'embryographie systématique des Cyperaceae – Cyperoideae. *Bulletin du Jardin Botanique de l'Etat, Bruxelles*, 35, 219–354.

Velde, G. van der, 1986. Developmental stages in the floral biology s.l. of Dutch Nymphaeaceae (*Nymphaea alba* L., *Nymphaea candida* Presl, *Nuphar lutea* (L.) Sm.). *Acta Botanica Neerlandica* 35, 111–113.

Velde, G. van der & Heijden, L. A. van der, 1981. The floral biology and seed production of *Nymphoides peltata* (Gmel.) O. Kuntze (Menyanthaceae). *Aquatic Botany* 10, 261–293.

Verhoeven, J. T. A., 1979. The ecology of *Ruppia*-dominated communities in western Europe. I. Distribution of *Ruppia* representatives in relation to their autecology. *Aquatic Botany* 6, 197–268.

Verhoeven, J. T. A., 1980a. The ecology of *Ruppia*-dominated communities in western Europe. II. Synecological classification. Structure and dynamics of the macroflora and macrofauna communities. *Aquatic Botany* 8, 1–85.

Verhoeven, J. T. A., 1980b. The ecology of *Ruppia*-dominated communities in western Europe. III. Aspects of production, consumption and decomposition. *Aquatic Botany* 8, 209–253.

Verhoeven, J. T. A., Schmitz, M. B. & Pons, T. L., 1988. Comparative demographic study of *Carex rostrata* Stokes, *C. diandra* Schrank and *C. acutiformis* Ehrh. in fens of different nutrient status. *Aquatic Botany* 30, 95–108.

Verhoeven, J. T. A. & Van Vierssen, W., 1978. Distribution and structure of communities dominated by *Ruppia, Zostera* and *Potamogeton* species in the inland waters of 'De Bol', Texel, The Netherlands. *Estuarine and Coastal Marine Science* 6, 417–428.

Verkleij, J. A. C., Pieterse, A. H., Horneman, G. J. T. & Torenbeek, M., 1983. A comparative study of the morphology and isoenzyme patterns of *Hydrilla verticillata* (L. f.) Royle. *Aquatic Botany* 17, 43–59.

Vickery, R., 1983. *Lemna minor* and Jenny Greenteeth. *Folklore* 94, 247–250.

Viinikka, Y., 1976. *Najas marina* L. (Najadaceae). Karyotypes, cultivation and morphological variation. *Annales Botanici Fennici* 13, 119–131.

Vose, P. B., 1962. Delayed germination in Reed Canary-Grass *Phalaris arundinacea* L. *Annals of Botany* new series 26, 197–206.

Vuille, F.-L., 1987. Reproductive biology of the genus *Damasonium* (Alismataceae). *Plant Systematics and Evolution* **157**, 63–71.

Vuille, F.-L., 1988. The reproductive biology of the genus *Baldellia* (Alismataceae). *Plant Systematics and Evolution* **159**, 173–183.

Wade, W., 1802. Catalogus plantarum rariorum in comitatu Gallovidiae, praecipue Cunnemara inventarum. *Transactions of the Dublin Society* **2**, 103–127.

Walker, A. O., 1912. The distribution of *Elodea canadensis*, Michaux, in the British Isles in 1909. *Proceedings of the Linnean Society of London* **1911–1912**, 71–77.

Wallace, E. C., 1975. *Carex* L. *In*: C. A. Stace (ed.), *Hybridization and the Flora of the British Isles*, pp. 513–540. London: Academic Press.

Walsh, H., 1944. *Myriophyllum heterophyllum* Michx. and *M. spicatum* L. in the Halifax canal. *Naturalist, Hull* **1944**, 143–144.

Walters, S. M., 1949. Biological Flora of the British Isles. *Eleocharis* R. Br. *Journal of Ecology* **37**, 192–206.

Walters, S. M., 1975. *Veronica* L. *In*: C. A. Stace (ed.), *Hybridization and the Flora of the British Isles*, pp. 371–372. London: Academic Press.

Watson, L. & Dallwitz, M. J., 1992. *The grass genera of the world*. Wallingford: CAB International.

Watson, K. J. & Murphy, K. J., 1988. *The Aquatic Vegetation of the Forth and Clyde Canal 1988*. Unpublished report to Nature Conservancy Council.

Webb, D. A., 1967. Generic limits in European Lythraceae. *Feddes Repertorium* **74**, 10–13.

Webb, D. A. & Scannell, M. J. P., 1983. *Flora of Connemara and the Burren*. Cambridge: Royal Dublin Society & Cambridge University Press.

Weber, J. A. & Noodén, L. D., 1974. Turion formation and germination in *Myriophyllum verticillatum*: phenology and its interpretation. *Michigan Botanist* **13**, 151–158.

Weber, J. A. & Noodén, L. D., 1976. Environmental and hormonal control of turion germination in *Myriophyllum verticillatum*. *American Journal of Botany* **63**, 936–944.

Webster, S. D., 1988a. *Ranunculus penicillatus* (Dumort.) Bab. in Great Britain and Ireland. *Watsonia* **17**, 1–22.

Webster, S. D. 1988b. *Ranunculus* L. subgenus *Batrachium*. *In*: T. C. G. & M. D. B. Rich (comps), *Plant Crib*, pp. 8–17. London: Botanical Society of the British Isles.

Webster, S. D., 1990. Three natural hybrids in *Ranunculus* L. subgenus *Batrachium* (DC.) A. Gray. *Watsonia* **18**, 139–146.

Webster, S. D., 1991. *Ranunculus penicillatus* (Dumort.) Bab. in Ireland. *Irish Naturalists' Journal* **23**, 346–354.

Weeda, E. J., 1976. Over het optreden van *Potamogeton praelongus* Wulf., o.a. bij Buinen (Dr.). *Gorteria* **8**, 89–98.

Weeda, E. J., Meijden, R. van der & Bakker, P. A., 1990. FLORON – Rode Lijst 1990. *Gorteria* **16**, 1–26.

Welch, D., 1961. Water Forget-me-nots in Cambridgeshire. *Nature in Cambridgeshire* no. 4, 18–27.

Welch, D., 1993. *Flora of North Aberdeenshire*. Banchory: privately published.

Wells, R. V., 1968. *Veronica anagallis-aquatica* × *catenata* – at Meonstoke, Hampshire. *Proceedings of the Botanical Society of the British Isles* **7**, 391.

Wentz, W. A. & Stuckey, R. L., 1971. The changing distribution of the genus *Najas* (Najadaceae) in Ohio. *Ohio Journal of Science* **71**, 292–302.

West, G., 1905. A comparative study of the dominant phanerogamic and higher cryptogamic flora of aquatic habit, in three lake areas of Scotland. *Proceedings of the Royal Society of Edinburgh* **25**, 967–1024.

West, G., 1910. A further contribution to a comparative study of the dominant phanerogamic and higher cryptogamic flora of aquatic habit in Scottish lakes. *Proceedings of the Royal Society of Edinburgh* **30**, 65–182.

West, R. G., 1953. The occurrence of *Azolla* in British interglacial deposits. *New Phytologist* **52**, 267–272.

Wheeler, B. D., 1978. The wetland plant communities of the River Ant valley, Norfolk. *Transactions of the Norfolk and Norwich Naturalists' Society* **24**, 153–187.

Wheeler, B. D., 1980a. Plant communities of rich-fen systems in England and Wales I. Introduction. Tall sedge and reed communities. *Journal of Ecology* **68**, 365–395.

Wheeler, B. D., 1980b. Plant communities of rich-fen systems in England and Wales II. Communities of calcareous mires. *Journal of Ecology* **68**, 405–420.

Wheeler, B. D., 1980c. Plant communities of rich-fen systems in England and Wales III. Fen meadow, fen grassland and fen woodland communities, and contact communities. *Journal of Ecology* **68**, 761–788.

Wheeler, B. D. & Giller, K. E., 1982. Status of aquatic macrophytes in an undrained area of fen in the Norfolk Broads, England. *Aquatic Botany* **12**, 277–296.

White, F. B. W., 1898. *The Flora of Perthshire*. Edinburgh: Perthshire Society of Natural Science.

White, J. W., 1912. *The Flora of Bristol*. Bristol: John Wright & Sons.

Whitehead, F. H., 1971. Comparative autecology as a guide to plant distribution. *In*: E. Duffey & A. S. Watt (eds), *The Scientific Management of Animal and Plant Communities for Conservation*, pp. 167–176. Oxford: Blackwell Scientific Publications.

Wiegleb, G., 1988. Notes on pondweeds – outlines for a monographical treatment of the genus *Potamogeton* L. *Feddes Repertorium* **99**, 249–266.

Wiersema, J. H. & Hellquist, C. B., 1994. Nomenclatural notes in Nymphaeaceae for the North American Flora. *Rhodora* **96**, 170–178.

Wiggers Nielsen, L., Nielsen, K. & Sand-Jensen, K., 1985. High rates of production and mortality of submerged *Sparganium emersum* Rehman during its short growth season in a eutrophic Danish stream. *Aquatic Botany* **22**, 325–334.

Wigginton, M. [J.], 1989. *Survey of Shropshire and Cheshire Meres 1987*. NCC England Field Unit Report no. 59. Unpublished report.

Wigginton, M. J. & Graham, G. G., 1981. *Guide to the Identification of Some of the More Difficult Vascular Plant Species*. England Field Unit Occasional Paper no. 1. Banbury: Nature Conservancy Council.

Willby, N. J. & Eaton, J. W., 1993. The distribution, ecology and conservation of *Luronium natans* (L.) Raf. in Britain. *Journal of Aquatic Plant Management* **31**, 70–76.

Williams, I. A., 1929. A British *Veronica* hybrid. *Journal of Botany* **67**, 23–24.

Wilmot-Dear, M., 1985. *Ceratophyllum* revised – a study in fruit and leaf variation. *Kew Bulletin* **40**, 243–271.

Wilson, A., 1938. *The Flora of Westmorland*. Arbroath: privately published.

Wilson, A., 1956. *The Altitudinal Range of British Plants*, edn 2. Arbroath: T. Buncle & Co.

Wilson, K. L., 1981. A synopsis of the genus *Scirpus* sens. lat. (Cyperaceae) in Australia. *Telopia* **2**, 153–172.

Witztum, A., 1977. An ecological niche for *Lemna gibba* L. that depends on seed formation. *Israel Journal of Botany* **26**, 36–38.

Witztum, A., 1986. Seed viability in *Lemna gibba* L. *Israel Journal of Botany* **35**, 279.

Wolfe-Murphy, S. A., Smith, S. J. & Preston, C. D., 1991. Irish pondweeds I. A recent record of *Potamogeton × cooperi* (Fryer) Fryer from Co Antrim. *Irish Naturalists' Journal* **23**, 457–458.

Wolley-Dod, A. H. (ed.), 1937. *Flora of Sussex*. Hastings: Kenneth Saville.

Woodhead, N., 1951. Biological Flora of the British Isles. *Subularia aquatica* L. *Journal of Ecology* **39**, 465–469.

Wooten, J. W., 1971. The monoecious and dioecious conditions in *Sagittaria latifolia* L. (Alismataceae). *Evolution* **25**, 549–553.

Wyse Jackson, P. & Sheehy Skeffington, M., 1984. *Flora of Inner Dublin*. Dublin: Royal Dublin Society.

Yeo, R. R., 1964. Life history of Common Cattail. *Weeds* **12**, 284–288.

You, J., Sun, X.-Z. & Wang, H.-Q., 1991. A preliminary study on the polyploid series and cytogeography of *Najas graminea*. *Acta Phytotaxonomica Sinica* **29**, 230–234.

Young, C. P. L. & Stewart, N. F., 1986. *A Pilot Survey of Fife and Kinross Lochs to Assess Changes in the Aquatic Flora Since Surveys Conducted by G. T. West between 1905 and 1909*. Unpublished report, Nature Conservancy Council.

Zákravský, P. & Hroudová, Z., 1994. The effect of submergence on tuber production and dormancy in two subspecies of *Bolboschoenus maritimus*. *Folia Geobotanica et Phytotaxonomica* **29**, 217–226.

Ziegler, P., 1969. *The Black Death*. London: Collins.

INDEX TO ACCOUNTS OF SPECIES, SUBSPECIES AND HYBRIDS

Printed in the United States
by Baker & Taylor Publisher Services